STP 1093

Buried Plastic Pipe Technology

George S. Buczala and Michael J. Cassady, editors

ASTM
1916 Race Street
Philadelphia, PA 19103

Library of Congress Cataloging-in-Publication Data

Buried plastic pipe technology / George S. Buczala and Michael J.
Cassady, editors.
    (STP ; 1093)
    Papers from a symposium held in Dallas, Tex., Sept. 10-13, 1990
and sponsored by the ASTM Committee F-17 on Plastic Systems and
Subcommittee D20.23 on Reinforced Plastic Piping Systems and
Chemical Equipment.
    "ASTM publication code number (PCN) 04-010930-58."
    Includes bibliographical references and index.
    ISBN 0-8031-1395-1
    1. Pipe, Plastic--Underground--Congresses.   I. Buczala, George
S., 1932-   . II. Cassady, Michael J., 1942-   . III. ASTM
Committee F-17 on Plastic Piping Systems.   IV. ASTM Committee F-17
on Plastic Piping Systems.   Subcommittee D20.23 on Reinforced
Plastic Piping Systems and Chemical Equipment.   V. Series: ASTM
special technical publication ; 1093.
TJ930.B873   1990
668.4'9--dc20   90-14395
                                                                      CIP

NOTE

The Society is not responsible, as a body,
for the statements and opinions
advanced in this publication.

Peer Review Policy

Each paper published in this volume was evaluated by three peer reviewers. The authors
addressed all of the reviewers' comments to the satisfaction of both the technical editor(s)
and the ASTM Committee on Publications.

The quality of the papers in this publication reflects not only the obvious efforts of the
authors and the technical editor(s), but also the work of these peer reviewers. The ASTM
Committee on Publications acknowledges with appreciation their dedication and contribution of time and effort on behalf of ASTM.

Printed in Baltimore
October 1990

Foreword

The symposium on Buried Plastic Pipe Technology was held in Dallas, Texas, on 10–13 Sept. 1990. ASTM Committee F-17 on Plastic Piping Systems and Subcommittee D20.23 on Reinforced Plastic Piping Systems and Chemical Equipment sponsored the symposium in cooperation with ASTM Committees D-20 on Plastics and D-18 on Soil and Rock, the Uni-Bell PVC Pipe Association, the Fiberglass Pipe Institute, and the Plastics Pipe Institute. George S. Buczala, Philadelphia Electric Company, and Michael J. Cassady, Battelle Columbus Laboratories, served as co-chairmen of the symposium and co-editors of the resulting publication.

Acknowledgment

On behalf of the symposium planning committee, we thank the speakers for their preparation and participation, the peer reviewers for their time, and the ASTM Staff for their patience.

George S. Buczala and Michael J. Cassady

Contents

CONTENTS

Overview

This symposium on Buried Plastic Pipe Technology was organized to provide the many constructors of water, sewer, drainage, waste managment, irrigation, and gas projects with state of the art engineering data and techniques for the successful use of plastic piping materials.

The workshop on the day prior to the formal symposium set forth the basic properties of the various plastic piping materials-both thermoplastic and thermoset. The workshop gave the participants an introduction to the plastic pipe.

Session one deals with Testing and Standards. Ronald Bishop described a new application for ASTM D2412 Determination of External Loading Characteristics of Plastic Pipe by Parallel-Plate Loading. This method is typically used to describe and establish pipe stiffness values for plastic pipe. Mr. Bishop extends the method over an extended time and increasing deflection to measure the retention of pipe stiffness for PVC pipe samples exposed to various environments.

Polyethylene (PE) piping has become a material of choice for many applications that benefit from this material's special ability to endure fatigue, impact loading, large deformations, abrasive materials and very agressive environments. The durability of PE's may be compromised by any one of the following causes: chemical and physical aging, weathering, creep under load, and fracture under load. Stanley Mruk in his paper reviews each of these limits and the measures taken to ensure, by modern standards, that only suitable durable materials are used for piping. Particular attention is given to the characterization and testing used to ensure that PE materials sensitive to slow crack growth are not used in piping applications.

"Widespread use of a piping product will only be achieved when there are detailed comprehensive product performance standards that can be confidently utilized by the specifying engineer." said L. E. Pearson. Mr. Pearson in his presentation, on Recent Changes in Fiberglass Pipe Specifications, describes the changes in fiberglass pipe standards issued by the American Water Works Association (AWWA) and ASTM. Among these changes have been the expansion to multible stiffness ranges, a 50 year design criteria, increased deflection to crack-damage requirements, establishment of long term ring bending strength test method, and updating and modification of test methods and performance criteria for strain corrosion and hydrostatic design basis. AWWA and ASTM product standards have been made consistent.

The German Specification ATV-127 is appropriate for static calculations of buried gravity and pressure pipelines. H. Schneider relates this specification to plastic pipe design. The specification is based on experience and allows pipe installations to be analyzed for various pipe stiffnesses, backfill and bedding conditions. Various inputs to the system such as pipe properties, soil properties and traffic loads result in a vertical deflection, a wall stress and/or strain, and a buckling calculation. Combined with minimum requirments for factors of safety, the calculated wall stress is used as a design basis for thermoplastic materials. Thermoset materials use strain as a design basis. The analysis for thermosets may be extended to a multiple layer stress or strain calculation.

D. A. Gregorka, etal., National Sanitation Foundation (NSF), Health Effects Standard and Certification of Plastics Pipe, addresses the impact NSF standards 14 and 61 on specifiers, users, and designers of plastic piping for potable water systems. Special emphasis is placed in extraction testing and toxicology requirements. NSF Standard 61, covers the health effects of indirect additives to drinking water for all types of piping materials.

In the Design part of the Symposium, the presentations centered on controlled tests on buried pipe, evaluation testing of pipe installed up to 20 years, and measurment of buried pipe deflections.

Amster Howard in Fullerton PVC Pipe Test Section describes a test section of 27-inch poly (vinyl chloride) (PVC) pipe installed in 1987. Initial measurements included pipe deflections, pipe invert elevations, soil properties, and in-place unit weights. Periodic measurments were made during the first two years to establish pipe deflection time-lag factor. The pipe was installed with three different bedding and backfill conditions.

A. P. Moser, etal., ask the question "Is PVC strain limited after all these years?". Over the years that PVC pipe has been used in buried non-pressure applications, a debate has continued over the right way to design products that are stress relaxation life dependant; but are subject to fixed strain and stress relaxation conditions over their useful life. Data from notched and unnotched pipe rings under fixed circumferential deflections of 30% to 40% is included.

Many of the presentations dealt with soil properties, installation techniques, and their effects on service performance. Kennedy, etal., describe the design of undergrund thrust restrained systems for PVC pipe. Direct shear tests were made to study the pipe-to-soil friction. The resulting data were used to formulate design parameters for PVC pipe thrust restrained systems in a wide range of soil types. Selig discussed the basic soil property requirements for basic trench and embankment requirements. He described the characteristics of compacted soils and gave representative stress-strain parameters. Greenwood and Lang introduced empirically-based modifications to the original Spangler approach to abtain a new calculation method for estimating vertical deflection of flexible pipe. These modifications are based on recent research results. Along thee lines, K. G. Leondaris describes several installations in the Middle East. These installations of GRP pipes were in areas of prevailing high temperatures, high and saline ground water tables, and corrosive soils.

Plastic pipes have long life because of their resistance to corrosion and erosion. This makes them attractive for use under long-term landfills and in aggresive environments such as sanitary landfills. R. K. Watkins report on tests at Utah State University on the performance of plastic pipes under high landfills. Plastic pipe can perform under enormous soil loads -- hundreds of feet -- if an envelope of carefully selected soil is carefully placed about the pipe. The creep of plastic materials allows the pipe to relax and so conform with the soil in a mutually supportive pipe-soil interaction.

Lars-Eric Jansen reported on the use of flexible thermoplastic pipes such as polyethylene and polypropylene for submarine outfall systems. These pipes are well suited for this use because they can be extruded in long sections, towed fully equipped with anchoring weights to the outfall site and sunk directly on to a seabed with a minimum of underwater work.

Moore and Selig describe a buckling theory for design of buried plastic pipes which combines linear shell stability theory for the structure with elastic continuum analysis for the assessment of the ground support. The theory provides stability estimates which are superior to those generated using 'spring' models for the soil, predictions of phenomena such as long-wavelength crown buckling without the need to pre-guess the deflected shape, and rational assessment of the influence of shallow cover and the quality and quantity of backfill material. Buckling as a performance limit for buried plastic pipe is discussed, and the selection of appropriate soil and polymer moduli for use in the theory is also considered.

Collins and Svetlik describe techniques to rehabilitate existing piping facilities. Collins reports on the use of centrifugally cast fiberglass pipe to renew reinforced concrete pipe by sliplining. Svetlik describes four generic processes for insert renewal of existing piping systems. These processes are linear expansion, rolldown reduction, hot swage reduction, and visco-elastic reduction.

The goal of the symposium and ASTM STP 1093 is to provide the many constructors of water, sewer, drainage, waste management, irrigation, and gas projects with state of the art engineering data and techniques for the successful use of plastic piping materials. The planning committee thinks we have done this; however, the opinions of the attendees at the symposium and the users of this volume are welcomed and solicited. Comments on needed technology and standards should be relayed to any of the planning committee members.

Planning Committee Members:

Robert Bailey
(513) 226-8706

George S. Buczala
(215) 841-4881

Michael J. Cassady
(614) 424-5568

Jayme Kerr
(215) 299-5518

Robert Morrison
(419) 248-6162

Stanley Mruk
(201) 812-9076

Ernest Selig
(413) 545-2862

Dennis Bauer
(214) 243-3902

George S. Buczala

Michael J. Cassady

Philadelphia Electric Company
2301 Market St.
Philadelphia, PA 19101;
symposium chairman and
editor.

Battelle, Columbus Labs
505 King Ave.
Columbus, OH 43201;
symposium chairman and
editor.

Testing and Standards

Ronald R. Bishop

RETENTION OF PIPE STIFFNESS FOR POLYVINYL CHLORIDE (PVC) PIPE
SAMPLES EXPOSED TO VARIOUS ENVIRONMENTS AND CONSTANT STRAIN

REFERENCE: Bishop, Ronald R., "Retention of Pipe Stiffness for
Polyvinyl Chloride (PVC) Pipe Samples Exposed to Various
Environments and Constant Strain" Buried Plastic Pipe Technol-
ogy, ASTM STP 1093, George S. Buczala and Michael J. Cassady,
Eds., American Society for Testing and Materials, Philadelphia,
PA 1990.

ABSTRACT: The key physical design parameter for flexible buried
non-pressure pipe is Pipe Stiffness. The ever broadening
recognition of the inherent chemical resistance of polyvinyl
chloride (PVC) pipe has led to a wide range of possible new
applications. The effect of environment on PVC and other
plastic pipe is typically measured by weight change or strength
change on unstressed samples exposed to various environments
such as described in ASTM D-1784, Standard Specification for
Rigid Polyvinyl Chloride (PVC) Compounds or by stress or strain
crack resistance. Herein a new method of measuring the long
term pipe stiffness of samples exposed to various environments
is proposed. This is accomplished by monitoring the instantaneous slope of the load deflection curve as a function of
time. Six-inch (150mm) long pipe stiffness samples are tested
to determine the initial stiffness by the method of ASTM D-2412-
87, Standard Test Method for Determination of External Loading
Characteristics of Plastic Pipe by Parallel Plate Loading. The
sample is then clamped in a position of fixed deflection of
either 5% or 7.5%. At time intervals of 1 day, 1 week, 2 weeks,
4 weeks, 8 weeks, 16 weeks and 32 weeks the load increment is
increased to produce an added deflection increment to a total
deflection of 7.5% or 10%. The slope of the load-deflection
curve for this new load increment is calculated for each new
time interval. Data from various PVC pipe samples in environments consisting of 5% sulfuric acid, 5% sodium hydroxide, tap
water and air are presented. Results for up to two years
exposure are included.

KEYWORDS: Polyvinyl Chloride (PVC) Pipe, Sewer Pipe, Pipe
Stiffness, Stress Relaxation, Constant Strain.

*Director of Technical Services, Carlon Division of
 Lamson & Sessions, 25701 Science Park Drive,
 Beachwood, OH 44122.

INTRODUCTION

The key pipe property which defines a plastic pipe's response to a
buried loading condition is pipe stiffness, PS.   This property is
analogous to the stiffness constant of a steel spring and defines
its resistance to applied external forces.   This property is defined
in ASTM D-2412, Standard Test Method for Determination of External
Loading Characteristics of Plastic Pipe by Parallel Plate Loading,
as the load per unit length of conduit required to produce a 5%
deflection (or decrease in dimension of the internal pipe diameter).   The results of this test for typical thermoplastic
materials are dependent upon several test conditions, including:
the rate of loading, temperature of the samples, deflection or
strain level, etc.   For the ASTM test, these parameters are
quantified at a crosshead rate of 0.5 ± 0.02 inches (12.5 ± 0.5mm)
per minute, a temperature of 73.2 ± 3.6°F (23° ± 2°C) and a
deflection of 5% of the inside diameter.   When this test is
conducted by a test apparatus that continuously monitors and records
load and deflection simultaneously, the result is plotted as a load
vs. deflection curve as shown in Figure 1.

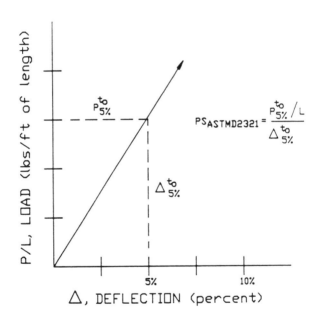

FIGURE 1:   TYPICAL LOAD-DEFLECTION CURVE FOR PVC PIPE

From Figure 1, the ASTM pipe stiffness is defined as:

$$PS = \frac{P_{5\%}/L}{\triangle_{5\%}} \tag{1}$$

Where:

PS = Pipe Stiffness (lb/in^2) (KPa)
$F_{5\%}$ = Load to produce a 5% deflection (pounds) (Newtons)
L = Sample length (inches) (mm)
$\triangle_{5\%}$ = Inside diameter decrease (inches) (mm)

These properties can also be defined as a function of the pipe material property, modulus of elasticity, E as follows:

$$PS = 0.149 \ EI/R^3 \tag{2}$$

Where:

E = Material Modulus of Elasticity (lb/in^2) (KPa)
I = Moment of Inertia of pipe wall cross section per unit length (in^4/in) (mm4/mm)
R = Mean radius (in) (mm)

This pipe stiffness quantity then defines the secant slope between 0% and 5% deflection on the load per unit length vs. deflection curve based on a sample tested in air as produced after temperature conditioning at 73.2°F (23°C). ASTM product standards for plastic pipe used in non-pressure applications define minimum pipe stiffness product requirements based on this test.

The foregoing test and property determinations are restricted to a relatively short term (minutes) duration. Thermoplastic materials are generally very resistant to chemicals at concentrations normally found in domestic sewer and storm drain systems. However, no formal commonly accepted test procedure for determining the influence of common environments on pipe stiffness for extended periods of time are available. Herein, the results of a new proposed test method extending for a period in excess of one year and for environments consisting of air, water, 5% sulfuric acid and 5% sodium hydroxide are reported. This new test method allows determination of a true pipe stiffness (instantaneous slope of the load-deflection curve) at any time increment and is described herein. This test has been applied to three PVC sewer pipes made to ASTM standards with three distinctively different PVC compounds.

TEST PROCEDURE

Pipe Retention Test Procedure

The proposed test procedure for the plastic pipe stiffness retention closely follows the ASTM D-2412 test procedure. The first step is to establish the average sample length by taking four equally spaced measurements taken to the nearest 1/32 inches (0.8mm) (ASTM D-2122, Test Method for Determining Dimensions of Thermoplastic Pipe and Fittings). D2412 requires a length of 6 inches± 1/8 in (150mm).

The second step establishes the average thickness by taking eight equally spaced measurements using a ball anviled micrometer (ASTM D-2122).

The third step measures the average outside diameter by taking four equally spaced measurements using calipers accurate to .001 in. (.025mm), or a vernier circumferential wrap tape (ASTM D-2122).

Fourth, the average inside diameter is established by subtracting the average outside diameter by two times the average thickness. This I.D. will be used throughout the test for computing the percent deflection for all test samples.

Fifth, the testing apparatus is required to be a properly calibrated compression testing machine with a constant rate of crosshead movement. For results reported herein, the testing machine is a MTS 810 Material Testing System. The pipe sample must be compressed at a constant rate of deflection of 0.5 + .02 inches (12.5mm)/min. Loading is applied through two parallel flat loading plates with a length equal to, or exceeding the sample length, and a width not less than the pipe contact width at the maximum deflection. Each pipe sample will have an initial pipe stiffness test recording the load-deflection measurements continuously up to 10% of the original average I.D. The initial pipe stiffness is calculated as per equation (1) above.

Sixth, the pipe samples will be conditioned undeflected in the specified environment for one week.

Seventh, after the one week environmental conditioning the pipe stiffness test is performed (this provides the effects on environment exposed unstrained samples) the pipe samples are secured in the fixture at 5% deflection (Fig. 2).

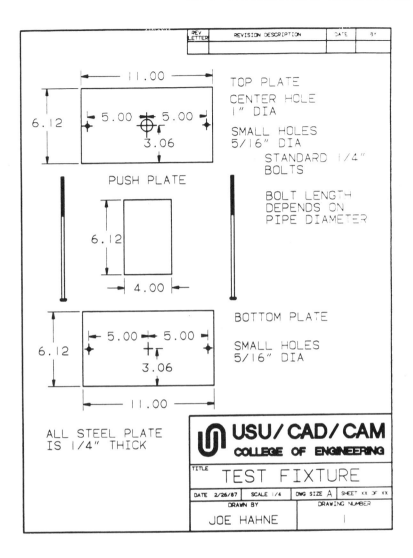

FIGURE 2: TEST FIXTURE DIMENSIONS

Eighth, the deflected pipe samples are returned to the environment and subsequent stiffness tests (based on a 5% through 10% deflection increment) are performed to the following schedule: One day, seven days, fourteen days, twenty-eight days, and every thirty days thereafter for a period of one year or more (Fig. 3).

Ninth, the pipe stiffness and load relaxation of the samples are monitored by extracting the values from the load deflection curve taken during each pipe stiffness test over the time period.

FIGURE 3: PIPE STIFFNESS TEST IN PROCESS

A note about the test environment; certain considerations in testing
pipe in chemical environments must be resolved. The procedures for
handling hazardous materials are clearly established by government
standards and must be strictly followed (Fig. 4). The containment
device should be surrounded by an adequate spillage area, and both
must be resistant to the chemical environment. Set up the environment in a well ventilated area and use proper safety equipment. The
test fixture may be subject to corrosion in the particular environment therefore protection must be provided. In this case, the
sulfuric acid corroded the fixture. The best corrosion protection
found for the sulfuric acid environment was acid resistant vinyl
paint, manufactured by Carlon, reinforced with PVC plastic sheet
glued to the steel plate by contact cement, the edges were caulked
with silicon caulking. The steel rods were painted and placed
inside small diameter PVC tubing, then filled with silicon caulking,
and the ends were heavily caulked with the silicon and allowed to
dry. Alternately, acid resistant stainless steel fixtures could be
used.

FIGURE 4:   SAFETY PRECAUTIONS FOR HANDLING
             ENVIRONMENT EXPOSED SAMPLES

Specific Gravity Determination

Each test was followed by a specific gravity test, and an ignition
"burn out" test.   ASTM D-792, Standard Test Method for Specific
Gravity and Density of Plastics by Displacement describes the
specific gravity test that was used.   An analytic balance and
distilled water at a temperature of 72°F (23°C) were used.   The
sample was cut such that each sample had a dry weight greater than
50 grams; therefore, method A-3 of the standard was used.   First, a
dry weight of the sample was recorded, then a weight of the sample
immersed in the water and suspended from the balance by light wire
and a wire hook.   Specific gravity of the sample is calculated by
the relation.

$$Sp\ gr\ (@\ 23°C) = a/\ (a+w-b) \qquad (3)$$

Where:

a = dry weight (lb) (grams)
b = sample weight (lb) (grams) immersed in the water
w = weight of the wire and hook immersed in the water (lbs) (grams)

Ash Content Determination

The ash content of the sample was determined by the ignition burn out test, described in Sections 54 through 56 of ASTM D-229 Method of Testing Rigid Sheet and Plate Materials used for electrical insulation. The sample is cut into small pieces and dried in an oven for 2 hours at 105° to 110°C. After weighing the sample, it is placed in a crucible and then burned to a constant weight.   The percent ash is based on the ratio of ash weight to dry weight.

DEFINITION OF TERMS

A complete understanding of results obtained from the foregoing test procedure is dependent upon the complete understanding of some basic definitions which are listed below and given in Figure 5.

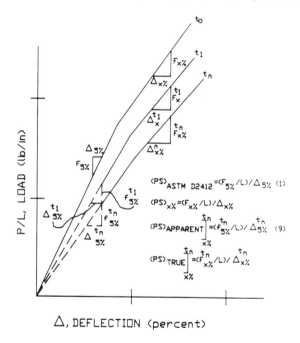

$$(PS)_{ASTM\ D2412} = (F_{5\%}/L)/\Delta_{5\%} \quad (1)$$

$$(PS)_{x\%} = (F_{x\%}/L)/\Delta_{x\%}$$

$$(PS)_{APPARENT}\Big|_{x\%}^{t_n} = (F_{5\%}^{t_n}/L)/\Delta_{5\%}^{t_n} \quad (9)$$

$$(PS)_{TRUE}\Big|_{x\%}^{t_n} = (F_{x\%}^{t_n}/L)/\Delta_{x\%}^{t_n}$$

Δ, DEFLECTION (percent)

FIGURE 5:   TRUE PIPE STIFFNESS VS. APPARENT

The ASTM D-2412 test described in equation (1) above can be written in more complete terms by defining the time frame of the test, the deflection range over which the load and deflection are taken and the environment in which the sample is exposed as follows:

$$(PS_{0-5\%})_{t=0} = \frac{[F_{0-5\%}/L]^{\text{Air}}}{[\triangle\ 0\text{-}5\%]\ t=0} \qquad (4)$$

Note: "Air" labels appear above both numerator terms.

Similarly, additional definitions for stiffness may be described by using different portions of the deflection curve (though for the deflection range indicated the stiffness is nearly linear, the relationship is non-linear) as given in equations (5) and (6) and shown of Figure 5.

$$(PS_{0-10\%})_{t=0} = \frac{F_{0-10\%}/L}{[\triangle\ 0\text{-}10\%]\ t=0} \qquad (5)$$

$$(PS_{7.5-10\%})_{t=0} = \frac{F_{7.5-10\%}/L}{[\triangle\ 7.5\text{-}10\%]\ t=0} \qquad (6)$$

Additionally, the pipe's response to a new increment of load and deflection may be characterized through the newly described procedure. This calls for periodic deflection tests to be made from a fixed deflection level (5% for these tests) to a new deflection level (7-1/2% and 10% for these tests). By this means a new time dependent but true (slope of an in process deflection and load increment) pipe stiffness can be determined by equation (7) and (8).

$$(PS_{5-7.5\%})_{t=n} = \frac{[F_{5-7.5\%}/L]^{\text{envir}}}{[\triangle\ 5\text{-}7.5\%]\ t=n} \qquad (7)$$

$$(PS_{5-10\%})_{t=n} = \frac{[F_{5-10\%}/L]^{\text{envir}}}{[\triangle\ 5\text{-}10\%]\ t=n} \qquad (8)$$

Note: "envir" labels appear above the numerator terms in (7) and (8).

Each of the foregoing defines the true slope of the load-deflection curve and at a particular period of time for the additional deflection increment 5 to 7.5% or 5 to 10%. This is significant because the time dependent properties of plastics are most often described by creep or stress-relaxation constants. These utilize an accumulated strain/deflection combined with a fixed load for creep properties or a decayed load with a fixed deflection for stress relaxation. They are apparent properties which describe a mathematical relationship but do not describe the behavior of the material or its ability to respond to a new load or deflection increment. For the specific case of stress relaxation (fixed deflection or strain) as demonstrated with this new test method, this apparent pipe stiffness is defined as:

$$(PS_{apparent})_{t=n} = \frac{f_{5\%}^{tn}/L}{\triangle\ _{5\%}^{t=n}} \qquad (9)$$

Note: "envir" labels appear above the terms in (9).

This stiffness or slope is determined by the load required to maintain the deflection in the test fixture at time, n. It is shown as a dotted line in Figure 5. The value of this apparent stiffness reduces with time for all materials (plastics, metals at high temperature, etc.) that exhibit visco-elastic properties. It is not an indication that the material is softening or that its ability to withstand a new load increment has been decreased.

TEST PROGRAM

The test program itself consisted of monitoring the results of the pipe stiffness retention series for a time period exceeding one year. PVC sewer pipe samples have been chosen as representative of those currently commercially available through municipal distribution channels. They represent products made by three different domestic pipe manufacturers. Each of the pipe products utilized a distinctly different PVC compound as characterized by its ash content. A basic description of the different samples is given in Table 1. Four different environments were chosen to represent extremes of conditions found in typical domestic sewer system. Air environment has been chosen as a control with water, acid and caustic solutions used for the other cases. All tests were conducted at the laboratory at Utah State University in Logan, Utah under the direction of Dr. Owen K. Shupe and A. P. Moser.

TABLE 1:   TEST PROGRAM DESCRIPTION
PIPE DESCRIPTION

	Pipe A	Pipe B	Pipe C
Manufacturer	A	B	C
ASTM Product Standard	D-3034	F-789	D-3034
SDR (-)	35	_b	35
$t_{min.}$ (in)[a]	.240-.253	.212-.227	.251-.254
t_{ave} (in)[a]	.248-.254	.235-.242	.257-.260
OD (in)	8.390-8.400	8.390-8.400	8.390-8.42
PVC Cell Class (ASTM D-1784)	12454B	12154A	13364B
Specific Gravity (-) (ASTM D-792)	1.41-1.42	1.64-1.66	1.53-1.54
Pipe Stiffness (PSI)[a] (ASTM D-2412)	54.1-60.95	65.7-70.79	44.91-48.34
Ash Content (%) in burnout (ASTM D-229)	14.5-16.3	39.7-41.4	29.7-30.7

[a] Ranges of results are based on measurements of 20 samples for each pipe.

[b] SDR = $OD/t_{min.}$ is not calculated by ASTM F-789.

TEST RESULTS

Pipe Stiffness Retention

Results plotting pipe stiffness ratio defined as the time dependent true stiffness from equation (7) divided by the ASTM D-2412 initial stiffness at the same deflection defined by equation (4) are given in Figure 6 as a function in time. For air, this ratio for each of the pipe samples is nearly the same and has remained constant for the time periods exceeding one year or 10,000 hours.

AIR

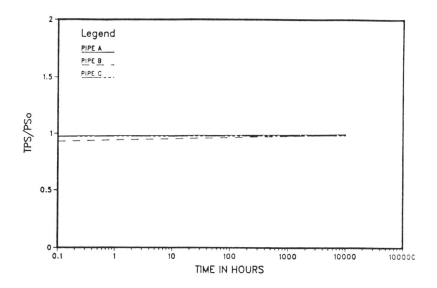

FIGURE 6: The true Pipe Stiffness (TPS) does not significantly change from the Initial Pipe Stiffness (PSo) when subjected to constant strain in the air environment.

The same data for samples exposed to tap water are given in the same format in Figure 7. Again, pipes designated A, B and C perform similarly and note a small decrease in the ratio of initial pipe stiffness to pipe stiffness at 10,000 hours is evident. Results for the sulfuric acid environment are given in Figure 8 and for the sodium hydroxide environment, in Figure 9. Again, for both of these extremely different environments, the slope of the load-deflection curve as shown by the pipe stiffness ratio remains relatively unchanged for the 10,000 hours time duration shown and there is only slight variation among the PVC pipe formulations tested.

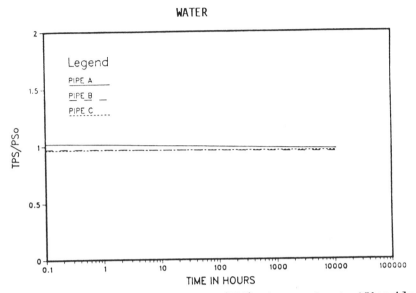

FIGURE 7: The True Pipe Stiffness (TPS) does not significantly change from the Initial Pipe Stiffness (PSo) when subjected to constant strain as demonstrated by the figure in the water environment.

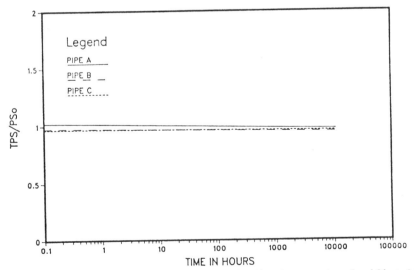

FIGURE 8: The True Pipe Stiffness (TPS) does not significantly change from the Initial Pipe Stiffness (PSo) in the H_2SO_4 environment.

SODIUM HYDROXIDE

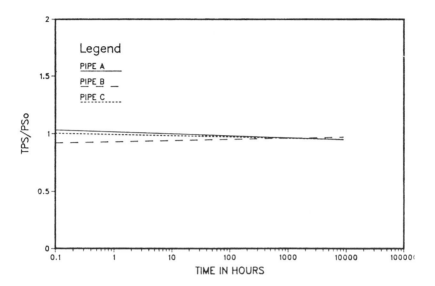

FIGURE 9: The True Pipe Stiffness (TPS) does not change signifi-
            cantly from the Initial Pipe Stiffness in the NaOH
            environment, as shown by the figure.

CONCLUSIONS

A new means for evaluating the plastic pipe property of pipe
stiffness has been presented, which provides several advantages over
that method historically used and described in ASTM D-2412. A
method of simulating the incremental deflection process actually
seen by non-pressure pipe in underground installations has been
presented. This technique provides a laboratory means of (1)
determining a true pipe stiffness at a given time, (2) evaluating
the effects of various environments on the long term stiffness of
plastic pipes and conduits subject to fixed strain conditions and
(3) gathering data on rate of stress-relaxation of various pipe
exposed to environmental conditions.

The specific results of these long term pipe stiffness tests
conducted on a variety of PVC pipe having PVC compound formulations
with ash contents of 15%, 30% and 40% by weight. Samples were made
by three different manufacturers have shown only minor variations in
true slope of the load-deflection curve for the test periods
exceeding one year. Environments of air, water, 5% sulfuric acid
and 5% sodium hydroxide demonstrate the same patterns.

RECOMMENDATIONS

The long term pipe stiffness test method described in this article is useful for screening pipe materials, products and processing to ensure the most critical design property is known in the "long term." Even though no reduction in true stiffness is revealed in these tests, it is suggested that an ASTM standardized test method based on this description be developed as a means of characterizing long term performance of thermoplastic and other pipe for buried non-pressure applications. The suitability of plastic pipe in a number of specific industrial environments will be best evaluated by such long term testing.

Stanley A. Mruk

THE DURABILITY OF POLYETHYLENE PIPING

REFERENCE: Mruk, S. A., "The Durability of Polyethylene Piping", Buried Plastic Pipe Technology, ASTM STP 1093, George S. Buczala and Michael J. Cassady, Eds., American Society for Testing and Materials, Philadephia, 1990.

ABSTRACT: A review of the literature on the physical and chemical aging of polyethylene piping has been undertaken. In typical uses the only types of aging mechanism that may limit polyethylene (PE) pipe's durability are photodegradation, oxidative degradation, and slow crack growth under tensile stressing.

Current generation PE pipe grade materials have superior resistance to these forms of aging. This resistance has been confirmed by accelerated aging testing, by fracture mechanics testing, and by evaluating pipes after long service histories.

Appropriate requirements, test methods, and protocols have been incorporated into product standards to adequately protect commercial grade PE piping against aging over the long service life expected in public utility and similar applications.

KEYWORDS: polyethylene piping, stress-rupture, long-term strength, aging, durability, slow-crack growth.

INTRODUCTION

Polyethylene piping has become a material of choice for many applications that benefit from its unique combination of properties: heat-fusibility; excellent ductility; immunity to corrosion; and very high chemical resistance. Most of the gas distribution piping being installed in the U.S., Canada, and many other coun-

Stanley A. Mruk is the Executive Director of the Plastics Pipe Institute, Wayne Interchange Plaza II, 155 Route 46 West, Wayne, NJ, 07470.

tries is made from PE. This pipe is increasingly used in industry and in municipal applications conveying water, sewer, and waste-water. And a number of methods for the trenchless installation of new and for the rehabilitation of old pipe are based on PE pipe.

The growing acceptance of PE piping for engineering applica-tions has led to an increased interest and need to broaden our understanding of its engineering behavior and durability limits. In particular, designers and users of piping systems intended for long-term service wish to know how and under what conditions the life of PE piping may be limited by some mechanism analogous to corrosion of metal piping. Furthermore, these persons want to also know how to select and to specify an appropriately durable system.

While the vast majority of the PE piping that has been in-stalled--some of it over 25 years ago--has been trouble-free, some field failures have occurred in service as a result of aging. This paper reviews the various potential aging mechanisms for polyethylene and describes the measures that have been taken to ensure that they will not compromise the potential excellent durability of this material.

AGING

The durability of plastics may be limited by either chemical or physical aging. Chemical aging refers to the loss of perfor-mance properties caused by the gradual breakdown of polymer mole-cules into smaller units. Physical aging is the result of gradual adverse change in the physical state and order of a plastic. For a plastic to be durable it must resist both forms of aging under the anticipated service conditions.

Chemical Aging

As with all materials, the molecular structure of polymers may be altered by certain chemical reactions. Being non-conductors, plastics are immune to the galvanic and electrochemical effects which result in the corrosion of metals. Plastics are unaffected by most electrolytes such as acids, bases and salts. They are susceptible only to these chemicals when they are of sufficient concentration or are of such chemical nature to induce other than electrochemical activity. Outstanding corrosion resistance is, of course, a principal reason for the broad acceptance of plastics piping, particularly for underground applications.

Some chemicals, generally strong oxidizing agents such as ozone, nitric acid, sulfuric oxides, and concentrated sulfuric acids, will attack many plastics. Plastics produced by condensa-tion-type polymerization may also be subject to the hydrolitic action of water. A plastic's susceptibility to a particular form of chemical attack primarily depends on the base polymer. It is also affected by the nature of any additives, such as a property modifier which may be blended in with the base polymer, and on the type and quantity of antioxidant, or other additives, that have been incorporated into the plastics composition to protect the base polymer and additives against the fabrication and end use conditions.

All polymers can be chemically degraded by the application of sufficient heat. Heat alone causes thermal degradation. But heat combined with an oxidizing agent, such as the oxygen in ordinary air, can dramatically accelerate oxidative degradation. Moreover, subjecting a polymer melt to excessive shear action can break polymer chains. High shear can be caused by too narrow a clearance between the cylinder walls and the screw of an extruder or of an injection molding machine. For these reasons manufacturers of plastic pipe and fittings use only properly stabilized plastic compounds and maintain very close tolerances on processing conditions, particularly those that affect thermal and mechanical energy history.

Some polymer chains can be broken by micro-organisms, notably those made by nature, like the long cellulosic chains in cotton. Plastics piping is not made from any of these materials.

Radiation can also chemically degrade a polymer just as it does vegetable and animal matter. The only radiation that is of practical concern for most plastic pipe applications is the ultra-violet (UV) segment in sunlight. Over sufficiently long-term exposure to sunlight, unprotected plastics can have their properties adversely affected by photodegradation. Manufacturers of thermoplastic piping for underground service add sufficient levels of UV stabilizers, or UV blocks such as finely divided carbon black or titanium dioxide, to protect pipe and fittings during prolonged outdoor storage--generally, for at least two years. As has been demonstrated by over three decades of outdoor experience with polyethylene jacketed telephone cable, the addition of about 2 percent finely divided carbon black to polyethylene results in more than 25 years of protection against sunlight.

Of all these possible ways of chemical aging that can be encountered in natural gas, sewer and water service, polyethylene is susceptible only to photo, thermal and oxidative degradation. Polyethylene pipe material specifications require that proper levels of effective UV blocks or stabilizers, thermal stabilizers, and antioxidants be added to control these potential chemical aging processes.

A number of studies attest to the excellent chemical-aging resistance of PE when it has been protected against photo, thermal, and oxidative degradation in accordance with the state-of-the-art and as prescribed by current piping standards. In a paper presented at the 1983 American Gas Association (AGA) Distribution Conference, Palermo and DeBlieu [1] reviewed the results of evaluations of buried PE gas pipes with 18 years service at Philadelphia Electric and 20 years at Wisconsin Public Service. In both cases no significant change was observed in any of the physical or performance properties.

Moreover, in a paper presented at the 1985 AGA Distribution Conference, Toll [2] reported on the weathering performance of a colored (i.e. non-carbon black containing) PE gas pipe protected for above-ground storage by the incorporation of a U.V. stabilizer. Two years of outdoor exposure in Florida resulted in no adverse effect on any of the performance properties.

In addition, evaluations conducted on PE cable jacketing and PE sheet have yielded similar results. In a Bell Laboratories study of various PE wire and cable jacketing which had been buried in soil for eight years, it was found that the PE was completely intact as a coating [3]. There were no color changes and no discernible reaction with the metallic conductors and the soil. No significant change in physical properties was observed except in one case where embrittlement and other loss of properties were noted in a PE which did not contain anti-oxidant. Properly protected PE has now been successfully used in this application for over 30 years.

Polyethylene film has become a preferred encasement for the protection against corrosion of buried gray and ductile cast iron pipe. The introduction to AWWA C 105, the American Water Works Association standard for PE encasement, states the following:

> Tests of polyethylene used to protect gray and ductile cast iron pipe have shown that, after 25 years of exposure to severely corrosive soils, strength loss and elongation reduction are insignificant. U.S. Bureau of Reclamation (BUREC) studies of polyethylene film used underground show that tensile strength was nearly constant and elongation was only slightly affected during a seven-year test period. BUREC'S accelerated soil-burial testing (acceleration estimated to be 5 to 10 times that of field conditions) shows polyethylene to be highly resistant to deterioration.

The problems that can be created by the use of an improperly protected PE unfortunately have visited a few water utilities. These utilities experienced many failures in PE service lines which seemed to be linked to an uncharacteristic embrittlement after a few years of service. In a study conducted for the Plastics Pipe Institute, the engineering consulting firm Simpson, Gumpertz & Heger determined that all of these failures were traceable to one defective material [4]. The defect was the lack of sufficient stabilizer: It probably was never added; or if it was, it was consumed during improper extrusion. That this problem could occur was unanticipated for it has been a long standing practice to make durable products, such as cable jacketing, pipe encasement, and pipe and fittings, only from PE material with adequate thermal antioxidant protection. To ensure that the lack of this protection will not recur, appropriate revisions have been made in AWWA and ASTM standards for PE pressure piping. A minimum quality of thermal stabilizer and antioxidant protection is now required not only for the starting material, but also for the finished product. The requirement on the finished product is sufficient to also protect PE against the added thermal exposure during the heat fusion process.

Furthermore, two studies recently reported by the Gas Research Institute (GRI) attest to the long-term durability of PE gas piping. In one study conducted for GRI by Battelle Memorial Institute [5], the chemical and physical properties of different plastic gas pipes installed in 1963 and succeeding years in a service yard of Columbia Gas of Ohio were evaluated. In the other, conducted by L. J. Broutman & Associates [6], more than 40 gas companies submitted samples of different plastic pipes that were removed from service after from one to 20 years operation under very differing soil conditions. The results of both projects, which were conducted independently, show that commercially available PE gas pipe materials did not experience statistically significant changes in their chemical and physical properties for periods of up to 20 years.

Overall, chemical aging of buried polyethylene pipe is not a concern as long as it has been properly stabilized. Current standards adequately cover this requirement.

Physical Aging

Some thermoplastics exist as entanglements of randomly coiled, interpenetrating molecular chains that form a relatively unordered amorphous structure. Examples of amorphous polymers include polyvinyl chloride (PVC), cellulose acetate butyrate (CAB) and acrylonitrile-butadiene-styrene (ABS). Other thermoplastics have a partly crystalline structure consisting of portions of molecular chains that lie beside portions of neighboring chains, thereby forming regular arrays of compact and very well ordered regions. In between these crystalline regions lie amorphous regions of disordered portions of polymer chains. Examples of crystalline polymers are polyethylene (PE), polybutylene (PB), polypropylene (PP), and nylon.

Amorphous polymers are frequently formulated with property modifiers to improve some property, such as impact or flexibility, or with processing aids to facilitate molding or extrusion. If these additives are not carefully selected or are not properly incorporated, they could migrate from or coalesce within the plastic which would cause its properties to revert to their original unmodified state or to suffer in some other regard. This transition is a form of physical aging. Crystalline polymers, such as PE, generally do not contain property modifiers or extrusion aids. Therefore, they do not age by migration of additives.

However, semi-crystalline polymers can very gradually increase in their crystalline order by a process akin to the annealing of

metals. In fact, minor increases in density, which reflect the
degree of crystal order, have been noted in some of the aged PE's
recovered for the previously referenced aging studies. But none of
these changes led to a significant change in any of the physical
and performance properties.

Certain plastics, when subjected for a long time to tensile
stresses substantially lower than those necessary to bring about
short-time rupture, will develop crazes and small cracks which grow
ever so slowly until eventually rupture occurs. This extended
time-scale formation and growth of crazes and cracks is not caused
by any chemical degradation of the polymer; it is the result of
purely mechanical and/or thermal forces. The formation of cracks
is initiated by the action of stress on defects in the plastic.
Crack growth rate is accelerated by stress intensity, by cycling
the stress (fatigue), by elevating the temperature, and often also
by exposure to certain environments. The latter observation has
led to the name "environmental stress-cracking". When no stressing
is present or when it is present below a certain threshold value,
the crack-producing agents have no discernible effect on the
polymer. The sensitivity of a polymer to crack formation and
growth under stress is greatly dependent on molecular structure
parameters, such as molecular weight and the nature and frequency
of polymer branching.

Polyethylene is one of the plastics potentially vulnerable to
reduced durability by the development and growth under tensile
stressing of very slowly propagating slits, or cracks. When this
mechanism--commonly referred to as slow crack growth (SCG)--is in
play, the durability of PE is delimited by the time for the first
slowly growing crack to run through the entire wall thickness of a
product and, thereby, cause failure.

Since the beginning of the industry, the importance of making
pipe from high stress crack resistant polyethylene has been
recognized and heeded. One of the earliest means for evaluating
the stress crack sensitivity of PE is by the use of an environmental stress-crack resistance (ESCR) test, such as ASTM D 1693, "Test
Method for Environmental Stress-Cracking of Ethylene Plastics". In
this test the time to fail by crack growth is greatly accelerated
by subjecting a highly strained (i.e. stressed) specimen to the
combination of an initial flaw (a razor produced notch), elevated
temperature, and a powerful stress-cracking liquid. By empirical
correlation with field performance, minimum ESCR requirements were
established for PE pressure piping materials. The use of high ESCR
materials greatly contributed to the generally very good performance record and the broad acceptance achieved by PE piping.

O'Donoghue, et al., estimate that about 350,000 miles (560,000
km) of plastic pipe are in gas distribution service in the United
States, a substantial portion of which is PE [7]. Much more PE
pipe has been installed through the years for other buried pressure
uses, including water, sewer, and wastewater. The vast majority of
these PE pipe installations have been trouble-free.

However, in some pipes failures have occurred after many years of service through the 'brittle' SCG mechanism. All these failures have been associated with the presence of external forces arising from rock impingement, excessive bending, differential settlement, and other causes. These seem to have acted in concert with internal pressure and residual stresses on defects contained in the pipe or fitting wall. This experience pointed out the need for improved methodology for material selection to ensure that none of the PE materials used for pressure pipe would be susceptible to SCG in properly installed pipe under the conditions typically encountered in gas, water, wastewater, and sewer services.

This objective has been largely accomplished. The new tests and material requirements which have been put in place in the applicable product standards ensure superior durability and reliability of PE piping.

The new methodology has also fostered the development of a new generation of pipe materials with outstanding SCG resistance. The next section reviews these developments and reports on some of the continuing work which promises to simplify future material selection and quality assurance testing for optimum SCG resistance.

HIGH RESISTANCE TO SLOW CRACK GROWTH: KEY TO DURABILITY

One consequence of the viscoelastic nature of thermoplastic materials is that their breaking strength is significantly dependent on duration of loading and temperature. For trouble-free, long-term service the pressure rating of a thermoplastic pipe must be established based on the pipe material's long-term strength under the anticipated service conditions.

In 1961 the Plastics Pipe Institute proposed a new method for forecasting the long-term strength of thermoplastic pressure pipe materials. Soon after industry adopted this method to stress rate its materials. In 1967, after the addition of some refinements, ASTM adopted the PPI proposal as D 2837, "Standard Method for Obtaining Hydrostatic Design Basis (HDB) for Thermoplastic Pipe Materials". This ASTM method, which has undergone a number of additional refinements through the years, is the backbone of the successful field performance history that has been achieved for over the past 30 years with all the major thermoplastic pipes, including PVC as well as PE.

Referring to Fig. 1, method D 2837 establishes a pipe material's hydrostatic design basis (HDB) by essentially the following
steps:

1. Hoop stress versus time-to-fail data covering a time span from
   about 10 to at least 10,000 hours are developed by conducting
   sustained pressure tests on pipe specimens made from the material under evaluation. The required test procedure is ASTM
   method D 1598, "Time-to-Failure of Plastic Pipe Under Constant
   Internal Pressure". The test is conducted under specified
   conditions of external and internal environment (usually water,
   air, or natural gas inside and outside the pipe) and temperature (generally 73°F (23°C) for ambient temperature design);

2. The resultant data are plotted on log hoop stress versus log
   time-to-fail coordinates, and the 'best-fit straight line'
   running through these points is determined by the method of
   least squares;

3. Provided the data meet certain tests for quality of correlation, the least squares line is extrapolated mathematically to
   the 100,000 hour intercept. The hoop stress value at this
   intercept is called the long-term hydrostatic strength (LTHS);

4. Depending on its LTHS, a material is categorized into one of a
   finite number of HDB categories. For example, if a material
   has an LTHS between 1,200 and 1,520 psi (8.27 and 10.48 MPa),
   it is assigned to the 1,250 (8.62 MPa) psi HDB category. If
   its LTHS is between 1,530 and 1,910 (10.55 and 13.17 MPa)
   psi, it is placed in the next higher HDB category, 1600 psi

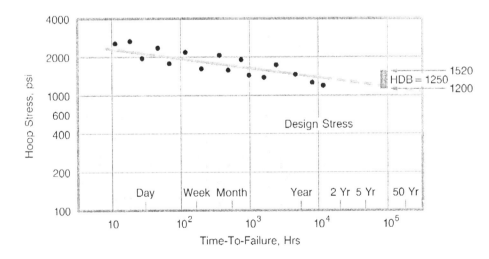

Figure 1 – The forecasting of the hydrostatic design basis
in accordance with ASTM D 2837.

(11.03 MPa). By the D 2837 system, the value of each higher HDB category is 25 percent above the preceding one. This preferred number categorization was selected to reduce the number of material strength categories and, thereby, simplify pressure rating standardization.

The HDB is the accepted basis for pressure rating thermoplastic pipe. To safely pressure rate pipe, a hydrostatic design stress (HDS) is established from the HDB by multiplying the latter by an appropriate design factor (DF). In selecting the DF, due consideration is given to all variables in piping design and installation that result in other than pressure generated stresses. Consideration is also given to those variables that affect the material's capacity to safely resist these stresses. The convention in the U.S. is to use a DF of 0.5 or less for thermoplastics pipe. Smaller design factors than 0.5 (that is, larger 'safety' factors than 2) are specified for certain applications. For example, the Federal Code of Regulations mandates a design factor of 0.32 for thermoplastic pipe in natural gas distribution.

In applying ASTM D 2837 to forecast a material's HDB the fundamental assumption of this method must be kept in mind; that is, the straight stress versus time-to-fail line depicted by the first 10,000 hours of loading will continue through at least 100,000 hours. If it does not, and if the departure from linearity takes a steep downturn around the end of the D 2837 mandated test period of 10,000 hours, then method D 2837 will yield an overestimate of a material's actual long-term strength (See. Fig. 2). In such cases the design factor may not be adequate to offset the unanticipated downturn in strength, and failure could occur after considerably less time than projected. This scenario is believed to have happened in the few cases of premature service failures with certain polyethylene pipes.

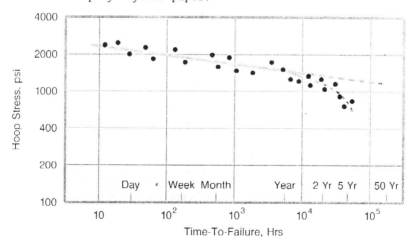

Figure 2 – Illustration of the possibility that ASTM D 2837 may overestimate the actual long-term strength in a case where downturning begins at, or after, the mandated 10,000 hour test period.

To be sure, the assumption that when a stress versus time-to-fail line is straight for 10,000 hours it will continue to be straight considerably beyond this period has proved valid for most pipe grade thermoplastics, including PVC and CPVC. However, as demonstrated by some field experience, D 2837 left an open *window* to certain poorer long-term performing PE materials.

About 10 years ago, two essentially different approaches were undertaken in the United States to close this *window*. The Plastics Pipe Institute (PPI) undertook the evaluation of elevated tempera-ture pressure testing as a means of determining in considerably shorter time a PE material's resistance to a downturning in its long-term strength at ambient temperature. Having noted that pre-mature field failures in PE invariably occur through the brittle-like SCG mechanism, the Gas Research Institute (GRI) initiated a series of research investigations directed at elucidating the fracture-mechanics principles behind this form of failure. Both avenues of exploration have yielded very fruitful and practical information.

Fracture Mechanics Evaluation

Along with other work done in the U.S. and abroad, the GRI work indicates that the SCG brittle-like failure process occurs in two stages [8]. First is crack initiation. In this stage a sustained tensile stress induces a micro-damage zone around an included flaw or degraded polymer or other defect which acts as a stress intensifier. This zone slowly grows until it attains a critical crack dimension at which very slow stable crack growth commences. Increasing flaw size has been noted to decrease initiation time. The period required for crack initiation can sometimes be substantial--times as long as one-half the time for complete failure by SCG have been noted [9].

The second stage is the propagation of the slow-moving, stable crack through the pipe or fitting wall. The crack grows in a direction perpendicular to the maximum tensile stress. Crack growth rate is raised by the increase of the applied stress or temperature. Polymer molecular weight and other molecular struc-tural parameters, such as type and frequency of branches exert a powerful influence on the time for initiation and on the rate of growth of slow moving cracks.

The analytical evaluation of the kinetics of slowly propagat-ing cracks has been fraught with a number of challenges, among which are:

- In some PE piping materials crack initiation and SCG at ambient temperatures proceed at extremely slow rates making their study not only difficult but very time consuming;

- Rates may be accelerated by increasing temperature or stress, but care has to be exercised not to change the fundamental mechanism being evaluated. For example, high

stress can induce blunting at the tip of a crack which
dramatically slows its growth rate;

- The samples being evaluated should adequately represent the
extrusion, molding, and other fabrication variables
including the interface in heat-fused joints that can exist
in installed piping systems.

Notwithstanding these and other challenges, several SCG tests
have been developed which are useful for evaluating the SCG resistance of PE materials. One of the most suitable is the three-point
bend SCG test developed for GRI by Battelle [7]. In this test (See
Fig. 3) a 120° sector of a pipe section is centrally notched to a
specified depth and placed in a three-point bending configuration.
A predetermined load is applied and the crack growth is measured
with a calibrated microscope. The measured crack depth is then
plotted against time. These data are then interpreted on the basis
of linear elastic fracture mechanics (LEFM) principles. At the heart
of LEFM methodology is the following basic relationship:

$$\frac{da}{dt} = AK^m$$

where: a = crack length
        K = stress intensity factor (depends on the stress and
            the geometry of the flaw)
        t = time
        A and m = material constants

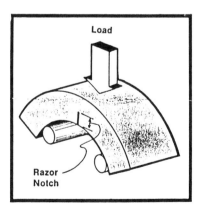

Figure 3 – Schematic of the loading configuration for a
pipe section in the slow crack growth test.

By analyzing SCG data obtained on certain older PE materials
O'Donoghue, et al. derived their LEFM constants [7]. Using this
information, the authors applied the above relationship to forecast
the failure time for pipes made of these materials when exposed to

a particular combination of initial damage and stress intensification conditions which had been observed to result in premature
failure in actual service. Because of the presence of service
induced damage, the extra time for crack initiation could be
discounted. Under this condition, the authors obtained reasonably
good agreement between the predicted and the actual failure times.

However, this reference also reports that some of the new
generation PE materials, which began to displace the older materials in the early eighties, exhibit such tenacity against SCG that
they do not lend themselves to LEFM principles and computational
methods. To interpret SCG data for these materials, damage (failure)
methods beyond LEFM remain to be formulated and validated. Nevertheless, this fracture mechanics approach produces slow crack growth
data that are very useful for evaluating and comparing PE pipe
materials. But the most significant finding is that fracture
mechanics studies attest that the current generation PE's are highly
resistant to SCG.

Elevated Temperature Pressure Testing

Parallel to the work in fracture mechanics PPI undertook an
exploration to determine if the long-term ambient temperature
strength properties of PE could be quantitatively forecasted from
shorter-term elevated temperature pressure testing of pipe. It has
been recognized for some time that PE pipe under long-term pressure
tests can fail by one of three distinct failure modes (See Fig. 4):

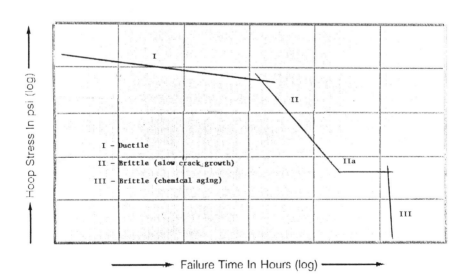

Figure 4 - Potential modes of failure of PE pipe under long-
term pressure testing.

- The first (Mode I) is a ductile failure whereby the specimen ruptures as a consequence of a mechanism initiated by the gross yielding at some location, usually in a region of maximum tensile stress. A decreased test pressure results in a decreased tensile stress which inhibits circumferential creep and delays yielding [10]. No irreversible structural damage occurs prior to yielding.

- The second (Mode II) is by the slowly propagating slit mechanism which has been observed under service conditions and which is the subject of the fracture mechanics studies previously discussed. In the SCG zone, the failure stress regresses much more rapidly with time under load than in the ductile mode--hence the label the downturn region. Near the intersection of Modes I and II, commonly called the 'knee', there is competition between failure mechanisms which is denoted by a scatter of failures by both modes.

- Below a certain threshold stress (denoted by IIa), there is insufficient energy to initiate and propagate a slit.

- The third (Mode III) is the result of the chemical aging or molecular breakdown of the polymer. The regression of strength with time under load occurs most rapidly after a polymer has sufficiently degraded.

Because ductile failures (Mode I) occur by gross yielding, they are relatively unaffected by very localized stress concentrations which tend to be relieved through deformation. The brittle-like slit failures (Mode II), on the other hand, are initiated and propagated in response to the maximum net stress, including the effect of localized stress risers. Accordingly, design for ductile materials is based on average stress; but for brittle behaving materials, design must consider the maximum tensile stress at any point along the pipe that could be generated by all potential loads.

Thermoplastics piping design presumes ductile behavior. Clearly, if a pipe material's durability is delimited by its "brittle" strength, ductile design may result in premature failure. As previously pointed out, most in-service failures of PE piping systems have been by the brittle-like SCG mode and are in response to localized stress concentrations. Ductile or chemical-aging failures are very rare events. The inference is clear: For maximum durability PE piping materials should be so selected and used to ensure that over their entire design lifetime they will retain their ductile quality under the anticipated service conditions. In other words, optimum durability is attained by precluding failure by Modes II and III.

As previously pointed out, suitable protection against chemical aging (Mode III) is effected through proper polymer stabilization. To exclude from pressure pipe applications those materials with inadequate resistance to SCG, PPI proposed in 1985 that D 2837 be only used for the forecasting of a PE's ambient temperature long-term strength when independent pressure tests at

elevated temperatures validate the inherent assumption of D 2837:
That the ductile performance exhibited by the first 10,000 hours of
required testing shall continue through the extrapolation period up
to at least 100,000 hours. PE materials that do not validate would
be excluded from long-term pressure service.

The fundamental relationship behind the PPI proposed validation procedure is the following activated rate-process equation
which has been found to relate the effects of temperature and
stress on the SCG rate of many solids, including certain forms of
silver, platinum, zinc, aluminum as well as various plastics [11]:

$$t = Ae^{-\frac{U}{KT}}$$

where   t = time-to-fail under load
         A = constant
         U = activation energy for the SCG mechanism (a
             function of stress)
         K = Boltzman's constant
         T = absolute temperature

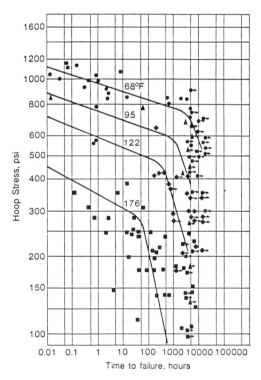

← Arrows indicate test in progress.

Figure 5 – Effect of increasing test temperature on the time
          required to fail by the brittle slow–crack growth
          mode.

The possibility that rate-process principles may also be
applicable to the evaluation of SCG failures in PE pipe was suggested by the observation that in pipe pressure testing elevating
the test temperature greatly reduces the time required to reach the
SCG, brittle-like failure region (Fig. 5). An evaluation of many
sets of such elevated temperature data has shown that in this
brittle region and at a condition of constant stress, the log
time-to-fail is directly proportional to the reciprocal of the
absolute temperature [12].

Based on further evaluation of such data, PPI determined that
the following equation, which was derived from rate process theory,
gave the best general correlation between stress, temperature, and
time-to-fail in the SCG mode [13]:

$$\log t = A + \frac{B}{T} + \frac{C}{T} \log S$$

where   t = time-to-fail
        T = absolute temperature
        S = hoop stress
   A,B,C = coefficients

Based on this rate process method (RPM) equation, PPI adopted
a test method for validating the use of ASTM D 2837 [14]. With
reference to Fig. 6, this method is as follows:

1.   The log stress versus log time-to-fail line for the <u>ductile</u>
     failure zone for 23°C is established (line aa^1) by applying
     method D 2837 on stress rupture data collected through 10,000
     hours;

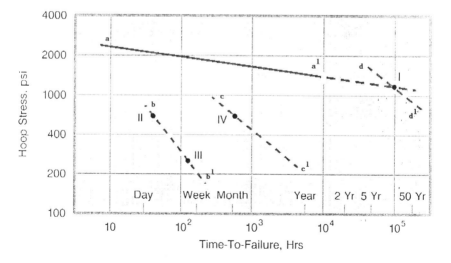

Figure 6 – Methodology adopted by ASTM D 2837 to validate the
assumption of the continuance through at least
100,000 hours of the ductile line established by
the data obtained through 10,000 hours.

2. Line aa^1 is extrapolated in accordance with D 2837 to yield the LTHS, the ductile strength intercept at 100,000 hours (point I);

3. The line for brittle-like failure at some higher temperature, for example 90°C, is determined (line bb^1) by conducting a series of tests at two stress levels and by averaging the log-failure times for each level (points II and III);

4. To test the assumption inherent to D 2837, namely that line aa^1 continues straight from 10,000 through at least 100,000 hours to point I, we make the hypothesis that, as a worst case, point I marks the location at which ductile failure line aa^1 downturns into brittle failure line dd^1;

5. This hypothesis is tested by obtaining brittle data at the same stress as for point II, but at some intermediate temperature, for example 75°C. The average of the log time-to-fail for these tests (point IV) is then compared to that predicted for this stress and temperature by the RPM equation in which coefficients A, B, and C are calculated from points I, II, and III;

6. If the experimental log time-to-fail for point IV equals or exceeds that predicted by the RPM equation, the above hypothesis is considered validated and method ASTM D 2837 may be used to calculate the HDB for 23°C. If the experimental log time-to-fail for point IV is less than predicted, the use of D 2837 is disqualified, and the material is not considered adequate for pressure pipe.

To test the efficacy of this proposed validation test requirement, PPI applied it on elevated temperature data obtained on six pipe materials with varying histories of field experience. The materials that exhibited some problem with premature SCG failures did not validate. Those that did validate had as long as 20 years of satisfactory field performance [14].

In 1988 this validation requirement was added to ASTM D 2837. For the benefit of the user and industry, PPI publishes a listing of the HDB's of all thermoplastic pipe materials which have been established in accordance with ASTM D 2837 and the additional requirements in PPI TR-3, "Policies and Procedures for Developing Recommended Hydrostatic Strengths and Design Stresses for Thermoplastic Pipe Materials". Since January 1986, the only PE's that have been included in this list are those that satisfy the above validation requirement.

By the adoption of the validation requirement the *window* in D 2837, which allowed the selection of PE materials with less than satisfactory resistance to SCG, was closed. As demonstrated by fracture mechanics, elevated temperature, and other testing, current generation PE materials have outstanding resistance to SCG.

Q. C. TESTING FOR SLOW CRACK GROWTH RESISTANCE

Assurance of excellent long-term durability requires more than just selecting an appropriate quality material. One of the most important benefits of the research that has been conducted on PE pipe materials is its reduction to better and faster quality control and quality assurance testing. For example, all current ASTM standards for PE pipe include elevated temperature require-ments to test for minimum SCG resistance in the finished product. In addition, the AWWA standards for PE pressure pipe have also been updated to include this and other tests that help ensure superior chemical aging and SCG resistance [15]. Additional, more effective Q. C. tests are under consideration by ASTM, AWWA and other standardization bodies. A discussion of these tests is beyond the scope of this report.

CONCLUDING REMARKS

Several types of chemical and physical aging can occur in polymers. For polyethylene the only type of chemical aging that could be encountered in gas, water, sewer, wastewater, and similar service are by photodegradation (when exposed to sunlight) and oxidative degradation. Current standards require that PE piping system materials include sufficient ultraviolet stabilizers and antioxidants to prevent, or delay these forms of aging. The adequacy of this protection has been documented by evaluation of pipes with long-term service history.

The only type of physical aging to which PE compositions may be susceptible is the formation and very slow growth of cracks when PE is subjected to tension for very long periods. The extent of susceptibility to this form of stress-cracking is very much materi-al dependent. The current generation pipe grade PE's have out-standing resistance to SCG. New requirements have recently been added to standards which adequately ensure that pressure PE pipe and fittings are made only from such materials. In addition, elevated temperature pressure tests have been added to PE standards to confirm that the finished product has the anticipated resistance to SCG.

No equivalent theoretically based protocol is yet in effect for establishing the SCG resistance requirements for PE materials used for non-pressure buried applications. As SCG only occurs in response to net tensile forces, it is less likely to afflict buried non-pressure pipe which, generally, is subject to net compressive stresses. However, even in non-pressure pipe tensile stresses may be generated by pipe bending, by diametrical deformation, by rock impingement, and by other loads. But unlike the case of pressure pipe where the tensile stress generated by internal pressure is persistently present, the tensile stresses induced by pipe reaction to external loads tend, generally, to decrease because of stress relaxation. However, since this relaxation is never complete, there will always remain some level of tensile stress even in a non-pressure pipe. Because of this, some standards for non-pressure PE pipe take the very prudent course of requiring that the pipe be made only from ASTM D 2837 rated materials.

Other standards rely on minimum stress crack resistance (ESCR) requirements for specifying PE materials with suitable resistance to SCG. The minimum acceptable levels for ESCR have been empirically established based on field experience. The progress that has been made in the understanding of the fracture mechanics behavior of PE materials will be most useful for the future development of more theoretically based criteria for the SCG resistance requirements for PE materials for non-pressure applications.

REFERENCES

[1]   Palermo, E. F. and Ivan DeBlieu, "Aging of Polyethylene Pipe in Gas Distribution Service," presented at the 1983 American Gas Association, Operating Section, Distribution Conference.

[2]   Toll, K. G., "Polyethylene Piping Systems Gas Distribution Service Retention of Performance Properties," presented at the 1985 American Gas Association, Operating Section, Distribution Conference.

[3]   De Coste, J. B., "Effects of Soil Burial Exposure on the Properties of Plastics for Wire and Cable," The Bell System Technical Journal, Vol. 51, No. 1, 63 (1972).

[4]   Chambers, R. E., Performance of Polyolefin Plastic Pipe and Tubing in the Water Service Application, a report prepared for the Plastics Pipe Institute by Simpson, Gumpertz & Heger, Consulting Engineers (1984).

[5]   Battelle-Columbus Division, Material Property Changes in Aged Plastic Gas Pipe, Gas Research Institute Report, GRI-88/0286, 1989.

[6]   L. J. Broutman & Associates, Aging of Plastics Pipe Used for Gas Distribution, Gas Research Institute Report, GRI-88/0285, 1989.

[7]   O'Donoghue, R. S., Kanninen, M. F. and Mamoun, M. M., "Predicting the Useful Service Life of PE Gas Pipes by the Use of Recent Technology Advances," presented at the 1989 American Gas Association, Operating Section, Distribution Conference.

[8]   Bowman, J., "Can Dynamic Fatigue Loading Be a Valuable Tool to Assess MDPE Pipe System Quality?," presented at the American Gas Association Eleventh Plastic Fuel Gas Pipe Symposium, 1989.

[9]   X. Lu and Brown, N., "Predicting Failure in Gas Pipe Resin," presented at the American Gas Association Eleventh Plastic Fuel Gas Pipe Symposium, 1989.

[10] Mruk, S., "The Ductile Failure of Polyethylene Pipe," SPE Journal, Vol. 19, No. 1, January, 1963.

[11] Bartenev, G.M. and Zuyev, V. S., "Strength and Failure of Viscoelastic Materials," First English Translation, Pergamon Press, Oxford, 1968.

[12] Szpak, E. and Rice, F. G., "A Procedure for Confirming the
      ASTM Extrapolation of the Strength Regression of
      Polyethylene Pipe," presented at the American Gas
      Association Sixth Plastic Fuel Gas Pipe Symposium, 1978.
[13] Palermo, E. F. and DeBlieu, I. K., "Rate Process Concepts
      Applied to Hydrostatically Rating Polyethylene Pipe,"
      American Gas Association Ninth Plastic Fuel Gas Pipe
      Symposium, 1985.
[14] Mruk, S., "Validating the Hydrostatic Design Basis of PE
      Piping Materials," American Gas Association Ninth Plastic
      Fuel Gas Pipe Symposium, 1985.
[15] Chambers, R. E., "Standards Herald Wider Role for Plastic
      Water Pipe," American City and County, November, 1989.

Lee E. Pearson

FIBERGLASS PIPE PRODUCT STANDARDS - AN UPDATE ON PERFORMANCE

REFERENCE: Pearson, L. E., "Fiberglass Pipe Product
Standards - An Update on Performance", Buried Plastic
Pipe Technology, ASTM STP 1093, George S. Buczala and
Michael J. Cassady, Eds., American Society for Testing
and Materials, Philadephia, 1990.

ABSTRACT: Widespread use of a piping product will only
be achieved when there are detailed comprehensive
product performance standards that can be confidently
utilized by the specifying engineer. These standards
should establish the performance criteria and
requirements so that products made by different
processes and materials can be validly evaluated and
compared on the most important basis -- performance.

Fiberglass pipe product standards issued by ASTM and the
American Water Works Association have undergone
substantial changes over the past few years. Among
these significant changes have been the expansion to
multiple stiffness ranges, a 50 year design basis
criteria, increased deflection to crack- damage
requirements, establishment of a long term ring bending
strength test method, and updating and modification of
test methods and performance criteria for strain
corrosion and hydrostatic design basis. A major
revision of the design appendix of AWWA C950 has been
completed. AWWA and ASTM product standard requirements
have been made consistent.

This review of fiberglass pipe product standards, the
major revisions, and the upgrading of performance
requirements will demonstrate that fiberglass pipe can
be confidently evaluated, specified and used in a wide
variety of applications.

KEYWORDS: fiberglass pipe, performance standards, pipe
design, shape factors, combined loading.

Lee E. Pearson is Manager, Product Design & Process
Technology, International Pipe Operations, at Owens-Corning
Fiberglas Corporation, Fiberglas Tower, Toledo, Ohio 43659.

FIBERGLASS PIPE PRODUCT STANDARDS

Is fiberglass pipe RTRP (reinforced thermosetting resin pipe) or RPMP (reinforced plastic mortar pipe)? Are the thermosetting resins used epoxy, polyester, or vinyl ester? Do you produce fiberglass pipe by filament winding or centrifugal casting?

All are fiberglass pipe -- a very diverse and versatile class of engineering materials. Because of the variety in materials and manufacturing processes possible with fiberglass pipe, it is most important to have standards that address performance, the proper basis on which to compare products.

There are four key fiberglass pipe product standards that address performance regardless of material, process, or diameter. These are:

. ASTM D3262-88, Standard Specification for "Fiberglass" (Glass-Fiber- Reinforced Thermosetting-Resin) Sewer Pipe.
. ASTM D3517-88, Standard Specification for "Fiberglass" (Glass-Fiber- Reinforced Thermosetting-Resin) Pressure Pipe.
. ASTM D3754-88, Standard Specification for "Fiberglass" (Glass-Fiber- Reinforced Thermosetting-Resin) Sewer and Industrial Pressure Pipe.
. ANSI/AWWA C950-88, AWWA Standard for Fiberglass Pressure Pipe.

Each of these standards has recently been substantially revised, with design basis and product performance requirements strengthened. Comparable requirements have been made consistent. There are also a number of other significant ASTM fiberglass pipe standards and fiberglass pipe test methods and practices that have undergone revision and are continually being reviewed and revised based on current state-of-the-art. The space limitation of this paper does not allow addressing all of these supporting documents. However, a complete listing of ASTM fiberglass pipe standards is given in Appendix A.

In addition to performance requirements, AWWA C950-88 includes a very comprehensive design appendix for fiberglass pipe. A number of very significant changes in design philosophy and design approach have been introduced in a major revision of this appendix. These important design changes will be reviewed.

All of these standards are primarily directed at pipes used in buried municipal water and sewer applications. These pipes, however, are also used in other installations, such as slip lining, pipeline rehabilitation, and aboveground.

Product Range - Requirements

The range of fiberglass pipe products addressed by these four product standards is quite large. A listing of the classification, designation, and performance requirements includes consideration of the following:

. Types: RTRP, RPMP
            filament wound, centrifugally cast

. Liner: reinforced, unreinforced

. Grade: polyester, epoxy (note that for standardization
            purposes, vinyl esters are considered polyesters)

. Pressure Classes: gravity, 50, 75, 100, 125, 150, 175, 200,
                     225, 250 psi (90, 135, 180, 225, 270, 315,
                     360, 405, 450 kPa)

. Pipe Stiffness: 9, 18, 36, 72 psi (62, 124, 248, 496 kPa)

. Workmanship: visual inspection criteria

. Diameters: 1" to 144" (25 mm to 3658 mm)

. Lengths: 10, 20, 30, 40, 60 ft. (3.05, 6.10, 9.15, 12.19 and
            18.29 m)

. Wall Thickness: minimum - average and single point

. End Squareness: ± 1/4" (± 6 mm) or 0.5% diameter

. Strain Corrosion: long term - 50 year
                     annual control

. Hydrostatic Leak Testing: all pipe to 54" (1372 mm) to
                             twice rated pressure

. Hydrostatic Design Basis: long term - 50 year
                             reconfirmation

. Deflection: Level A - No damage
               Level B - No failure

. Hoop Tensile Load Capacity: minimum levels

. Axial Tensile Load Capacity: minimum levels

. Joint Tightness: twice rated pressure
                    type of joint

. Long Term Ring Bending Strength: long term - 50 year

This paper will focus on several of the requirements which have
undergone substantial change and strengthening in the areas which
relate to long term design and performance -- pipe stiffness,
deflection to damage/ failure, strain corrosion, hydrostatic design
basis, and long term ring bending.

Pipe Stiffness

Early versions of these standards were based on a single pipe stiffness category, reflecting general product usage at the time--10 psi. As products of higher stiffness became used and specified, in poor soil areas for example, accommodation of higher stiffnesses was necessary.

Standardized pipe stiffness classes were established, each giving a doubling of stiffness over the previous category to reflect recognizable changes in product performance. Standard pipe stiffness classes are 9, 18, 36 and 72 psi (62, 124, 248, 496 kPa).

In addition to meeting the specified stiffness category, fiberglass pipes must now pass increased deflection without damage (Level A) and deflection without structural failure (Level B) requirements. These levels were established so that all pipes, regardless of stiffness, are required to exhibit a minimum level of strain tolerance (approximately 0.8% for Level A and 1.2% for Level B) consistent with practical handling and installation considerations. These requirements are shown in Table 1.

TABLE 1 -- Deflection Without Damage/Failure Requirements

Pipe Stiffness psi (kPa)	Deflection Without Damage (Level A)	Deflection Without Failure (Level B)
9 (62)	18%	30%
18 (124)	15%	25%
36 (248)	12%	20%
72 (496)	9%	15%

Strain Corrosion

While fiberglass pipe is an inherently very corrosion resistant material, it is not totally immune to corrosive attack under strained conditions. In sanitary sewer applications, it is possible to generate sulfuric acid droplets on the pipe crown. The effect of such acid is more severe at high strain levels. Fiberglass sewer pipes (D3262 and D3754) must demonstrate long term resistance to the possible acid strain corrosion effects of 1.0 N (5%) sulfuric acid. This concentration is representative of the most adverse conditions found in sanitary sewers.[1] In such an environment, conventional materials are rapidly deteriorated.

To establish the proper strain corrosion performance requirements for a broad range of stiffness classes, it was necessary to examine the influence of pipe stiffness on pipe behavior. Of primary interest is the strain level induced in a pipe when deflected in a buried condition. The general expression is:

$$\varepsilon_b = Df\ (T/D)\ \left(\frac{\Delta D}{D}\right)$$

where:

ε_b = bending strain, in/in (mm/mm)

T = total wall thickness, in. (mm)

D = diameter, in. (mm)

ΔD = vertical deflection, in. (mm)

Df = shape factor (dimensionless)

The shape factor (Df) has been emirically found to be a function of pipe deflection level and stiffness as well as the installation (for example, backfill material and density, compaction method, haunching, trench configuration, native soil, and vertical loading).[2, 3, 4]   While the Df value will be larger at low pipe deflections, the product of Df and deflection (strain) will be highest at higher deflections. Therefore, assuming, conservatively, that installations are achieved by tamped compaction with inconsistent haunching, and that long term deflections are in the order of 5%, the following values of Df were selected as realistic, representative, and limiting for the establishment of standardized performance requirements.

TABLE 2 -- Df, Tamped compacted sand

PIPE STIFFNESS psi (kPa)	Df
9 (62)	8.0
18 (124)	6.5
36 (248)	5.5
72 (496)	4.5

With Df defined as a function of stiffness class and with the common acceptance that pipes should be capable of 5% long term (50 year) deflection, the maximum installed bending strain is expressed as:

$$\varepsilon_b\ max\ =\ 0.05\ (Df)\ (T/D)$$

Using a long term safety factor of 1.50, the minimum 50 year strain corrosion performance must be:

$$\varepsilon_{scv}\ \geq\ 0.075\ (Df)\ (T/D)$$

The minimum strain corrosion performance levels for the various pipe stiffness categories are then:

TABLE 3 -- Minimum strain corrosion performance

Pipe Stiffnes psi (kPa)	Minimum 50 Year εscv Performance
9 (62)	0.60 (t/D)
18 (124)	0.49 (t/D)
36 (248)	0.41 (t/D)
72 (496)	0.34 (t/D)

Strain corrosion performance is established by conducting a series of long term deflected corrosion tests with 5% sulfuric acid, according to ASTM D3681 - Test Method for Chemical Resistance of Fiberglass Pipe in a Deflected Condition.(Figure 1)   A minimum of 18 tests is required.

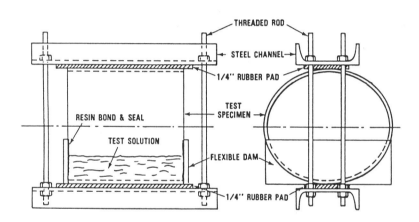

Figure 1 - Strain Corrosion Test Apparatus

Test data (tests must be conducted beyond 10,000 hours) are extrapolated to 50 years to establish long term strain corrosion performance.(Figure 2)   Previously, long term performance was defined at 100,000 hours.   The data are statistically analyzed for acceptability.   An alternative test procedure allows establishment of a minimum 50 year strain corrosion performance value even for products where actual test failures are difficult to obtain.   Again, a minimum of 18 samples must be tested and data must be generated to over 10,000 hours.

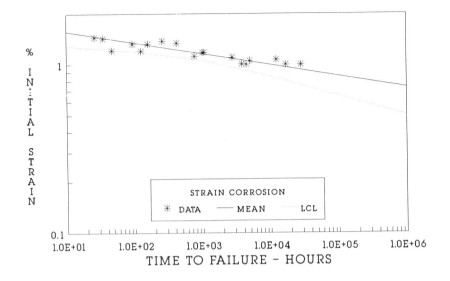

TIME TO FAILURE - HOURS

Figure 2 - Strain Corrosion Data Analysis

Because of the long term nature of the strain corrosion test, it is a qualification test -- that is, a test used to prove product performance. However, standards D3262 and D3754 also include requirements for annual control or reconfirmation tests to demonstrate continued product compliance. Test Method D3681 has been revised to include standard reconfirmation testing and analysis procedures.

Hydrostatic Design Basis

Fiberglass pressure pipes (D3517, D3754 and ANSI/AWWA C950) require establishment of a long term hydrostatic design basis in accordance with ASTM D2992 - Standard Practice for Obtaining Hydrostatic or Pressure Design Basis for "Fiberglass" (Glass-Fiber-Re-inforced Thermosetting-Resin) Pipe and Fittings. This qualification test requirement involves the testing of at least 18 samples to over 10,000 hours and extrapolating the results to determine long term hydrostatic design stress or strain from which pressure ratings may be established.

The long term rating point for hydrostatic design basis, like strain corrosion, has been increased from 100,000 hours to 50 years reflecting a more typical project life.

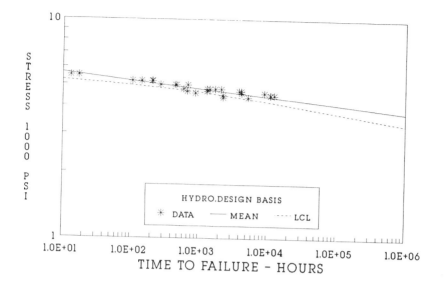

Figure 3 - HDB Data Analysis

Figure 3 shows a typical set of HDB data and the analysis required to establish a 50 year HDB. This example analysis is on a
stress basis, but it could also be done on the strain basis that is
becoming common practice, as it more lends itself to a range of
compositions and is felt to better reflect material limits.

The pipe pressure rating or pressure class is determined by
application of a design factor (inverse of service factor) to the long
term, 50 year value. The minimum design factor at 50 years as given
by AWWA C950-88 is 1.8. Pressure classes may be developed on either a
stress or strain basis as follows:

$$Pc \leq (\frac{HDB_{50}}{F.S.}) (\frac{2t}{D}) \qquad \text{(stress basis)}$$

or

$$Pc \leq (\frac{HDB_{50}}{F.S.}) (\frac{2t \ E_H}{D}) \qquad \text{(strain basis)}$$

Pc = pressure class - psi (kPa)

HDB_{50} = hydrostatic design basis at 50 years in psi (kPa) for
stress or in/in (mm/mm) for strain

t = reinforced wall thickness - inch (mm)

D = mean pipe diameter - inch (mm)

E_H = hoop tensile modulus - psi (kPa)

Standard Practice D2992 now includes detailed test and analysis procedures for the reconfirmation of the hydrostatic design basis. Standards D3517 and D3754 require reconfirmation at least once every two years to demonstrate continued product compliance. Significant changes in materials or the manufacturing process should also warrant a reconfirmation test.

Long Term Ring Bending

AWWA C950-88 includes the qualification test requirement to develop a long term ring bending strength (Sb). The long term ring bending strength is used in conjunction with HDB to evaluate the combined stress (strain) behavior of buried fiberglass pipe. Consistent with the requirements for strain corrosion and hydrostatic design basis, long term ring bending strength is established at 50 years.

Presented in AWWA C950-88 is a new test method to establish Sb. A series of constant load, creep to failure tests are conducted with the test samples totally immersed in environments controlled at pH4 or pH10 (to simulate the range of soil conditions). Change in deflection with time is monitored to failure or to an abrupt slope change as shown in Figure 4.

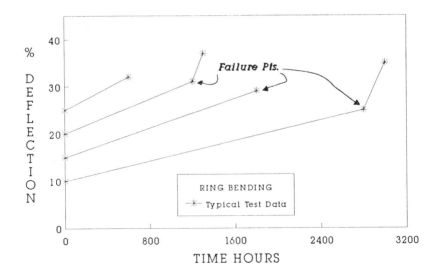

Figure 4 - Long Term Ring Bending Test Data

At least 18 tests are required to establish a regression line. The data must be distributed and analyzed in accordance with ASTM D3681 as shown in Figure 5. A minimum design factor of 1.5 is applied to the 50 year ring bending strength value.

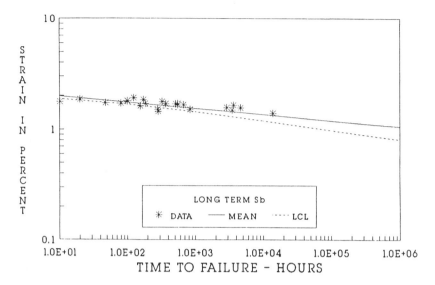

TIME TO FAILURE - HOURS

Figure 5 - Sb Determination

STRUCTURAL ANALYSIS

In addition to performance requirements, ANSI/AWWA C950-88 includes a very comprehensive appendix on the structural design and analysis of buried fiberglass pipe. It is widely referenced and used around the world and has been substantially revised and upgraded.

Deflection

The prediction of buried pipe deflection is addressed by a form of the Iowa Formula, first published by Spangler[5] in 1941, the most widely recognized deflection prediction method. As used in AWWA C950-88, modifications from the work by Howard[6] are incorporated to improve reliability and accuracy.

A very significant change in design philosophy has been made in the use of the predicted deflection values for product design and analysis. Deflection is calculated to demonstrate that the selection of pipe and installation procedures are such that the allowable or limiting deflection of the product (usually 5%) is not exceeded. However, it is the manufacturer's maximum allowable deflection for the product that is used in all subsequent design calculations. For example, vertical pipe deflection might be predicted to be 2.4% long term. If, however, the product's allowable long term deflection is the more typical 5%, then 5% must be used in all design calculations.

This has a substantial impact on product analysis and design. It is a conservative position as it assumes that the pipe is always used at the maximum limit, generally not the case, but it ensures that the product is analyzed at the allowable limits.

Buckling

The design appendix of AWWA C950-88 contains a buckling analysis method that is based on a modification of Luscher[7], as a result of extensive testing and analysis of fiberglass pipe.[8] The effects of vacuum, ground-water and soil loading are related to the stiffness provided by the pipe-soil system. This buckling analysis method, first developed in AWWA C950-81, has been widely adopted by other piping systems -- for example, steel pipe in the AWWA M-11 manual.

Combined Loading

The combined effects of internal pressure and ring bending from pipe deflection is a most important design consideration for buried pipe. AWWA C950-88 introduces several modifications in the analysis of combined strain.

The pressure portion of combined loading is compared to the long term (50 year) hydrostatic design basis (HDB) and the bending portion of the loading is compared to the long term (50 year) ring bending (Sb) strength. A design factor is then applied to the combination.

The following illustrates the combined loading approach. Only the more common strain basis is covered, however, an analogous stress basis is included in AWWA C950-88.

The strain due to internal pressure is:

$$\varepsilon_p = \frac{PD}{2 E_H t}$$

where:

ε_p = pressure strain (mm/mm)

P = pressure - psi (kPa)

D = diameter - inch (mm)

t = reinforced wall thickness - inch (mm)

E_H = hoop tensile modulus (kPa)

The maximum allowable strain is related to hydrostatic design basis by:

$$\frac{\varepsilon_p}{HDB_{50}} \leq \frac{1}{1.8}$$

It is assumed that the pipe is initially deflected to its maximum allowable or limiting deflection. As the pipe is then pressurized, this deflection is reduced as the internal pressure tends to reround the pipe and in turn reduce bending strain.

The effects of rerounding are addressed by the introduction of a rerounding coefficient. This approach is based on rerounding tests conducted on fiberglass pipe.[9]

$$r_c = (1 - \frac{P}{435}) \quad \text{or} \quad r_c = (1 - \frac{PN}{30})$$

where:

r_c = rerounding coefficient (dimensionless)

P = pressure - psi

PN = pressure - bars

The limiting deflection is multiplied by the rerounding coefficient to obtain rerounded deflection. The bending strain is then:

$$\varepsilon_b = Df \; r_c \; (T/D)(^{\Delta ya}/D)$$

where:

ε_b = bending strain, in/in (mm/mm)

Df = shape factor

r_c = rerounding coefficient

T = total wall thickness, in. (mm)

D = diameter, in. (mm)

Δya = limiting deflection, in. (mm)

AWWA C950-88 presents a comprehensive listing of shape factors. In addition to being a function of pipe stiffness, the shape factor is also influenced by the type of backfill and backfill compaction level. This recognizes that gravel backfills generally require less compaction effort resulting in a more uniform pipe shape (lower Df). For a given backfill material, higher compaction forces will give a less uniform shape and higher Df values.

This detailed tabulation of Df (Table 4) is desirable to allow analysis of a wide range of pipe projects and conditions. However, for the setting of a product standard performance level, as was described earlier for strain corrosion resistance, the maximum or limiting Df value is used.

Table 4 -- Shape Factors

		Pipe Zone Backfill Material & Compaction			
		Gravel		Sand	
		Dumped to Slight	Moderate to High	Dumped to Slight	Moderate to High
Pipe Stiffness psi (kPa)		Shape Factor Df (dimensionless)			
9	(62)	5.5	7.0	6.0	8.0
18	(124)	4.5	5.5	5.0	6.5
36	(248)	3.8	4.5	4.0	5.5
72	(496)	3.3	3.8	3.5	4.5

The maximum allowable bending strain is related to long term ring bending strength by:

$$\frac{\varepsilon_b}{Sb_{50}} \overset{<}{=} \frac{1}{1.5}$$

The combination of the two strains is further limited by an overall combined safety factor of 1.5 as follows:

$$\frac{\varepsilon_p}{HDB} + \frac{\varepsilon_b}{Sb_{50}} \overset{\leq}{=} 1/1.5$$

Figure 6 graphically represents the interaction of pressure and bending strain. A straight line relationship has been used. Recent published work by Bar-Shlomo[10] strongly indicates this straight line relationship to be conservative and that fiberglass pipe more likely behaves in a manner similar to that developed by Schlick[11] for cast iron pipes.

COMBINED STRAIN INTERACTION

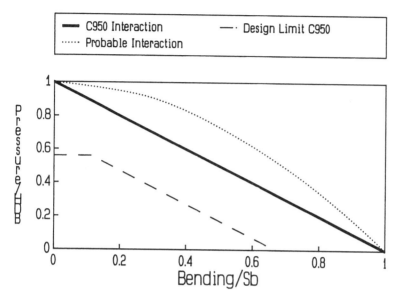

Figure 6 - Combined Strain Design

SUMMARY

The standardization process is an ongoing activity with existing documents under continual review and new needs being addressed. As new information and technology is available, it must be evaluated and incorporated into product standards.

The recent modification of these four significant ASTM and AWWA fiberglass pipe standards is a prime example of the standardization process at work. Product performance requirements were substantially upgraded and strengthened. Included were multiple stiffness ranges, a 50 year design basis, increased deflection to crack-damage, increased strain corrosion levels, and establishment of a long term ring bending strength. A major revision of the AWWA C950-88 design appendix addressed design at limiting deflection, accommodation of multiple pipe stiffness, rational pressure rerounding, and combined strain revisions.

Fiberglass pipe can be confidently specified and used in a wide variety of municipal and industrial applications. These major product standards allow this versatile, highly corrosion resistant material to be evaluated, compared and designed on the one important basis -- product performance.

REFERENCES

[1]   Marsh, H. N., "Middle East Sewer Conditions", Owens-Corning
      Fiberglas Publication No. INT-10847, September 1981.
[2]   Bishop, R. R, Lang, D. C., "Design and Performance of Buried
      Fiberglass Pipes -- A New Perspective", presented at American
      Society for Civil Engineering Conference, October, 1984.
[3]   Lang, D.C., Howard, A.K., "Buried Fiberglass Pipe Response to
      Field Installation Methods", presented at the American Society
      for Civil Engineering Conference, August, 1985.
[4]   Moser, A. P., et al, "Deflection and Strains in Buried FRP Pipes
      Subjected to Various Installation Conditions", Transportation
      Research Board, January, 1985.
[5]   Spangler, M. G., Handy, R.L., Soil Engineering, 4th Ed. 1982,
      Harper & Row, New York, NY.
[6]   Howard, H. K, "Modulus of Soil Reaction Values for Buried
      Flexible Pipe", American Society for Civil Engineering Journal
      of Geotechnical Engineering Division, January, 1977.
[7]   Luscher, U., "Buckling of Soil Surrounded Tubes", American
      Society for Civil Engineering Journal of Soil Mech. & Found.
      Div., Sect. 92:6:213, November, 1966.
(8)   Glascock, B. C., Cagle, L.L., "Recommended Design Requirement for
      Elastic Buckling of Buried Flexible Pipe", presented 39th
      Annual SPI Conference, January, 1989.
[9]   Moser, A. P., Clark, J., Bair, D. R., "Strains Induced by
      Combined Loading in Buried Pressurized Fiberglass Pipe",
      presented at American Society for Civil Engineering Confer-
      ence, August, 1985.
[10] Bar-Shlomo, S., "Stresses and Strains in GRP Pipes Due to
      Internal and External Pressures", 3R International, October,
      1987.
[11] Schlick, W. J., "Supporting Strength of Cast-Iron Pipe for Water
      and Gas Service", Iowa College Bulletin, June, 1940.

APPENDIX A

American Society for Testing Materials (ASTM)
Fiberglass Pipe Standards

Pipe Product Standards	Title
D2310	Standard Classification for Machine-Made Reinforced Thermosetting-Resin Pipe
D2517	Specification for Reinforced Epoxy Resin Gas Pressure Pipe and Fittings
D2996	Specification for Filament-Wound "Fiberglass" (Glass-Fiber-Reinforced-Thermosetting Resin) Pipe
D2997	Specification for Centrifugally Cast "Fiberglass" (Glass-Fiber-Reinforced-Thermosetting Resin) Pipe
D3262	Specification for "Fiberglass" (Glass-Fiber-Reinforced-Thermosetting Resin) Sewer Pipe
D3517	Specification for "Fiberglass" (Glass-Fiber-Reinforced-Thermosetting Resin) Pressure Pipe
D3754	Specification for "Fiberglass" (Glass-Fiber-Reinforced-Thermosetting Resin) Sewer and Industrial Pressure Pipe

Fittings & Joints

D1694	Specification for Threads (60° Stub) for "Fiberglass" (Glass Reinforced Thermosetting Resin)
D3840	Specification for "Fiberglass" (Glass-Fiber-Reinforced-Thermosetting Resin) Pipe Fittings for Non-Pressure Applications
D4024	Specification for Reinforced Thermosetting Resin (RTR) Flanges
D4161	Specification for "Fiberglass" (Glass-Fiber-Reinforced-Thermosetting Resin) Pipe Joints Using Flexible Elastomeric Seals

Test Methods & Practices

D2105   Test Method for Longitudinal Tensile Properties of Reinforced Thermosetting Plastic Pipe and Resin Tubes

D2143   Test Method for Cyclic Pressure Strength of Reinforced Thermosetting Plastic Pipe

D2924   Test Method for External Pressure Resistance of Reinforced Thermosetting Resin Pipe

D2925   Test Method for Beam Deflection of Reinforced Thermosetting Plastic Pipe Under Full Bore Flow

D2992   Practice for Obtaining Hydrostatic or Pressure Design Basis for "Fiberglass" (Glass-Fiber-Reinforced Thermosetting-Resin) Pipe and Fittings

D3567   Practice for Determining Dimensions of Reinforced Thermosetting Resin Pipe (RTRP) and Fittings

D3681   Test Method for Chemical Resistance of Fiberglass (Glass-Fiber-Reinforced Thermosetting-Resin) Pipe in a Deflected Condition

D3839   Practice for Underground Installation of Fiberglass (Glass-Fiber-Reinforced Thermosetting-Resin) Pipe

Hubert Schneider

ATV A 127 -- AS IT RELATES TO PLASTIC PIPE DESIGN

REFERENCE: Schneider, H., "ATV A 127 -- As it Relates to Plastic Pipe Design", Buried Plastic Pipe Technology, ASTM STP 1093, George S. Buczala and Michael J. Cassady, Eds., American Society for Testing and Materials, Philadelphia, 1990

ABSTRACT: The German specification ATV A 127 for static calculation of buried gravity pipes made of rigid and flexible materials is introduced. To show the handling of this regulation only for flexible pipes the set of equations is reduced to what is needed for those materials. Rules are given to work out the different types of loading and support conditions. Finally it is stated where the weak parts are and what should be done to modify it for flexible gravity and pressure pipes.

KEYWORDS: flexible pipe, static calculation, load condition, design, buckling

ABBREVIATIONS

The abbreviations used in the report are listed in the order as they appear in the following text:

E_B	=	soil modulus $[N/mm^2]$
D_{Pr}	=	Standard Proctor Density [%]
G	=	number of soil group/type [-]
p_E	=	vertical pressure due to soil load $[kN/m^2]$
æ	=	soil load reduction factor, Silotheory [-]
$æ_\beta$	=	æ recalculated to trench angle ß
γ_B	=	specific gravity of the soil $[kN/m^3]$
h	=	cover depth [m]
p_O	=	vertical pressure due to area load $[kN/m^2]$
$æ_O$	=	area load reduction factor, Silotheory [-]
$æ_{O\beta}$	=	$æ_O$ recalculated to trench angle ß
b	=	trench width [m]
φ^*	=	internal friction angle [^{O}]
K_1	=	ratio of lateral to vertical soil pressure [-]
δ	=	wall friction angle (see Eq. 6) [^{O}]
ß	=	trench angle (see Fig. 2) [^{O}]

Mr. Schneider is Managing Director of ComTec Ingenieurbüro für Verbund-werkstoffe GmbH, Feldchen 8, D-5100 Aachen, Federal Republic of Germany

57

a_F	=	correction factor due to a pressure distribution of approximately 2 : 1 [-]
d_a	=	external diameter of the pipe [m]
d_i	=	internal diameter of the pipe [m]
d_m	=	mean diameter of the pipe [m]
r_a	=	external radius of the pipe [mm]
r_i	=	internal radius of the pipe [mm]
r_m	=	mean radius of the pipe [mm]
p_F	=	pressure according to Boussinesq [kN/m^2]
F_A, F_E	=	design loads [kN]
r_A, r_E	=	design radii [m]
p	=	soil pressure on top of the pipe due to road traffic loading [kN/m^2]
φ	=	impact factor [-]
p_V	=	relevant soil pressure on top of the pipe due to road traffic loading [kN/m^2]
E_R	=	elasticity modulus of pipe material [N/mm^2]
σ_R	=	flexural strength of pipe material [N/mm^2]
γ_R	=	specific gravity of pipe material [kN/m^3]
ε_R	=	extreme fiber bending strain [%]
e	=	wall thickness of the pipe [mm]
y_{URD}/d_m	=	ultimate ring deflection [%]
f	=	reduction factor due to ground water [-]
α_B	=	reduction factor due to narrow trenching resp. to removal of sheeting [-]
E_2	=	effective soil stiffness/modulus to the side of the pipe (backfill surround) [N/mm^2]
E_{20}	=	deformation modulus [N/mm^2]
a'	=	active relative projection [-]
E_1	=	soil modulus of fill over the pipe crown [N/mm^2]
maxλ	=	maximum value of concentration factor for load distribution [-]
λ_R	=	concentration factor for load distribution (pipe) [-]
V_S	=	stiffness ratio [-]
S_R	=	ring stiffness of the pipe [N/mm^2]
I	=	moment of inertia, $I = e^3/12$ [mm^4/mm]
c_v^*	=	coefficient for vertical deflection [-]
c_{v1}	=	coefficient for vertical deflection as a result of q_v [-]
K^*	=	coefficient for reaction pressure [-]
c_{h1}	=	coefficient for horizontal deflection as a result of q_v [-]
2α	=	support angle (see Table 4, Fig. 7) [^{O}]
V_{RB}	=	system stiffness [-]
S_{Bh}	=	horizontal bedding stiffness [N/mm^2]
ζ	=	Leonhardt factor [-]
E_3	=	modulus of native soil to the side of the pipe [N/mm^2]
λ_{fu}, λ_{fo}	=	lower, upper limits of concentration factors for load distribution [-]
λ_{RG}	=	concentration factor for load distribution (pipe/trench) [-]
λ_B	=	concentration factor for load distribution (soil) [-]
q_v	=	total vertical load [N/mm^2]
q_h	=	total lateral load [N/mm^2]
q_h^*	=	total lateral reaction pressure [N/mm^2]
2β	=	angle of lateral bedding reaction pressure (see Fig. 9; estimated to be 120°) [^{O}]

m_{xx}	=	bending moment coefficients [-]
M_{xx}	=	bending moments due to various loads [kNm/m]
n_{xx}	=	normal force coefficients [-]
N_{xx}	=	normal forces due to various loads [kN/m]
$\sigma_{i,a}$	=	pipe wall stress [N/mm^2]
$\alpha_{ki,ka}$	=	correction factor for extreme fiber behaviour [-]
A	=	cross section of pipe wall [mm^2/mm]
W	=	resistance of pipe wall to bending [mm^3/mm]
$\varepsilon_{i,a}$	=	pipe wall strain [%]
Δd_v	=	vertical deflection [mm]
δ_v	=	relative vertical deflection [%]
α_D	=	tightness coefficient for critical external water pressure [-]
krit q_v	=	critical vertical load against buckling [N/mm^2]
FoS$_{qv}$	=	safety factor against buckling due to vertical load [-]
krit p_a	=	critical external (water) pressure against buckling [N/mm^2]
FoS$_{qv}$	=	safety factor against buckling due to vertical load [-]
p_i	=	internal pressure [N/mm^2]
p_a	=	external (water) pressure [N/mm^2]
γ_W	=	specific gravity of fluid [kN/m^3]
h_W	=	level of water table [m]
FoS$_{comb.}$	=	combined safety factor against buckling [-]

THE CALCULATION SYSTEM

The German specification for static calculation and design of buried pipes ATV A 127 [1] is founded upon experience, studies on soil behaviour and pipe material properties. Based on this work, the analytical model on the pipe/soil interaction was then worked out mainly by Leonhardt. The evaluation of this method started in the early sixties and the first issue was published in 1984. People started to work with the specification immediately and fed back their results and experiences. New and better knowledge on various materials was evaluated. All this led to the second edition, which was published in 1988. One of the most important modifications is the inclusion of fiberglass pipes. Minor modifications within the calculation system were necessary to cover the behaviour and the mode of design of GRP pipes. The determination of load concentration factors λ, is more precise now compared to the first edition. The original ATV specification is valid for rigid and flexible materials, which necessitates a wide variability of the system to cover all the different behaviours. Those variability problems are solved by a lot of tables and other conditions such as various load cases, bedding angles etc.

The title of this paper only refers to plastics pipes. In it the author reduces the amount of tables and formulas and simplifies the application of the specification demonstrated.

Knowing that a static calculation can only be done correctly if the loads on the pipe, the ambient conditions and the material behaviour is well known the ATV system can basically be divided into three steps:

1. Consideration and specification of loading and ambient conditions

2. Determination of deformation and stability behaviour

3. Determination of actual stress and/or strain data and stress/strain analysis

These calculation steps have to be worked out for plastics piping systems on short term and long term behaviour.

This mandatorily encludes, that the various steps of analysis have to be done using short term material/pipe properties for short term behaviour and using long term material/pipe properties for long term behaviour (i.e. stress/strain, deflection, buckling).

All loads have to be taken into account initially as well as long term. Only for the short term strain and deflection analysis the traffic load could be assumed to be "0".

For all cases where no soil expertise for a specific project is available ATV requires to distinguish the soil properties from the four following soil groups/types.

Group 1: Non Cohesive Soils

Group 2: Slightly Cohesive Soils

Group 3: Cohesive Mixed Soils, Silty Clays, Cohesive Sand and Fine Gravel, Cohesive Stony Weathered Soil

Group 4: Cohesive Soils (Clay, Loam)

The soil moduli are dependant on the Standard Proctor Density (SPD) which can vary between 85 % and 100 % (according to ATV A 127) and on the soil group/type. The reference values shall be calculated according to Eq 1.

$$E_B = \frac{2,74 \cdot 10^{-7}}{G} \cdot e^{0,188 \ D_{Pr}} \qquad (\ 1 \ )$$

This results for group 1 in a range of modulus between 2 N/mm^2 and 40 N/mm^2, for group 2 between 1,2 N/mm^2 and 20 N/mm^2, for group 3 between 0,8 N/mm^2 and 13 N/mm^2 and for group 4 between 0,6 N/mm^2 and 10 N/mm^2.

These soil moduli are confined moduli and serve as base values for a stress range between 0 N/mm^2 and approximately 0,1 N/mm^2.

The specific gravity γ_B, of the soil is assumed to be constant and to result in a value of 20 kN/m^3. (Author's recommendation: In cases of water table assumptions could be made to reduce the value of soil specific gravity from 20 kN/m^3 to a lower value depending on the level of water table. ATV only permits this, if measured data from a soil expertise are available.) The angle of internal friction, φ^*, ranges between 35° (Group 1) and 20° (Group 4) in 5°-steps.

LOADING
Soil Load

Pipes independent of the material type are subject to different modes of circumferential loading, such as

- soil load
- traffic load
- other area loads
- dead load (own weight)
- liquid filling
- internal pressure

The three loading modes mentioned first are handled to create pipe vertical deflection, pipe wall stress/strain and buckling (see Fig. 7, Fig. 8 and Fig. 9).

The others are only taken into account in case of the determination of pipe wall stress/strain.

The pressure in the soil is determined independently of the pipe material in the first steps. In a later stage it will be shown that flexible pipes do activate a positive load distributing behaviour of the soil which may decrease the pressure over the top of the pipe. Frictional forces against the trench walls may lead to a reduction of soil pressure and in those cases justify the use of the so called SILOTHEORY. In the actual issue of ATV A 127 it is assumed that the trench walls are maintained even long term. Taking the silotheory into account the mean vertical pressure due to the soil load may be determined according to Eq 2:

$$P_E = æ_\beta \cdot \gamma_B \cdot h \qquad\qquad (\ 2 \ )$$

For a uniformly distributed limited area load p_O the mean vertical pressure is to be calculated by Eq 3:

$$P_E = æ_{O\beta} \cdot p_O \qquad\qquad (\ 3 \ )$$

Further assumptions for the use of these reduction factors are:

$$E_1 \leq E_3 \text{ (for æ)}$$
$$E_1 < E_3 \text{ (for æ}_O)$$

If one or both of these assumptions are not fulfilled or under embankment conditions the reduction factors æ and $æ_O$ approach "1".

The reduction factors can be calculated according to Eq 4 resp. Eq 5

$$æ = \frac{1 - e^{(-2.h/b.K_1.\tan \delta)}}{2 \cdot h/b \cdot K_1 \cdot \tan \delta} \qquad (\ 4 \ )$$

$$æ_O = e^{(-2.h/b.K_1.\tan \delta)} \qquad\qquad (\ 5 \ )$$

Four conditions for fill above the pipe zone are discerned:

A1: Compacted fill undisturbed against native soil without analysis of degree of compaction. Condition A1 also applies to supporting sheet piling

A2: Vertical sheeting or lightweight sheet piles
or
installation sheeting to be removed after fill in stages
or
uncompacted fill
or
sluicing of the fill (valid for soil groups 1)

A3: Vertical sheeting to be removed after filling

A4: Equal to A1 but with analysis of degree of compaction. This condition must not be used with soil group G4.

For these standardized conditions the soil pressure ratio K_1 is assumed to be "0,5". Therefore the Eq 4 and Eq 5 can be reduced as follows

$$ æ = \frac{1 - e^{(-h/b.\tan \delta)}}{h/b \cdot \tan \delta} \qquad (4a) $$

$$ æ_O = e^{(-h/b.\tan \delta)} \qquad (5a) $$

The appropriate relationship between internal friction angle φ^* and wall friction angle δ is defined as follows depending on the fill condition:

$$ \begin{array}{ll} A1: & \delta = 2/3 \cdot \varphi^* \qquad (6a) \\ A2: & \delta = 1/3 \cdot \varphi^* \qquad (6b) \\ A3: & \delta = 0 \qquad (6c) \\ A4: & \delta = \varphi^* \qquad (6d) \end{array} $$

In case $\delta = 0$ the silo reduction factors are

$$ æ = æ_O = 1 $$

Several trench shapes are described in ATV A 127 (for example Fig. 1 and Fig. 2) and the silo factors have to be adapted to the trench angle ß according to Eq 7 resp. Eq 8

$$ æ_\beta = 1 - \beta/90 + æ \cdot \beta/90 \qquad (7) $$
$$ æ_{O\beta} = 1 - \beta/90 + æ_O \cdot \beta/90 \qquad (8) $$

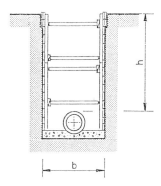

FIG. 1 -- Trench with parallel walls

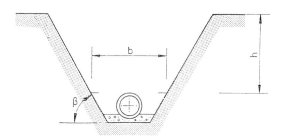

FIG. 2 -- Trench with sloping walls

Various standardized traffic loads are defined in ATV, such as:

Road Traffic Load

with three different vehicles:

SLW 60: wheel load 100 kN
SLW 30: wheel load 50 kN
LKW 12: wheel loads 20 kN, front
                    40 kN, rear

The soil loads resulting from road traffic loading can be cal-culated according to Boussinesq as an approximation with Eq 9, Eq 10 and Eq 11:

$$a_F = 1 - \frac{0,9}{0,9 + (4 \cdot h^2 + h^6)/1,1 \cdot d_m^{2/3}} \qquad (9)$$

$$p_F = \frac{F_A}{r_A^2 \cdot \pi} \cdot \left\{ 1 - \left[\frac{1}{1 + (r_A/h)^2} \right]^{3/2} \right\} + \frac{3 \cdot F_E}{2 \cdot \pi \cdot h^2} \cdot \left[\frac{1}{1 + (r_E/h)^2} \right]^{5/2} \qquad (10)$$

$$p = a_F \cdot p_F \qquad (11)$$

Depending on the road traffic load class the soil pressure p has to be multiplied by impact factors φ (see table 1 and Eq 12)

$$p_U = \varphi \cdot p \qquad (12)$$

TABLE 1 -- Design Loads, Radii and Impact Factors for Standard Vehicles

Standard Vehicle	F_A kN	F_E kN	r_A m	r_E m	-
SLW 60	100	500	0,25	1,82	1,2
SLW 30	50	250	0,18	1,82	1,4
LKW 12	40	80	0,15	2,26	1,5

Railway Traffic Loads

The rail load is based on the UIC 71 load pattern describe in specifications of the German Federal Railway. The soil pressure depends on the number of tracks, the cover depth and an impact factor (see Eq 13) which is related to the cover depth. The soil presure p shall be taken from table 2 where a linear interpolation can be done between two given cover depth values.

TABLE 2 -- Soil pressure p due to railway traffic load

h m	p in kN/m^2	
	1 track	2 or more tracks
1,50	48	48
2,75	39	39
5,50	20	26
≥ 10,00	10	15

The relevant impact factor φ shall be calculated according to Eq 13:

$$\varphi = 1,4 - 0,1 \cdot (h - 0,5) \geq 1,0 \qquad (13)$$

The relevant soil pressure p_v on top of the pipe due to railway traffic loading then is to be calculated according to Eq 14:

$$p_U = \varphi \cdot p$$

Aircraft Traffic Loads

The soil pressure p_v resulting from a design aircraft may directly be obtained from Fig. 3.

Using these values of soil pressure an impact factor of $\varphi = 1,5$ is included for the main landing gear and the load distributing actions.

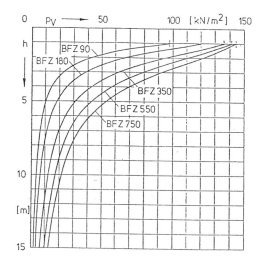

FIG. 3 -- Soil pressure p_V due to aircraft traffic loads

Any other area loads could be taken into account whenever some special regulations can be considered (for example: pressure propagation 2:1).

MATERIALS

In principle the ATV 127 worksheet is applicable for all plastics materials for which the long term behaviour is known. The characteristics of all pipe materials are influenced by aging, creep, fatigue loading and temperature.

This report only deals with plastics of which three are standardized in the worksheet, namely

Polyethylene, high density   (PE-HD)
Polyvinylchloride   (PVC-U)
Glass fiber reinforced plastic (UP-GF)

The properties mentioned below are taken from German application standards and valid for the following limiting conditions:

Aging	50 years
Long Term Behaviour	50 years
Cyclic Loading	$2 \cdot 10^6$ load cycles
Temperature	45 °C for DN ≤ 400
	35 °C for DN > 400

TABLE 3 a -- Properties of thermoplastic materials

Property	PE-HD		PVC-U	
	Short term	Long term	Short term	Long term
E_R in N/mm^2	1000	150	3600	1750
σ_R in N/mm^2	30	14,4	90	50
R in kN/m^3	9,5		13,8	

TABLE 3 b -- Properties of glass fiber reinforced plastics
(pipes centrifugally casted, filled)

Nominal Stiffness SN	Minimum Ringstiffness $S_R = E_R \cdot I/r_m^3$			Relative Ultimate Ring Deflection y_{URD}/d_m		Specific Gravity R
	Short Term	LT Gravity LT Pressure		Short Term	Long Term	
-	N/mm^2	N/mm^2		%	%	kN/m^3
2500	0,02	0,008	---	25	15	
5000	0,04	0,016	0,02	20	12	17,5
10000	0,08	0,032	0,04	15	9	

For GRP pipe materials the calculation values for the extreme fiber bending strain shall be determined according to Eq 15:

$$\varepsilon_R = \pm 4 \cdot \frac{e}{d_m} \cdot \frac{y_{URD}}{d_m} \qquad (15)$$

Note: The relationship between strain and deflection is depending on a deflection coefficient. For a two lines load condition and 0 % deflection the value is 4,28. For increasing deflection the coefficient decreases. By using the value of "4" in Eq 15 this decrease and a simplification is taken into consideration.

LOAD DISTRIBUTION

The load distribution on the pipe crown and to the side of the pipe is described by concentration factors λ. The different values for λ are depending on bedding requirements B1 through B4, the soil pressure ratio K_2 to the side of the pipe and the relative projection a' of the pipe shape. The bedding requirements are equal to the fill requirements which means: B1 is equivalent to A1, B2 to A2 etc. The soil moduli in the following equations are used with the below mentioned definitions (see Fig. 4)

E_1 = fill over the pipe crown
E_2 = backfill/soil to the side of the pipe
E_3 = native soil to the side of the pipe
E_4 = native soil under the pipe

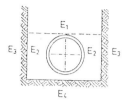

FIG. 4 -- Definition of soil moduli

Some special circumstances during or post installation have to be taken into account for the determination of E_2. This shall be done by reduction factors.

Ground water:

$$f = (D_{Pr} - 75)/20 \leq 1 \qquad (16)$$

Narrow trench:

$$\alpha_B = 1 - (4 - b/d_a) \cdot (1 - \alpha_{Bi})/3 \leq 1 \qquad (17)$$

with:

α_{Bi} = 2/3 for bedding requirement B1
α_{Bi} = 1/3 for bedding requirement B2
α_{Bi} = 0   for bedding requirement B3
α_{Bi} = 1   for bedding requirement B4

This results in an effective soil stiffness to the side of the pipe to be determined according to Eq 18:

$$E_2 = f \cdot \alpha_B \cdot E_{20} \qquad (18)$$

The soil pressure ratio K_2 in the soil to the side of the pipe is laid down taking into account that the system stiffness V_{RB} (see Eq 26) for plastics pipes (normally) is less than "0,1" as follows:

For soil group G1: K_2 = 0,4
                G2: K_2 = 0,3
                G3: K_2 = 0,2
                G4: K_2 = 0,1

For normal installation conditions all plastics pipes are laid according to Fig. 5. Therefore the relative projection a is equal to "1".

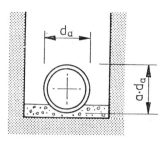

FIG. 5 -- Relative projection
for plastics pipes

The active relative projection a' can be calculated according to Eq 19:

$$a' = a \cdot E_1/E_2 = E_1/E_2 > 0,25 \qquad (19\ )$$

Note: For values a' < 0,251 use the value a' = 0,251

The calculation of the maximum concentration factor max λ is based on the consideration of a pipe of infinite ring stiffness on an absolute elastic soil in a wide embankment. Eq 20 reads as follows:

$$\text{max } \lambda = 1 + \cfrac{h/d_a}{\dfrac{3,5}{a'} + \dfrac{2,2}{E_4/E_1 \cdot (a' - 0,25)} + \left[\dfrac{0,62}{a'} + \dfrac{1,6}{E_4/E_1 \cdot (a' - 0,25)}\right] \cdot h/d_a} \qquad (20\ )$$

The relevant concentration factor over the pipe cross section λ_R is calculated according to Eq 21:

$$\lambda_R = \cfrac{\text{max } \lambda \cdot V_s + a' \cdot \dfrac{4 \cdot K_2}{3} \cdot \dfrac{\text{max } \lambda - 1}{a' - 0,25}}{V_s + a' \cdot \dfrac{3 + K_2}{3} \cdot \dfrac{\text{max } \lambda - 1}{a' - 0,25}} \qquad (21\ )$$

Stiffness ratio to be calculated according to Eq 22:

$$V_s = \frac{S_R}{|c_v{}^*| \cdot E_2} \qquad (22\ )$$

$$S_R = \frac{E_R \cdot e^3}{12 \cdot r_m^3} \qquad (23\ )$$

$$c_v{}^* = c_{v1} + 0,064 \cdot K^* \qquad (24\ )$$

$$K^* = \frac{c_{h1}}{V_{RB} + 0,0658} \qquad (25\ )$$

(for thermoplastic pipes E_R has to be taken from table 3 a)
(for thermosetting pipes S_R directly has to be taken from table 3 b)

TABLE 4 —— Coefficients for
vertical deflection

Support Angle 2α	c_{v1}	c_{h1}
60°	$-\,0,1053$	$+\,0,1026$
90°	$-\,0,0966$	$+\,0,0956$
120°	$-\,0,0893$	$+\,0,0891$
180°	$-\,0,0833$	$+\,0,0833$

By the system stiffness V_{RB} (according to Eq 26) the degree of demand of horizontal bedding reaction pressure is considered.

$$V_{RB} = S_R/S_{Bh} \qquad (26\ )$$

$$S_{Bh} = 0,6 \cdot \xi \cdot E_2 \qquad (27\ )$$

The Leonhardt factor (see Eq's 28a and 28b) takes into account the
difference in moduli of deformation between the backfill and the native
soil to the side of the pipe.

$$\triangle f = \frac{b/d_a - 1}{1,154 + 0,444 \cdot (b/d_a - 1)} \leq 1,44 \qquad (28a)$$

$$\xi = \frac{1,44}{\triangle f + (1,44 - \triangle f) \cdot E_2/E_3} \qquad (28b)$$

In cases of narrow trenches ($b/d_a \leq 4$) instead of the concentration
factor λ_R the factor λ_{RG} has to be used. Its value may be limited by
λ_{fu} (lower limit) or λ_{fo} (upper limit) (see Eq's 29a, 29b, 29c and 29d).

$$\lambda_{fu} \leq \lambda_{RG} \leq \lambda_{fo} \qquad (29a)$$

$$\lambda_{fu} = \frac{1 - e^{-h/d_a \cdot \tan \delta}}{h/d_a \cdot \tan \delta} \qquad (29b)$$

For $h \leq 10$ m: $\lambda_{fo} = 4,0 - 0,15 \cdot h \qquad (29c)$

For $h > 10$ m: $\lambda_{fo} = 2,5 =$ const. $\qquad (29d)$

For narrow trenches the concentration factor λ_{RG} shall be calculated according to Eq 30a resp. Eq 30b

For $1 < b/d_a \leq 4$: $\lambda_{RG} = \frac{\lambda_R - 1}{\cdot 3} \cdot b/d_a + \frac{4 - \lambda_R}{3} \qquad (30a)$

For $4 < b/d_a \leq \infty$: $\lambda_{RG} = \lambda_R \qquad (30b)$

The concentration factor λ_B is independent from the trench width and shall be calculated according to Eq 31.

$$\lambda_B = \frac{4 - \lambda_R}{3} \qquad (31\ )$$

In cases where λ_{RG} is limited by λ_{fu} or λ_{fo} the value of λ_B has to be calculated according to Eq 32a resp. Eq 32b.

$$\text{If } \lambda_{RG} = \lambda_{fu} \text{ then}$$
$$\lambda_B = \frac{b/d_a - \lambda_{fu}}{b/d_a - 1} \qquad (32a)$$

$$\text{If } \lambda_{RG} = \lambda_{fo} \text{ then}$$
$$\lambda_B = \frac{b/d_a - \lambda_{fo}}{b/d_a - 1} \qquad (32b)$$

The distribution of soil presures and relating concentration factors is demonstrated in Fig. 6

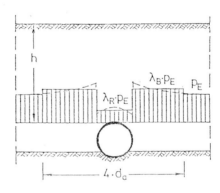

FIG. 6 -- Distribution of soil pressures

Now the total loads on the pipe can be calculated according to Eq 33

$$q_v = \lambda_{RG} \cdot (\ae_\beta \cdot \gamma_B \cdot h + \ae_{o\beta} \cdot p_0) + p_v \qquad (33\ )$$

Lateral (horizontal) pressure according to Eq 34

$$q_h = K_2 \cdot (\lambda_B \cdot p_E + \gamma_B \cdot d_a/2) \qquad (34\ )$$

Lateral (horizontal) reaction pressure according to Eq 35

$$q_h^* = (q_v - q_h) \cdot K^* \qquad (35\ )$$

In the ATV A 127 are two relevant support cases for plastics (flexible) pipes defined:

Support Case I: Supported in soil. Vertical reactions with rectangular distributions (see Fig. 7).

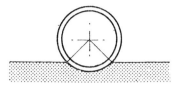

FIG. 7 -- Support case I

Support Case III: Support and bedded in soil with vertical and rectangular reaction distribution (see Fig. 8)

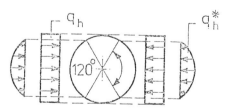

FIG. 8 -- Support case III

The lateral pressure on the pipe consists of a contribution q_h from the vertical load and the reaction pressure q_h^* due to pipe deflection (see Fig. 9).

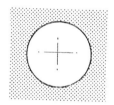

FIG. 9 -- Lateral pressure for support cases I and III

INTERNAL FORCES, STRESSES, STRAINS

The internal bending moments M and the internal normal forces N shall be calculated according to Eq's 36a, 36b, 37a, 37b, 38a, 39a, 39b, 40a, 41a and 41b whereby table 5 shall be taken into account.

Vertical Load

$$M_{qv} = m_{qv} \cdot q_v \cdot r_m^2 \tag{36a}$$

$$N_{qv} = n_{qv} \cdot q_v \cdot r_m \tag{36b}$$

Lateral Pressure due to q_v

$$M_{qh} = m_{qh} \cdot q_h \cdot r_m^2 \tag{37a}$$

$$N_{qh} = n_{qh} \cdot q_h \cdot r_m \tag{37b}$$

Lateral Reaction Pressure due to Deflection

$$M_{\bar{q}h} = m_{\bar{q}h} \cdot q_h^x \cdot r_m^2 \tag{38a}$$

$$N_{\bar{q}h} = n_{\bar{q}h} \cdot q_h^x \cdot r_m \tag{38b}$$

Own Weight

$$M_g = m_g \cdot \gamma_R \cdot e \cdot r_m^2 \tag{39a}$$

$$N_g = n_g \cdot \gamma_R \cdot e \cdot r_m \tag{39b}$$

Weight of Fluid

$$M_w = m_w \cdot \gamma_W \cdot r_m^3 \tag{40a}$$

$$N_w = n_w \cdot \gamma_W \cdot r_m^2 \tag{40b}$$

Internal/External (Water) Pressure

$$M_{pw} = (p_i - p_a) \cdot r_i \cdot r_a \cdot \left[0,5 - \frac{r_i \cdot r_a}{r_a^2 - r_i^2} \cdot \ln(r_a/r_i) \right] \tag{41a}$$

$$N_{pw} = p_i \cdot r_i - p_a \cdot r_a \tag{41b}$$

All coefficients "m_{xx}" and "n_{xx}" are support case and bedding angle dependent and have to be taken from table 5.

TABLE 5 -- Bending moment and normal force coefficients

Support Case/ Bedding Angle	Position at Circum- ference	Bending Moment Coefficients				
		m_{qv}	m_{qh}	$m_{qh}^{\star}$	m_g	m_w
I/60°	Crown	+0,286	-0,250	-0,181	+0,459	+0,229
	Springline	-0,293	+0,250	+0,208	-0,529	-0,264
	Bottom	+0,377	-0,250	-0,181	+0,840	+0,420
I/90°	Crown	+0,274	-0,250	-0,181	+0,419	+0,210
	Springline	+0,279	+0,250	+0,208	-0,485	-0,243
	Bottom	+0,314	-0,250	-0,181	+0,642	+0,321
I/120°	Crown	+0,261	-0,250	-0,181	+0,381	+0,190
	Springline	-0,265	+0,250	+0,208	-0,440	-0,220
	Bottom	+0,275	-0,250	-0,181	+0,520	+0,260
III/180°	Crown	+0,250	-0,250	-0,181	+0,345	+0,172
	Springline	-0,250	+0,250	+0,208	-0,393	-0,196
	Bottom	+0,250	-0,250	-0,181	+0,441	+0,220

		Normal Force Coefficients				
		n_{qv}	n_{qh}	$n_{qh}^{\star}$	n_g	n_w
I/60°	Crown	0,080	-1,000	-0,577	+0,417	+0,708
	Springline	-1,000	0	0	-1,571	+0,215
	Bottom	0,080	-1,000	-0,577	-0,417	+1,292
I/90°	Crown	0,053	-1,000	-0,577	+0,333	+0,667
	Springline	-1,000	0	0	-1,571	+0,215
	Bottom	0,053	-1,000	-0,577	-0,333	+1,333
I/120°	Crown	0,027	-1,000	-0,577	+0,250	+0,625
	Springline	-1,000	0		-1,571	+0,215
	Bottom	0,027	-1,000	-0,577	-0,250	+1,375
III/180°	Crown	0	-1,000	-0,577	+0,167	+0,583
	Springline	-1,000	0	0	-1,571	+0,215
	Bottom	0	-1,000	-0,577	-0,167	+1,417

Signs denote: Moment: + tension on inside,
         - tension on outside of pipe wall
  Normal Force: + tension,
         - compression in pipe wall

All coefficients apply only to circular pipes having a constant wall thickness over the circumference

The stress and/or strain determination shall be done as follows.

Stress determination only for thermoplastic materials has to be done with support case I and bedding angle 90°.

$$\sigma_i = N/A + \alpha_{ki} \cdot M/W \tag{42a}$$

$$\sigma_a = N/A - \alpha_{ka} \cdot M/W \tag{42b}$$

$\alpha_{ki} = 1 + 1/3 \cdot e/d_m$
= Correction factor for extreme fiber behaviour (inside) $\tag{43a}$

$\alpha_{ka} = 1 - 1/3 \cdot e/d_m$
= Correction factor for extreme fiber behaviour (outside) $\tag{43b}$

$A = e$
= Cross section of the pipe wall $\tag{44}$

$W = e^2/6$
= Resistance to bending of the pipe wall $\tag{45}$

Strain determination only for thermosetting materials has to be done with support case III and bedding angle 180°.

$$\varepsilon_i = \frac{e}{2 \cdot r_m^3 \cdot S_R} \cdot \left[\frac{e \cdot N}{6} + \alpha_{ki} \cdot M \right] \tag{46a}$$

$$\varepsilon_a = \frac{e}{2 \cdot r_m^3 \cdot S_R} \cdot \left[\frac{e \cdot N}{6} - \alpha_{ka} \cdot M \right] \tag{46b}$$

PIPE DESIGN

The calculated existing stresses or strains at any of the three positions of the pipe have to be compared with the applicable calculation values (see table 3a resp. table 3b and Eq 15)

Thermoplastic: $FoS = \sigma_R/\sigma \geq 2,5$ (Safety Class A) $\tag{47a}$
Thermosetting: $FoS = \varepsilon_R/\varepsilon \geq 2,0$ (Safety Class A) $\tag{47b}$

VERTICAL DEFLECTION, BUCKLING

The deflection and buckling analysis worked out with the backfill material properties to the side of the pipe (E_2) result in mean values for these behaviours. To take into account unavoidable dispersion in soil properties it is strongly recommended (mandatory) in ATV (see clause 8.4 of [1] to reduce the E_2-value by multiplying the result of Eq 1 with 2/3.

Therefore the values for

ξ, S_{Bh}, V_{RB}, a', max λ, S_{Bv}, K^*, c_v^*, V_S, λ_R, λ_{fo}, λ_{fu}, λ_B, λ_{RG}, q_v, q_h and q_h^*

have to be recalculated. Deflection and buckling analysis have to determined with support case III and bedding angle 180° for all plastic materials.

$$\triangle d_v = c_v^* \cdot \frac{q_v - q_h}{S_R} \cdot 2 \cdot r_m \qquad (48a)$$

$$\delta_v = c_v^* \cdot \frac{q_v - q_h}{S_R} \cdot 100 \qquad (48b)$$

The allowed vertical deflections are:

$$\text{Short Term:} \quad 4 \text{ \%}$$
$$\text{Long Term:} \quad 6 \text{ \%}$$

For buckling analysis the following calculations have to be done.

Critical Vertical Load

$$\text{krit } q_v = 2 \cdot \sqrt{S_R \cdot S_{Bh}} \qquad (49a)$$
$$\text{FoS}_{qv} = \text{krit } q_v / q_v \qquad (49b)$$

External Water Pressure

$$\text{krit } p_a = \alpha_D \cdot S_R \qquad (50a)$$

with α_D according to Fig. 10.

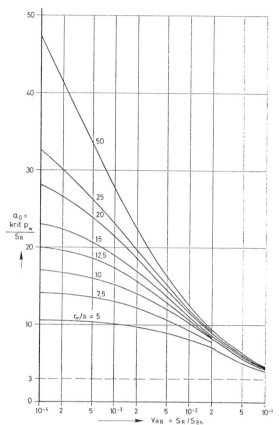

FIG. 10 -- Tightness Coefficient α_D for the
critical external water pressure

$$p_a = \gamma_{LI} \cdot (h_{LI} + d_a/2)$$

$$FoS_{pa} = krit \; p_a/p_a \qquad\qquad \begin{matrix}(50b)\\(50c)\end{matrix}$$

Combination of vertical soil load and external water pressure

$$FoS_{comb.} = \frac{1}{(1/FoS_{qv}) + (1/FoS_{pa})} \qquad (51\,)$$

Any single FoS and also the combined FoS has to be equal or greater
than 2,5 for all plastic materials.

CONCLUSION AND PROSPECTS

A lot of calculations compared to measured data show that the results of static calculations according to ATV A 127, especially on
plastics pipes, are higher than the real measured values. Compared to
some other calculation systems it easily could be demonstrated that
data determined by the ATV system are most close to practical observa-

tions. But even ATV A 127 is not perfect. Hereinafter some recommendations are given by the author to improve the system.

1. Equations should be incorporated into ATV A 127 to consider vertical deflections due to own weight before installation and horizontal
deflections due to compaction during installation to be taken into account not only for deflections but also for stresses/strains.

2. It is clearly mentioned in ATV A 127 that soil moduli determined
according to Eq 1 are only valid for a stress range between 0 N/mm^2 and
0,1 N/mm^2. Nowhere in the system an indication is given how to deal
with soil moduli in cases of higher stresses or cover depths.

3. The system used in ATV A 127 indicates that traffic loads are
continuously acting loads for 50 years life time. This is not very realistic. The load case "Traffic" should be reviewed and time periods
without any traffic load should be estimated and taken into account.

Here possibly assumptions could be made similar to those made for
the temperature "history". For example on roads with high traffic frequency the resulting traffic load could be as calculated according to
Eq 11. For medium traffic frequency the load could be rerated by x %
and for low frequency by y %.

4. Basically ATV A 127 is relevant only for solid wall pipes. But
now pipes with profiled wall constructions are getting more and more
applied. Therefore the system has to be expanded for those constructions.

5. ATV A 127 was elaborated for buried gravity piping systems.
Therefore minor efforts were made to determine the behaviour of buried
pressure pipes very carefully. No difference is made between flexural
stresses/strains due to life load and tensile stresses/strains due to
internal pressure. Nowhere in the document is mentioned that the values
for bending and tensile failure strains or for bending and tensile
strength for some materials might be different.

No "Rerounding"-Effect is taken into account although equations for
that benefit are laid down in several ISO documents. A combined safety
analysis seems to be necessary.

6. The effective angle, 2ß, of the bedding reaction pressure is
estimated in A 127 to be 120°. Many observations on buried pipelines
and a lot of soil box tests showed the necessity of a slight modification of this angle. The angle has to be increased to 170° through 180°
[2]. It could even be necessary to establish the angle depending on installation cases.

Various coefficients will be influenced by changing that angle 2ß.
The coefficient for horizontal deflection due to the bedding reaction
pressure c_{h2}, the moment and normal force coefficients due to bedding
reaction pressure m_{qh}^*/n_{qh}^* and the Leonhardt factor ζ has to be adapted.

7. More variations on support case/bedding angle combinations
should be incorporated in the system. The combination III/180° for

example seems not to be sufficient to cover all installation cases. This should be enlarged to III/120° and III/150° for example. In that case modifications on the coefficient for horizontal deflection due to the vertical load c_{h1} and the moment and normal force coefficients due to vertical load m_{qv}/n_{qv} have to be adapted.

REFERENCES

[1] Abwassertechnische Vereinigung, Richtlinie für die statische Berechnung von Entwässerungskanälen und -leitungen, St. Augustin, 1988

[2] Carlström, B., Leonhardt, G., Schneider, H., Statische Berechnung von erdverlegten Rohren aus glasfaserverstärktem Kunststoff, Internationaler Kongress Leitungsbau, Hamburg, 1987

David A. Gregorka, Peter F. Greiner, Stan S. Hazan, and Michael F. Kenel

THE ROLE OF ANSI/NSF STANDARD 61 AND THIRD-PARTY
CERTIFICATION IN PROVIDING SAFE DRINKING WATER

REFERENCE: Gregorka, D. A., Greiner, P. F., Hazan, S. S., and
Kenel, M.F., "The Role of ANSI/NSF Standard 61 and Third-Party
Certification in Providing Safe Drinking Water." Buried
Plastic Pipe Technology ASTM STP 1093, George S. Buczala and
Michael J. Cassady, Eds., American Society for Testing and
Materials, Philadelphia, 1990.

ABSTRACT: The NSF Drinking Water Additives Program is the
model for health effects evaluation of products and materials
in contact with drinking water, including buried plastics
pipe. One of the standards developed by the program, ANSI/NSF
Standard 61 (Drinking Water System Components - Health
Effects), covers the toxicology of indirect additives to
drinking water for all types of potable water contact products
and materials, including plastics pipe. For many years, NSF
Standard 14 (Plastics Piping Components and Related Materials)
has been the model toxicological and performance standard for
plastics pipe, serving regulators, users, industry, and the
public. Although it remains a viable standard, health effects
in NSF Standard 14 are now addressed by reference to ANSI/NSF
Standard 61.

How these Standards address potential health effects of
plastics pipe and their impacts on specifiers, users, and
designers is addressed. Extraction testing and toxicology
requirements are provided special emphasis. NSF's product
Certification (Listing) program for plastics is explained,
along with a discussion of the problems and opportunities
presented by consensus standards and third-party product
certification programs.

KEYWORDS: plastics, health effects, standards, certification,
toxicology, testing, additives, third-party, drinking water.

The National Sanitation Foundation's (NSF) Drinking Water
Additives program has drawn much interest since its inception in 1985.
This program addresses the health effects implications of water
treatment chemicals and other products used in conjunction with drinking
water treatment, storage, transmission, and distribution. One of two
standards developed as part of the program, ANSI/NSF Standard 61
(Drinking Water System Components - Health Effects), addresses the

toxicology of indirect additives to drinking water, including plastics and other types of pipe materials.

This Standard was originally adopted in October 1988, and revised in 1990. For over 25 years, NSF Standard 14 (Plastics Piping Components and Related Materials) has been the premier health and performance standard for plastics pipe, serving regulators, users, industry, and the public. Although NSF Standard 14 remains a viable standard, the health effects requirements of NSF Standard 14 are now addressed by reference to ANSI/NSF Standard 61.

This paper addresses how both Standards address health effects issues related to plastics pipe, and their impacts on specifiers, users, and designers. Special emphasis is placed on the extraction testing and toxicological requirements in ANSI/NSF Standard 61. NSF's third-party product certification program for plastics is also discussed, along with the problems and opportunities presented by consensus standards and product certification programs.

NSF STANDARD 14

NSF Standard 14 for plastics piping system components and related materials was originally adopted by NSF in 1965. For potable water applications, the Standard addresses both health effects implications and performance requirements. All components of plastics piping systems are covered by the Standard, including pipes, fittings, valves, lubricants, and solvent cements. In addition, the Standard has specific requirements for plastics materials, and for certain specified generic ingredients, including calcium carbonates, calcium stearates, hydrocarbon waxes, oxidized polyethylene waxes, and titanium dioxides.

Physical testing requirements are addressed by reference to American Society for Testing and Materials (ASTM), American Water Works Association (AWWA), and Plastics Pipe Institute (PPI) performance standards. In December 1988, Standard 14 was revised to reference ANSI/NSF Standard 61 for health effects requirements for covered potable water components.

ANSI/NSF STANDARD 61

ANSI/NSF Standard 61 was developed by a consortium of organizations and other interested parties under a cooperative agreement from the U.S. Environmental Protection Agency (EPA) to establish standards for all chemicals, materials, and other products used in conjunction with drinking water treatment, storage, transmission, and distribution.

For many years, EPA had in place an informal advisory program for evaluating various types of products and materials used in public drinking water systems. Under this program, manufacturers who marketed products intended for use in drinking water systems could submit information to EPA for its review. For products which passed the EPA review process, a letter of acceptance was issued to the manufacturer, and the products were placed on a list maintained by the Agency.

Although its official position was that this activity and the list were not regulatory functions but "technical advisory" functions, the process became the generally accepted "standard" for products used in drinking water contact.

One exception to the EPA review process was plastics pipe. For plastics products and materials, EPA relied entirely on NSF Standard 14 and directed inquiries for review to NSF's Plastics Listing Program.

The ultimate regulatory responsibility for accepting various chemicals, materials, and products for use in public drinking water systems rests with the state drinking water programs with primary enforcement authority (primacy). Most state drinking water programs, in the past, have relied upon the EPA list (and the NSF Listing of certified plastics products) as the basis for accepting or denying various products for use in public drinking water systems. Some states, such as New York and Ohio, developed their own evaluation criteria and programs for selected types of products.

In 1984, following a review of its program, the EPA determined that there were a number of serious deficiencies, and that it either had to invest substantial resources to correct the deficiencies, or it should discontinue the program. Neither option was a good one. The cost to totally revamp the program in terms of requirements and manpower was extremely high at a time when EPA's resources were becoming more and more scarce. Further, this was an area where the agency did not have a mandated responsibility to provide a service. However, if it simply dropped the program, it would have transferred a tremendous burden to the individual state drinking water programs. Under this scenario, each state program would be responsible for evaluating drinking water additives products and materials. This had the potential of fostering over 50 different sets of requirements across the country for drinking water additives products.

After rejecting these two options, EPA decided to foster development of national voluntary consensus standards for drinking water additives products, and to insure the availability of a third-party mechanism for product certification against the standards. The goal was to establish a single set of uniform national standards and a third-party certification program that would be acceptable to state drinking water programs, water utilities, and manufacturers.

In 1984, EPA issued a request for proposals (RFP) from organizations interested in developing national, voluntary, consensus standards to address the health effects of drinking water additives products, and to offer a program of third-party certification against those standards. It was really asking for a program equivalent to what NSF had provided for plastics piping system components for over 20 years! NSF viewed the proposed drinking water additives standards and certification program as a natural extension of its program for plastics. To appropriately respond, NSF formed a consortium of organizations which jointly developed a proposal in response to the EPA RFP. Included in the consortium at that time were the American Water Works Association Research Foundation (AWWARF), the Association of State Drinking Water Administrators (ASDWA), and the Conference of State Health and Environmental Managers (COSHEM).

EPA awarded the NSF-led Consortium a cooperative agreement, and a $185,000 grant of "seed money." The standards development activity alone cost over $1.6 million, and, in addition to EPA, was funded by NSF, manufacturers, and the consortium members.

About midway through the standards development activity, another organization, the American Water Works Association (AWWA), joined the Consortium. COSHEM has since ceased operations.

Using the established NSF consensus standards development process, the three year development activity resulted in the adoption of two new consensus standards. ANSI/NSF Standard 60 (Drinking Water Treatment Chemicals – Health Effects) addresses direct additives. Direct additives are water treatment chemicals, such as lime, chlorine, and alum, that are added directly to water.

On the indirect additives side, ANSI/NSF Standard 61 covers a diverse variety of products that have incidental contact with drinking water, and that may indirectly impart contaminants to the water. Included are products and materials such as pipes, coatings, gaskets, valves, and process media. Only this Standard, as it affects plastics pipe, is addressed in this paper.

Both Standards were developed consistent with American National Standards Institute (ANSI) voluntary standards development guidelines, and were adopted by NSF in October 1988 and approved as American National Standards in May 1989. Both are copyrighted, like most consensus standards, to protect their integrity. However, the copyright does not in any way restrict their use by any individual or organization.

Now that ANSI/NSF Standard 61 has been adopted and an NSF certification program implemented, all types of piping materials can now be evaluated for health effects implications under a single national, voluntary, consensus standard. Although the specific testing requirements vary from material to material, a single approach to pipe evaluation is used under the Standard. Specific requirements for pipe products, and plastics in particular, are addressed in the remainder of the paper.

ANSI/NSF STANDARD 61 REQUIREMENTS

Pipe Testing Requirements – Extraction, Analysis, and Normalization

ANSI/NSF Standard 61 requirements are based on health effects as they relate to the consumer "at the tap," and address two specific concerns:

1. Do any contaminants leach or migrate from the product into the drinking water?
2. If so, is the level of migration acceptable from a public health and toxicological viewpoint?

Plastics pipe and fittings are evaluated under Section 4, "Pipes and Related Products." Information on the size and intended use of the product is required, as is confidential information on the material formulation and ingredients used in water contact surfaces. A toxicological review of the information is required to determine potential contaminants of interest and to identify a specific testing regime. Testing is then performed, results analyzed, and a final toxicological assessment made.

The Standard also supports the evaluation of preblended potable water materials. Evaluation of these materials is done in the form of typical finished products, and they are evaluated fully to the requirements of the Standard. Certified materials can be used interchangeably in Certified finished products (pipe/fittings), so long as the alternate material is of the same generic type and meets designated end use requirements.

Chemical extraction is performed on the finished pipe or fitting using either an "in product" or "in vessel" exposure protocol. "In product" exposures involve filling the sample with extraction water, and are limited to products where this is practical. The intent is to expose only the water contact surfaces.

For "in vessel" exposures, a less costly option for homogeneous products (e.g., PVC pipe), product samples are cut to sizes that can be placed in exposure vessels and covered with extraction water. Under this option both the inside and outside of the product, as well as cut surfaces, are exposed to extraction water. In either case, the surface area to water volume ratio tested represents the smallest size produced. In some cases, for analytical sensitivity and convenience, the surface area to water volume ratio is exaggerated.

Once testing has been completed, extraction results are "normalized" to "at-the-tap" values based upon the intended use (e.g., water main, multiple user service line, service line, residential). Normalization mathematically adjusts measured laboratory contaminant concentrations to expected field use concentrations by factoring in surface area to water volume ratio differences and typical water flows.

The extraction protocol consists of three basic steps--washing, conditioning, and final exposure, followed by analysis and normalization.

Washing: To remove any extraneous debris or contamination that may have occurred from shipping and handling, the sample is first rinsed with cold tap water, followed by a deionized water rinse.

Conditioning: To simulate pre-use flushing and/or disinfection procedures, the sample is conditioned by exposing it at room temperature to pH 8 water for 14 days (or less if requested by the manufacturer). The water is changed 10 times over the 14 days, but no single exposure is less than 24 hours. During the first day of conditioning, the water contains 50 mg/L of available chlorine. During the remaining 13 days of conditioning, the water contains 2 mg/L of available chlorine. Hot application products are further conditioned with two, one-hour exposures at 82°C.

Final Exposure: The final exposure begins immediately after conditioning ends, and lasts for 16 hours. Based on the extraction program for contaminants of concern established during toxicological review, several final exposures are usually required, each under different exposure conditions (e.g., pH 5, 10, or 8; chlorinated or nonchlorinated). Only the final exposures are analyzed for contaminant extraction.

Analysis: Analyses are formulation dependent. The exposure waters are analyzed for the contaminants of interest using methods referenced in the Standard (typically EPA methods or methods from Standard Methods for the Analysis of Water and Wastewater) or by using alternate methods. The Standard specifies criteria for validating and using alternate methods.

For PVC and CPVC products, the Standard requires analysis for residual vinyl chloride monomer in the product. This method was adopted first in Standard 14. It involves dissolving products in a solvent and measuring vinyl chloride concentration in the headspace by gas chromatography. Prior studies have established a correlation between this measurement (ppm range) and the concentration of vinyl chloride extracted into water (ppb range) [1-6].

Normalization: The analytical results are mathematically adjusted to determine the maximum allowable levels of each contaminant. This step relates laboratory results to projected "at the tap" levels; i.e., the levels that would be experienced by the consumer under these conditions. The calculations and assumptions for pipe and related products are discussed in Appendix B, Table B.1, and Section 11 of the Standard.

Pipe Testing Requirements - Microbiological Growth Support

The Standard also requires that products not adversely affect water quality by supporting microbiological growth. Evaluation for the support of microbiological growth is generally not required on rigid plastics pipe and fitting products, but is generally required on products using plasticizers, solvent-containing products (e.g., cements), lubricants, gaskets, and similar materials.

The test method for the evaluation of microbiological growth support potential is detailed in Section D of the Standard. In brief, the protocol involves exposing a product sample to dechlorinated tap water inoculated with a fresh aliquot of water from a surface source (i.e., river) of suitable quality for treatment as drinking water. The uptake of dissolved oxygen (DO) is measured and compared with the uptake of an inert control (e.g., glass slide). Measurements are made during the fourth, fifth, and sixth weeks. The mean of DO values from the sample is subtracted from the mean DO value of the control. The result is the mean dissolved oxygen difference (MDOD), and is descriptive of the level of microbiological activity in the product.

In addition, during the final week of the test, samples are analyzed for the enumeration of pseudomonas species and total coliforms. At this time, there is no pass or fail level established for microbiological growth support testing. Results are required simply to be reported. However, the Joint Committee for Drinking Water Additives plans to revisit this issue in late 1990, once additional data is available, to determine whether or not a pass/fail level should be established.

Toxicology Requirements

Overview: The primary focus of the Standard is on potential contaminants extracted into drinking water from water contact surfaces of pipes, gaskets, coatings, and similar products. The first step in toxicology review is for product manufacturers to provide the reviewing toxicologists with detailed information about chemical composition, leachability, and toxicity. For the majority of plastics products, information on composition and toxicology is most often available from ingredient suppliers, occasionally available from formulators, and seldom available from end-product manufacturers. Ingredient supplier information is critical to the process.

The information is reviewed by qualified toxicologists, according to Appendix A of the Standard. Only contaminants of toxicological concern are identified for analysis. Desired limits of detection are specified based on the information needs of the reviewing toxicologist. For example, epichlorohydrin, a known carcinogen, ideally would be analyzed at a limit of detection below the target value (normalized for man's likely exposure), determined to be an acceptable level of risk for carcinogenicity. Not infrequently, artificially aggressive test systems (e.g., high surface to volume ratios) may be necessary to achieve required detection limits.

Selection of plastics ingredients or impurities for analytical testing is usually based on known toxicity, solubility, and concentration. Chemicals of unknown toxicity may be selected based on concentration, solubility, or knowledge of relationship to known toxic contaminants. Potential by-products, such as amines generated upon hydrolysis of isocyanates, may be selected for analysis when they are of toxicological concern. Plasticizers, solvents, dyes, and other components likely to extract may be chosen for analysis. Low molecular weight monomers typically receive priority attention. In addition, the Standard specifically requires that certain analytical tests be performed (Table 1).

TABLE 1--Testing Requirements for Plastics

PVC and CPVC	Regulated Metals[a] Tin Antimony Phenolics Volatile Organic Chemicals Residual Vinyl Chloride Monomer Plus Formulation Dependent Parameters
All Other Plastics	Regulated Metals[a] Phenolics Volatile Organic Chemicals Plus Formulation Dependent Parameters

[a]Arsenic, Barium, Cadmium, Chromium, Lead, Mercury, Selenium, Silver

Regulated Contaminants: EPA-regulated contaminants are evaluated by comparing the normalized analytical test result with one-tenth the EPA-specified maximum contaminant level (MCL). The one-tenth factor accounts for multiple sources of the contaminant in the water system.

Risk Assessment for Unregulated Contaminants: Table 2 details the minimum toxicological studies necessary to support certification of products leaching unregulated contaminants. They are determined by the normalized "at-the-tap" concentrations. Higher concentrations require more toxicology data. Concentrations less than 10 ppb are evaluated for potential mutagenicity; concentrations of 10-50 ppb are evaluated for both mutagenicity and subchronic toxicity. Similarly, reproduction and carcinogenicity data are required when normalized contaminant concentrations are 50-1000 ppb, and greater than 1000 ppb, respectively.

TABLE 2--Minimum Toxicity Studies

Normalized Leachate Concentrations			
0-10 ppb	10-50 ppb	50-1000 ppb	1000 ppb+

• Ames assay
 • Chromosomal
    aberration
	• Ninety-day
	    rodent study
		• Teratology Studies
		    (2 species)
		 • Multigeneration
		    Study
			• Two-year rodent
			    cancer bioassay

In addition, supplemental studies may be required at the discretion of the reviewing toxicologist. For example, in determining mutagenicity, a weight-of-evidence approach takes into account not just the core minimum Ames assay and chromosomal aberration study (Table 1), but can be based on supplemental studies such as unscheduled DNA synthesis, DNA adduct studies, and/or dominant lethal studies. Similarly, neurobehavioral studies, immunotoxicity studies, pharmacokinetic studies, etc., can be cited. Given the difficulties in extrapolating animal data to man, epidemiological or case control studies may be of obvious utility in making an informed scientific decision. Although virtually any study may be relevant, it is noteworthy that the Standard requires that core minimum studies be of design reflecting the most recent version acceptable to the EPA, the U.S. Food and Drug Administration, or the Organization for Economic Cooperation and Development, and conducted according to Good Laboratory Practices [7, 8].

Unregulated contaminants determined to be unequivocally mutagenic (a non-threshold response) are required to undergo a two-year rodent cancer bioassay. Tumor data is then extrapolated using a linear multistage mathematical model. Exposures to carcinogenic substances do not preclude certification, provided the level of exposure is associated with an acceptable level of risk.

Subchronic and reproduction studies (threshold phenomena) are commonly evaluated by identification of no-observable adverse effect levels in animal studies. An appropriate safety (uncertainty) factor is then applied to achieve a maximum drinking water level (MDWL). This process parallels EPA's procedures for developing proposed MCLs. Again, pass criteria are based on one-tenth this level (the Maximum Allowable Level or MAL) to allow for multiple sources of the same contaminant.

Toxicology Summary: To summarize the toxicological approach, potential contaminants extracted from water contact surfaces are first identified, then quantified. Complete formulation information, to the ingredient supplier level, is critical to the evaluation. Regulated contaminant concentrations are compared with EPA MCLs. Unregulated contaminants are evaluated against animal and human toxicity data. For unregulated contaminants, the approach used requires higher levels of toxicology data as contaminant concentrations increase. An MDWL is established. Following normalization, a product or material may not contribute more than the MAL to drinking water (i.e., one-tenth the MCL or one-tenth the MDWL).

NSF CERTIFICATION OF PLASTICS PIPE AND FITTINGS

Overview: The National Sanitation Foundation offers a product certification program for plastics pipe and fittings under ANSI/NSF Standard 61. The Certification (Listing) program provides for independent, third-party evaluation and includes provisions for:

- toxicological evaluation performed by degreed, experienced toxicologists.
- confidential business information safeguards.
- quality assurance audits of production facilities.
- sample collection by NSF's field staff.
- sample preparation and testing.
- Listings of Certified products.
    (Listing books or catalogs of Certified products are widely distributed to regulators, utilities, and industry. Listing information is also available on-line, or by phone.)
- registered Mark authorized for use on products, on packaging, and in advertising.
- ongoing follow-up (audit, testing) to ensure continued compliance with the Standard.

- new contracts for Certification Services are signed annually with each company.
- regular review and updating of the Standard.

The NSF Mark identifies products that have been evaluated by NSF and found to be in full compliance with the requirements of the Standard and with NSF's Certification policies.

Certification: Because it is not practical for public health agencies, water utilities, or consumer groups to evaluate the hundreds of health- and environmentally-related products and materials, they typically depend on the NSF Mark and Certification as evidence that the requirements of the Standard have been satisfied. The NSF Certification process for plastics pipe and fittings is illustrated in Figure 1 and proceeds as follows:

- a company applies to have products Certified under the Standard, providing information on the sizes, styles, end uses (e.g., use in watermain, multiple user service lines, single user service lines, and/or residential applications, as well as end use temperature). In addition, information is provided identifying the material (or formulation) used. All information is kept strictly confidential.
- formulation, processing, and manufacturing information on each material/ingredient used is provided directly from the material formulators and ingredient suppliers.
- the preliminary toxicology review examines every ingredient/material supplier. The review identifies contaminants of toxicological concern, and specifies the analytical testing required.
- a regional or program representative schedules an initial audit at the manufacturing plant. Subsequent audits may be unannounced or announced. During the audit, formulation information such as batch tickets, suppliers, lot numbers, shipping records, and other related documents are examined and confirmed. The representative examines QA/QC programs in place, and inspects for potential contamination and cross contamination problems.
- product is sampled and shipped to NSF's laboratories for testing. Chain-of-custody records are maintained. An inspection report is left with the plant contact, itemizing the items of noncompliance (if required) requiring a response within an established time period.
- the samples are tested for leveling of chemical and microbiological contaminants, as specified by the toxicologist from the initial toxicological review.
- the laboratory results are "normalized," based on end use information provided by the product manufacturer and the basic assumptions provided in the Standard.
- normalized concentrations are compared with MALs--generally one-tenth of EPA's MCLs--to account for multiple sources of contaminants.
- unregulated contaminants are reviewed in conjunction with the toxicological information provided to arrive at an MCL equivalent, the MDWL. The MAL for unregulated contaminants is set to one-tenth of the MDWL, again, to account for multiple sources of contaminants.

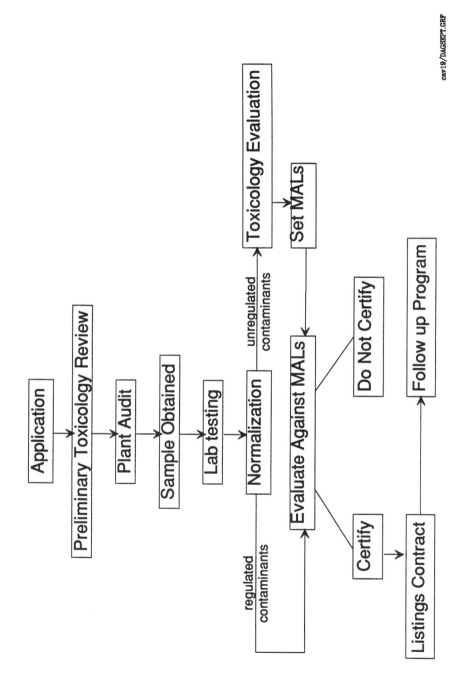

CERTIFICATION PATHWAY

Application → Preliminary Toxicology Review → Plant Audit → Sample Obtained → Lab testing → Normalization

Normalization → unregulated contaminants → Toxicology Evaluation → Set MALs

Normalization → regulated contaminants → Evaluate Against MALs

Set MALs → Evaluate Against MALs

Evaluate Against MALs → Do Not Certify

Evaluate Against MALs → Certify → Listings Contract → Follow up Program

car19/DAGSEPT.GRF

- when contaminant concentrations exceed the MAL, product rejection can result. When contaminant concentrations fall below the MALs and a product is eligible for Certification.

Once all the requirements of the Standard are satisfied when all the technical requirements have been satisfied and a contract executed, the product is Certified, and the Listing published. The Listing appears periodically in Listing books, is available electronically through NSF's on-line electronic Listings access service, or can be confirmed by directly by contacting NSF. Ongoing follow-up Certification procedures are implemented based on product types, the sources and types of materials used, and whether or not they are compounded/modified at the production plant.

Implications: The major perceived disadvantages to industry are the time and costs associated with Certification. Plastics products currently Certified under NSF Standard 14 are being reevaluated under the requirements of ANSI/NSF Standard 61 in order to maintain their NSF Certification. This effort will be completed by year end. New products, of course, must be evaluated at the outset for health effects under the ANSI/NSF Standard 61 requirements.

Once attained, however, the NSF Certification eliminates the need for multiple testing under various regulatory programs and provides for uniform interpretation of the Standard for equivalent types of products in the marketplace. All types of materials can be equally treated in terms of an independent health effects evaluation, demonstrating ongoing conformance with an established consensus Standard.

The NSF Mark adds credibility to products in the marketplace by demonstrating independent conformance with the Standard. Certified products generally meet the needs of water utilities' and other specifiers' contracts and specifications. By screening out products that are not appropriate for use in drinking water systems, the program may help utilities and other users avoid high removal and replacement costs.

Federal, state, and local regulatory agencies recognize and use NSF's Standard and Certification program as a basis for product/material acceptance in public drinking water systems. This third-party program frees regulatory agencies from the burden of maintaining redundant standards, testing, and evaluation programs. The Mark signifies to regulators that the audit, sampling, testing, and toxicological evaluation were performed by NSF, and that the product is part of a credible ongoing, third-party evaluation program.

For utilities, engineers, and other specifiers and users of the Standard, the NSF Certification program provides a means to clearly specify product requirements. For the first time, uniform health effects requirements are available for all types of pipe materials.

CONCLUSION

ANSI/NSF Standard 61 (and NSF Standard 14 by reference) and the NSF third-party Certification program provide a cost effective method for evaluating plastics and other types of piping materials. This Standard is the only widely accepted health effects consensus standard for pipe materials in the U.S., and is rapidly becoming incorporated into state regulations and water utility specifications.

It provides for comprehensive toxicological review and assessment, extraction testing, and microbiological growth support evaluation (when appropriate). Both regulated and unregulated contaminants are addressed.

REFERENCES

[1]   National Sanitation Foundation, Unpublished Data, 1976.
[2]   Bellen, G., Hammer, F., Dietz, E., and Kenyon, J., "Residual Vinyl Chloride Monomer Concentration (RVCM) In PVC and CPVC Products As A Measure of the Potential For Extraction of Vinyl Chloride Monomer in Water Exposed to Products," Proceedings of the Society of Plastics Engineers Regional Technical Conference, PVC: The Issues, Society of Plastics Engineers, September 1987.
[3]   Berens, A. R. and Daniels, C. A., "Prediction of Vinyl Chloride Monomer Migration from Rigid PVC Pipe," Polymer Engineering and Science, Vol. 16, No. 8, August 1976.
[4]   Crank, J. and Park G. S., Eds., Diffusion in Polymers, Academic Press, London, 1968.
[5]   Berens, A. R., "The Diffusion of Gases and Vapors in Rigid PVC," Journal of Vinyl Technology, Vol. 1, No. 1, March 1979.
[6]   Berens, A. R., "Diffusion and Relaxation in Glassy Polymer Powders: 1. Fickian diffusion of vinyl chloride in poly (vinyl chloride)," Polymer, Vol. 18, July 1977.
[7]   Code of Federal Regulations, Title 21, Part 58.
[8]   Code of Federal Regulations, Title 40, Part 792.

T.H. (Ted) Striplin

ULTRASONIC EVALUATION OF POLYETHYLENE BUTT FUSED JOINTS

REFERENCE: Striplin, T. H. (Ted), "Ultrasonic Evaluation of
Polyethylene Butt Fused Joints," Buried Plastic Pipe Technology, ASTM STP 1093 George S. Buczala and Michael J. Cassady,
Eds., American Society for Testing and Materials, Philadelphia, 1990.

ABSTRACT: Technology in ultrasonic inspection of polyethylene
heat fused joints is rapidly improving. Subtle flaws that
could not be detected previously can now be found. Sophisticated methods of destructively testing the fusions appear
to confirm these ultrasonic evaluations. Further study could
reveal the possibility of predicting long term fusion joint
reliability using state-of-the-art ultrasonics. It has been
proven that the use of ultrasonic testing of fusion joints
in the field and for requalification can greatly improve the
quality of workmanship.

KEYWORDS: Ultrasonic, high speed tensile, pulse-echo, pitch-
catch emulation, bend test.

Since its introduction in the 1960's the acceptance of
polyethylene piping in the gas industry has been phenomenal. Many
other industries such as municipal sewer and water, mining and
industrial plant piping have been using polyethylene piping in
increasing amounts. This growth in the usage of PE pipe has lead to
greater numbers of employees who are required to be proficient in
joining pipe materials.

The number of resin and pipe manufacturers has grown
accordingly in order to meet this ever increasing demand for PE
pipe. Unfortunately, nearly every pipe manufacturer has a different
set of parameters to join the pipe he manufactures.

The basic parameters are consistent with each manufacturer.
They are:
         - Clean
         - Face
         - Heat
         - Join
         - Hold

Mr. Striplin is Sales Manager for Fusion Products and Project
Manager for Ultrasonic Development at McElroy Manufacturing, Inc.,
833 North Fulton, Tulsa, Oklahoma 74115.

The critical variables in the process are heating and cooling times, heating temperature, and heating and joining forces. The recommended time, temperature and force values are seldom reproduced by different pipe manufacturers. Even those manufacturers using the same resins feel that their method produces slightly better results than the other manufacturers method. Because of these differences in procedure, plus the difference in design and performance of the fusion equipment a simple process has been made somewhat confusing.

Although failure in butt fused joints is not common, several sets of interacting circumstances can result in short term or long term failure. Some of these include poorly maintained joining equipment, dirty or poorly maintained heater plates, or failure to follow proper joining guidelines. Environmental conditions such as, wind or water preventing proper melt, cooling melted surfaces before joining, or dirt or debris falling on the prepared joint surfaces can also cause lack of integrity. The heat fusion process, although simple to perform and extremely reliable, like many other things we do frequently, can be thought to be infallible. Then we are shocked when we discover that it, and all processes, are not infallible.

These and other factors lead to the necessity of increased field inspection of fusions and qualification and requalification of operators. The only method available to us initially to nondestructively test butt joints was visual inspection of the outside diameter bead. If the bead appeared to be in accordance with the pipe manufacturer's recommendations it was assumed that the fusion was good. Generally this was the case. However, some of the joint failures that occurred had beads which passed visual inspection. Therefore, it was determined that there could be lack of bonding, voids or inclusions in the interfacial area of the joint which could not be visually detected.

The obvious solution to this dilemma was X-ray and ultrasonics. X-ray was expensive, bulky and failed to show the subtle flaws in the PE pipe joints. Ultrasonics, hopefully, is the answer.

Principles

Basic ultrasonic technology has not changed much in the last 50 years.

A short pulse of electric current excites a transducer (crystal) which vibrates at a high frequency such as 20 Mhz or 20,000,000 cycles per second. As this mechanical energy passes through the material being inspected (medium) the energy causes the individual particles in the medium to vibrate. The particles in the medium do not move as the sound wave does, but react to the energy of the sound wave. If the sound wave comes in contact with a discontinuity in the medium, the energy is disturbed and part of it will be rerouted in direction. The discontinuity can be voids or cracks, inclusions or any change in the basic mediums' acoustic impedance.

Methods

There are several methods of ultrasonic testing. Of them, Pitch-Catch and Pulse-Echo are the most popular. Pitch-Catch utilizes a sending transducer and a receiving transducer. The sending transducer transmits a sonic wave through the medium and expects to receive a certain amount of energy at the receiving transducer. If no flaws are found a strong signal is expected. If flaws are present, the signal received will be significantly reduced because the sound wave pattern is disturbed and fragmented.

Pulse-Echo utilizes one transducer which both sends and receives the sonic wave. The sonic wave is sent out in pulses and travels through the medium continuing uninterrupted unless flaws are present. If flaws are present a reflected sound wave will return to the transducer and is received between the pulses. The amount and strength of the reflected energy is dependant strictly on the size of the flaws and their geometric pattern.

Of these two methods, the most preferred is Pitch-Catch. Its ability to detect flaws is _not_ dependant on flaw size and geometry. In fact, small flaws are better detected by the disturbance they create than the reflection they give.

However, before selecting a method for inspecting a particular material two things must be considered:

Shape of Material: The shape of the material we are discussing (pipe) is obviously cylindrical. Since the area of concern is the fusion, we want to pass the sonic wave through the fusion zone.

Acoustic Properties of Material: Polyethylene is a p6or thermal, electrical and sound conductor. Because of this it is a difficult material to inspect ultrasonically.

Because the pipe is cylindrical, placement of transducers for Pitch-Catch poses a problem. (We obviously cannot place them against the pipe's inner wall.) Because of this and PE's poor conductivity Pitch-Catch becomes impossible. We are left with only one method, Pulse-Echo.

Traditionally Pulse-Echo inspection of PE pipe is accomplished by: Mounting a focused beam transducer on a lucite block which is at an angle to the centerline of the pipe. The bottom of the block is machined to the same radius as the pipe's outside diameter. A liquid couplant is placed between the transducer block and pipe, then the inspection begins.

The sonic wave is pulsed through the lucite block then refracts when it contacts the pipe, which has a different acoustic independence than the lucite. The sonic wave then travels through the pipe wall and eventually through the fusion zone. If no discontinuities are found, the sonic wave will continue through the pipe wall until it is absorbed. If discontinuities are found, a reflected sonic wave will return to the transducer. The energy which returns is then quantified. (The intensity is dependent on

(1) flaw size and geometry and (2) location in relation to the center of the sonic beams.) Once a value is given to the reflected energy, then the value is compared to a set threshold and a decision is made as to whether the flaws, which are detected, are too large to pass the inspection. Unfortunately, subtle flaws and sometimes obvious flaws may go undetected. Until recently, highly trained technicians were required to interpret the waveform which was generated from the reflections. Now products are available which automate this process.

A new concept has been introduced which operates primarily as a Pulse-Echo System, but emulates the Pitch-Catch method. Unlike conventional Pulse-Echo Systems, a sonic beam is transmitted through the pipe wall and spreads outward from the center. The transducer angle is such that not only the fusion zone is bombarded, but the inside diameter fusion bead is inspected as well. Each butt joint has an inside diameter bead and if properly formed, through proper joining practices, that bead will reflect a significant amount of energy which will return to the transducer. When the small flaws or discontinuities are present, the disrupted sound waves interfere with the sonic wave which would normally be reflected to the transducer. Therefore, small flaws or discontinuities can be detected, not from the reflections they give but from the disturbances they create. This technology provides more information about the joint which can be used to not only determine if the fusion is good or bad, but perhaps provide insight as to long term strength of the joint. This is probably the advantage this new technology has over standard technology - possible long term prediction.

Destructive Testing

The primary method of distructively testing butt joints in the field or for operator qualification is the bend back test. There are several sophisticated tests, although researchers do not agree on which is the best test to determine long term performance of the fusion joint. High temperature bath tests do not necessarily expose possible problems with impact or point loading. Tensile tests do not simulate impact or point loading stresses. Tensile-impact testing is probably the best of this group for determining fusion joint quality; however, it does not allow the tensile properties of the pipe to add to its strength. We have developed a machine which we feel gives an accurate evaluation of joint strength and reliability. "McSnapper" is a high speed tensile test with impact. While the standard tensile-impact tests are more impact than tensile, this test is more tensile than impact. Perhaps it should be called an Impact-Tensile Test.

Again, it appears that this test can be a useful tool in determining long term joint reliability. Figure 1 shows a typical load/time relationship that the McSnapper provides. The X-axis shows the time sequence with each plot point in milliseconds. The Y-axis shows the load in kilograms that the sample withstands. As shown, the sample withstands a load of 1605 kilograms and tapers off to 1206 kilograms. This pull shows good ductility. Several control samples (samples with no fusion) are pulled and used to compare with

CONTROL SAMPLE

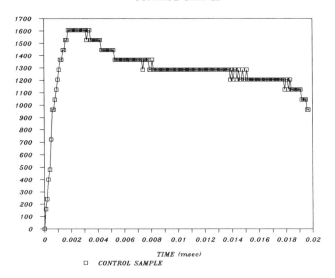

Fig. 1 - - Unfused Control Sample

GOOD FUSION

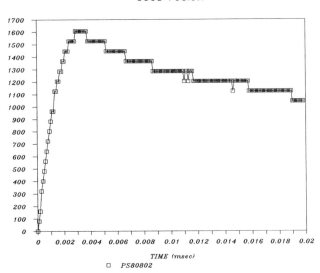

Fig. 2 - - Good Fusion

the fused samples. If the control samples do not exhibit good ductility the pipe lot is not used for the ultrasonic evaluation. Figure 2, represents a good fusion, and Figures 3, and 4 represent two bad fusions.

The good fusion (Figure 2) exhibits good ductility as all good fusions should. The initial load is 1605 kilograms and tapers down gradually to about 1043 kilograms in the 200 millisecond span.

Figure 3 shows a bad fusion which builds up to a load of 1285 kilograms and holds for only 1/10 of a millisecond before brittle failure. This failure could be readily detected using the bend back test. However, Figure 4 shows a bad fusion (cold joint) which illustrates some interesting characteristics. The sample withstands a load of 1605 kilograms initially and tapers off slightly for 3 milliseconds. Then in just over 6 milliseconds after the pull was initiated it fails in the brittle mode. Our testing has found this type of brittle failure to be the way a cold joint typically fails in this high speed tensile test. Notice that the control sample (Figure 1), good fusion (Figure 2), and the bad fusion - passed bend test (Figure 4) all had an initial load of 1605 kilograms. However, the bad fusion (Figure 4) began dropping rapidly until brittle failure at 6 milliseconds. Therefore, while not conclusive, it appears that the failure type in Figure 4 initially withstands the same load as the pipe or a good fusion joint, which is sufficient to pass a bend back test but not the high speed tensile test.

While it is a simple task to find gross flaws which will not pass a bend test, the challenge is to establish parameters in the ultrasonic test system's programming to detect minute flaws. Since detection of these subtle flaws is quite difficult, it is a painstaking and timely process. But, hopefully, the result of this task will be well worth the effort.

Field Use

Although ultrasonic testing of fusion joints has been performed for several years, highly trained technicians with considerable judgement were required to read the wave form. The process was slow, tedious, and open to interpretation. Recently, more automated versions of both old and new technologies have become available. These developments allow the average person with minimal training to perform field inspections of fusion joints on a productive basis.

Random Field Testing

Ultrasonic testing can be used to supplement visual field inspections currently being carried out by an inspection staff. Random inspections of field fusions will insure long term reliability of the PE pipeline. Most bad fusions are a result of worn or faulty equipment or use of incorrect fusion procedures. Random field testing can quickly expose these conditions because they will be consistent from joint to joint.

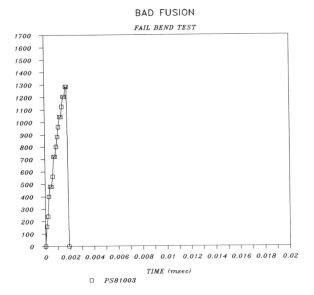

Fig. 3 - - Very Bad Fusion

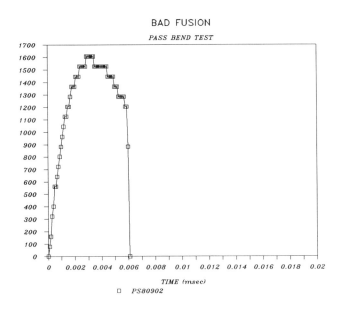

Fig. 4 - - Cold Joint

Total Inspections

100% ultrasonic inspection of fusion joints in pipes laid in more critical applications should be considered. Road crossings, insertions and installations in highly populous areas are some of the applications which apply.

Qualification of Operators

Qualification and requalification of fusion operators can be a very important function of ultrasonic testing. This task, as it is performed now, is a time consuming and costly operation. The time involved in cutting straps from a fusion joint then performing the bend test can be from 5 to 10 minutes. And then, only 40% (2" IPS pipe) or 14% (6" IPS pipe) of the joint area is required to be tested according to Pipeline Safety Regulations Part 192.285. If 50 operators were to be qualified in one day, this procedure alone could take over 8 hours. Using state-of-the-art ultrasonics, 100% of the fusion area can be inspected in about one hour and without the risk of injury related to band saw operation.

State-of-the-art ultrasonic testing equipment has the capability of recording the inspection so a permanent record can be maintained.

Improved Performance

A gas utility used ultrasonics, in January, 1989, to requalify the fusion operators. The results were reported as follows:

1. Only 57% of the fusion joints passed visual inspection.

2. Only 22% of the fusion joints passed ultrasonic inspection.

These figures probably do not represent the industry. Two factors were responsible for these staggering results:

1. Operators were using improper fusion procedures.

2. The fusion equipment was in need of repair.

The utility involved initiated a program to repair the equipment to the manufacturer's standards and began immediate retraining of their fusion personnel. After this program was implemented, retesting revealed an impressive 100% passing the visual inspection and over 98% passing the ultrasonic inspection on the first try. As a company official said "Overall quality of workmanship has improved since we began using ultrasonic testing".

Although fusion problems are rarely evident, ultrasonics can help provide quality and reliability to all industries that use polyolefin piping.

ACKNOWLEDGMENTS

The author acknowledges with appreciation the help of Sue Brown, Donna Dutton, and David Dutton in the preparation of this paper.

Jeremy A Bowman

THE FATIGUE RESPONSE OF POLYVINYL CHLORIDE AND POLYETHYLENE
PIPE SYSTEMS

REFERENCE: Bowman, J.A., "The Fatigue Response of Polyvinyl
Chloride and Polyethylene Pipe Systems," Buried Plastic Pipe
Technology, ASTM STP 1093, George S. Buczala and Michael J.
Cassady, Eds., American Society for Testing and Materials,
Philadelphia, 1990.

ABSTRACT: Fluctuating internal pressures induce fatigue stresses
in pipes and fittings. The influence of these fatigue loadings on
the strength of unplasticized polyvinyl chloride (UPVC) and
polyethylene (PE) pipe systems is examined. The fatigue response
of UPVC pipe is well defined. For PE pipe systems elevated
temperature fatigue tests identify fittings and joints as the
prefered failure sites. The literature indicates that PE pipe
systems, jointed by butt fusion in particular, have the best
projected fatigue lifetimes, and are capable of withstanding
significant surge fatigue stressing at 20/23°C.

KEYWORDS: Plastic pipes, plastic pipe fittings and joints,
polyvinyl chloride, polyethylene, fatigue behaviour.

INTRODUCTION

The low pressure and low temperature pipe markets have come to be
dominated by plastics, and in particular by unplasticised polyvinyl
chloride (UPVC) and the two linear polyethylenes, medium (MDPE) and
high (HDPE) density polyethylene. Both UPVC and MDPE/HDPE are used for
buried potable (drinking) water systems, for sewer pipes and pipe
systems in chemical plants. In these applications internal pressure
fluctuations and traffic loading can induce fatigue stresses. It is
important to assess if these fatigue loadings induce such damage that
pipe systems failure becomes either a possiblility or a reality.

The paper examines the fatigue behaviour of UPVC and MDPE/HDPE
pipe systems subject to fatigue arising from internal pressure

Dr. Jeremy Bowman is processing and development manager at Fusion
Plastics Ltd, Carrwood Road, Chesterfield, Derbyshire England.

fluctuations. The first section identifies and discusses the four different fatigue failure modes exhibited by thermoplastic materials. The second and third sections discuss the fatigue behaviour of UPVC, and MDPE/HDPE pipe systems respectively; consideration is given both to the response of pipes and the behaviour of joints and fittings. The final section attempts to quantify the fatigue damage pipe systems sustain in service. From the comparison of the laboratory studies and the expected in service fatigue damage, recommendations follow on the prefered materials and jointing methods for fatigue tollerant plastic pipe systems.

FATIGUE FAILURE MODES FOR THERMOPLASTIC MATERIALS

Definitions of Fatigue Test Variables

With dynamic fatigue it is possible to impose one of three well used loading profiles, see Figure 1. The variables associated with these different fatigue loading profiles can be divided into two groups, one associated with stress, the second with time.

The stress variables are defined for the sinusoidal loading profiles, see Figure 1 (a) and (b). From the minimum (σ_{min}) and maximum (σ_{max}) stresses, the mean stress, stress ratio, stress range and stress amplitude can be calculated:

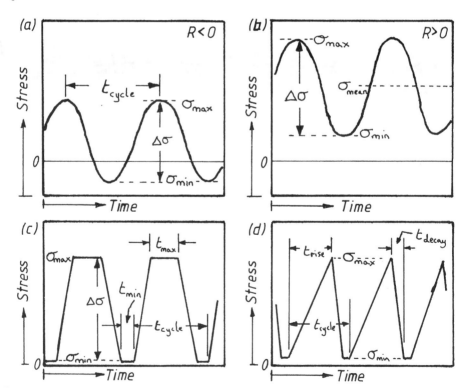

Figure 1. Schematic presentation of different loading profiles; (a) and (b) sinusoidal, (c) trapezoidal and (d) sawtooth.

Mean Stress (σ_{mean}); σ_{mean} = $\frac{1}{2}(\sigma_{max} + \sigma_{min})$

Stress Ratio   (R);   R   = $\sigma_{min} / \sigma_{max}$

Stress Range $(\Delta\sigma)$; $\Delta\sigma$ = $\sigma_{max} - \sigma_{min}$

Stress Amplitude (σ_a); σ_a = $\frac{1}{2}(\sigma_{max} - \sigma_{min})$

Convention regards tensile stresses as positive and compressive as negative, so if compressive stresses are present R < 0. With pipe systems the internal gage pressures are usually positive, so R > 0.

For pipe fatigue, stresses arise from varying internal pressures. The minimum (P_{min}) and maximum (P_{max}) pressures give ΔP, as $\Delta P = P_{max} - P_{min}$. Knowing ΔP, the hoop stress range in the pipe, $\Delta\sigma_H$, can be calculated, for thin walled pipes, by

$$\Delta\sigma_H = \frac{\Delta P . d_{OD}}{2t} \qquad\qquad 1\ (a)$$

and for thick walled pipes by

$$\Delta\sigma_H = \frac{\Delta P . (d_{OD} - t)}{2t} \qquad\qquad 1\ (b)$$

where d_{OD} is the outside diameter and t the wall thickness.

The time variables are annotated in Figure 1 (c) for the trapeziodal loading profile, and discussed individually below.

   (i)   Cycle time (t_{cycle}) is the time required for one complete cycle. The test frequency (f) is given by f = 1/t_{cycle}

   (ii)   Time at maximum stress (t_{max}) is the time at or close to the maximum stress.

   (iii)   Time at minimum stress (t_{min}) is the time when the applied stress is at the minimum.

   (iv)   Rise (t_{rise}) and Decay (t_{decay}) times are the times required to raise and lower the stress.

For sinusoidal loading only the cycle time is clearly identifiable. For trapezoidal profiles all times are identifiable. Note for the square wave $t_{rise} \to 0$ and $t_{decay} \to 0$, while for the sawtooth profile $t_{max} = 0$, see Figure 1 (d).

Having identified the fatigue loading variable's the fatigue failure modes will be discussed.

Thermal Failure

Fatigue thermal failure of thermoplastics is due to the accumulation of hysteretic energy and the poor heat transfer characteristics of polymers (1). The temperature of a plastics component subjected to high frequency fatigue can rise by up to 70°C.

For most thermoplastics strength declines with increasing temperature, so fatigue induced temperature increases precipitate the failure of the component, usually in a ductile mode.

The possibility of a buried plastics pipe system exhibiting thermal failure is remote. This is because the fluids within and around the pipes conduct away heat, the pressure fluctuations are generally of a low frequency, and the applied stress ranges small. However, laboratory tests should avoid the significant self-heating by keeping the frequency of loading to 2Hz or below.

Cumulative Creep Damage Failure

A plastics pipe system can experience a fatigue profile such that in each cycle the time at maximum stress (t_{max}) is both measureable yet significantly less (<0.001) than the static stress lifetime. If the number of loading cycles (n) required to induce failure is equal to N_f, then the time under maximum stress, τ_{FAT}, is given by

$$\tau_{FAT} = \sum_{n=1}^{N_f} (t_{max})_n \qquad \qquad 2\,(a)$$

For a loading profile with each cycle equivalent, equation 2 (a) simplifies to

$$\tau_{FAT} = N_f.t_{max} \qquad \qquad 2\,(b)$$

Equations 2 (a) and (b) define the creep damage introduced into the pipe system when subjected to fatigue. If sufficient creep damage is introduced, pipe system failure follows when

$$\tau_{FAT} \approx \tau_{SR}$$

where τ_{SR} is the creep or stress rupture lifetime for the same test temperature (T) and at $\sigma = \sigma_{max}$. This form of fatigue failure is thus creep controlled, and the test may be termed an interrupted creep or interrupted stress rupture test.

Figure 2 illustrates an example of creep controlled failure of a HDPE pressure pipe, tested at 80°C under fatigue with R = 0 and $\Delta\sigma$ = 4.9MPa. As t_{max} was varied, τ_{FAT} was largely constant, with the variable value for N_f predicted from a knowledge of the static stress lifetime (2). The fatigue failure of UPVC pipes, at high value of σ_{max}, can also arise from the accumulated creep damage (3).

Cycle Dependent Fatigue Failure

In cycle dependent (or true) fatigue failure the repeated application of the stress induces damage which manifests itself by the cycle controlled propagation of a crack (1). Since the application of the stress induces damage, the time at maximum stress, t_{max}, should not effect the number of cycles to failure, N_f.

For a small diameter, butt jointed, HDPE equal tee (4), Figure 3 shows N_f was essentially constant as the creep time per cycle was

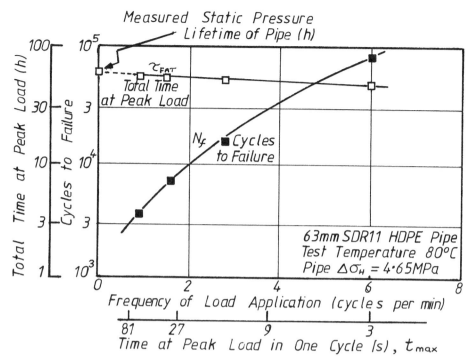

Figure 2. For a fixed temperature and pipe hoop stress range, the influence of t_{max} on N_f and τ_{FAT}. Note the latter remains constant as t_{max} varies, indicating cumulative creep damage failure.

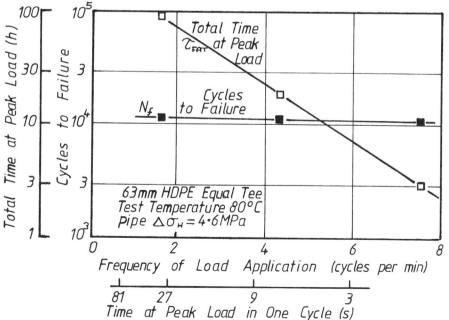

Figure 3. The influence of t_{max} on N_f and τ_{FAT}. Note N_f remains constant as t_{max} varies, indicating a true fatigue failure.

increased from about 1   to 30 seconds; the accumulated creep damage,
τ_{FAT} was not constant and changed with t $_{max}$, see Figure 3.

For cycle controlled fatigue failure it is usual to present data
as a double logarithmic plot of stress range ($\Delta\sigma$) against N_f, where
straight line plots are usually observed (1). This implies N_f, is
related to $\Delta\sigma$ by a power law relationship of the form

$$N_f \propto (\Delta\sigma)^{-b}$$ 3

where b>1. Decreases in $\Delta\sigma$ extend the fatigue lifetimes, with N_f
not influenced by t $_{max}$, unlike the case of creep controlled failure.

Combined Creep and Fatigue Damage

For metals, it is known that combined creep and fatigue loading
can introduce damage, either independantly or synergistically, to
induce failure (5). This topic must be considered since both constant
and fluctuating internal pressures are observed with pumped
pipelines. A code has been devised by the American Society of
Mechanical Engineers (ASME) to separate and quantify the creep and
fatigue damage (5). The measured fatigue life, N_f, is normalised
using N°_f the fatigue lifetime for a cycle with little or no creep
damage. The cumulative creep damage τ_{FAT}, is normalised using the
stress rupture lifetime, τ_{SR}, measured at the same temperature and
maximum stress used in the fatigue test. The normalised fatigue damage
(N_f/N°_f) and the normalised creep damage (τ_{FAT}/τ_{SR}) can be added to
give the combined damage:

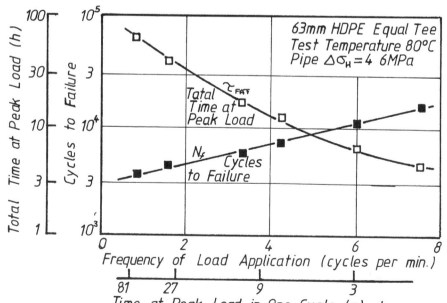

Figure 4. The failure of a different HDPE equal tee (from that in Figure
3) showing combined creep and fatigue failure mechanism. Both N_f and τ_{FAT}
varied as t_{max} was changed.

$$\text{Combined Damage} = \frac{N_f}{N_f} + \frac{\tau_{FAT}}{\tau_{SR}} \qquad \qquad 4$$

For the combined damage greater than 1, the mixing of creep and fatigue is beneficial. For the combined damage equal to unity the fatigue and creep loading introduce damage in direct proportion to their contribution. For the combined damage less than 1, mixing creep and fatigue is deleterious, with the lifetime shorter than that predicted from N_f and τ_{SR}. For the combined damage less than 1 a case case was seen for an injection molded HDPE equal tee fatigue tested at 80°C (6). Figure 4 shows both N_f and τ_{FAT} varying as t $_{max}$ was changed, this implying that both the creep and fatigue elements of loading contributed to the failure process.

This brief review of the fatigue failure modes for unreinforced thermoplastic resins is used to interpret the data on the fatigue performance of UPVC and MDPE/HDPE pipe systems. Separate consideration is given to the behaviour of pipes, fittings and joints.

FATIGUE BEHAVIOUR OF UPVC PIPE SYSTEMS

UPVC Pipe

The fatigue behaviour of UPVC pipes, loaded by internal pressure fluctuations, has been examined in depth by Joseph and Leevers (3)(7). Un-notched pipes, with an outside diameter of about 60mm and with a nominal wall thickness of either 3.4 or 4.2mm, were subjected to a trapezoidal loading profile, at a frequency of 1 Hz at 20°C, with a fixed minimum pressure of 0.4 MPa. Figure 5 presents data on the 60mm pipe as log $(\Delta \sigma_H)$ versus log (N_f), with Figure 6 replotting the data as log $(\Delta \sigma_H)$ against log (τ_{FAT}).

From Figures 5 and 6 it can be seen that
  (i) at high stress ranges the fatigue strength is controlled by the static stress lifetime; (creep controlled failures):
 (ii) for $\Delta \sigma_H \leqslant 30MPa$, the failures are "brittle", true fatigue failures;
(iii) a reasonable straight line plot was obtained when using double logarithmic axes for $\Delta \sigma_H$ against N_f (see equation 3);
 (iv) the fatigue strength of the thicker walled pipes was similar to the thinner walled;
  (v) the was no clear evidence of a fatigue limit.

In addition to the above, Joseph and Leevers (3) showed that 60mm UPVC pipe from different manufacturers had very similar fatigue strengths, that mean stress had little influence on fatigue strength (7), and that the performance of 168mm diameter pipe was very similar to the 60mm (7).

Two other studies on the fatigue strength of UPVC pipes are considered. Hucks (8) tested 2" and 4" pipes at 24°C using a base pressure of 0.35 MPa and a range of frequencies between 0.0017 and 0.38Hz. No clear influence of frequency on N_f was observed (8), despite fracture mechanics studies showing crack growth rates accelerating with decreasing frequency (1)(9). Vinson (10) tested 6"

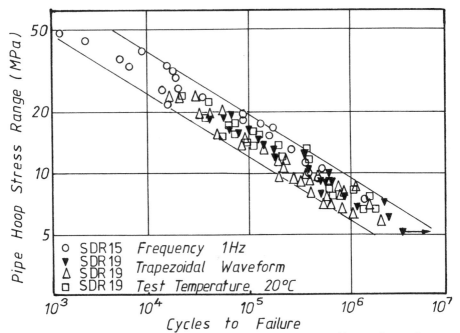

Figure 5. The influence of pipe hoop stress range on the number of cycles to failure for UPVC pipe. Note the different   symbols represent different pipe manufacturers.

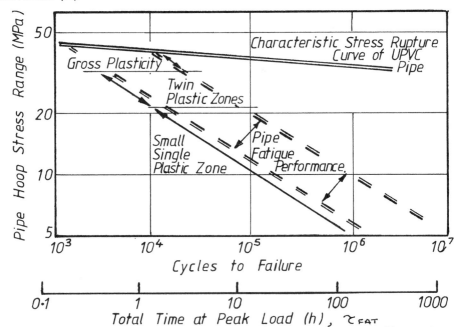

Figure 6. The data in Figure 5 represented schematically to illustrate the calculated total time at peak load. At low values of $\Delta\sigma_H$ the failures are brittle with τ_{FAT} very small compared to τ_{SR}.

SDR 17 UPVC pipe at 23°C, using a sawtooth profile, with frequencies between 0.1 and 0.17 Hz using a base pressure of between 0.35 and 0.48 MPa. When this data of Vinson (10) and Hucks (8) is compared to that of Joseph and Leevers (3), the agreement between the different studies is good, see Figure 7. The fatigue strength of conventional UPVC appears to be well defined and little influenced by the manufacturer of the pipe, the pipe diameter or the frequency of the test.

The good agreement of the data cited above allows the projection of a lower bound value (11), see Figure 7. This can define the influence of $\Delta\sigma_H$ on the fatigue lifetime (N_f) of UPVC pipe subjected to pulsating internal pressure loading at about 20°C;

$$N_f = 2.85 \times 10^8 \ (\Delta\sigma_H)^{-3.35} \qquad\qquad 5$$

Finally the behaviour of oriented UPVC pipe is commented on. Rings taken from pipe were subjected to fatigue to reveal an enhanced fatigue strength (12). Previous studies have shown pipe rings and complete pipes gave similar fatigue strengths (11). Molecular orientation is therefore seen as a route to enhance the fatigue strength of UPVC pipes;

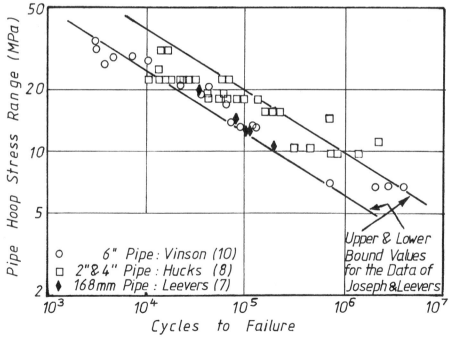

Figure 7. Data on the fatigue strength of UPVC pipes from Vinson (10), Hucks (8) and Leevers and Joseph (7). This is compared to the data from Joseph and Leevers (3) shown in Figure 5. The agreement between the different studies is good indicating a commonality of fatigue response for UPVC pipes.

Other Factors Influencing the Fatigue Strength of UPVC Pipe

In the transportation and laying of plastic pipe, surface
scratches can be introduced. Vinson (10) introduced a single long
(610mm) scratch at the minimum wall. Those scratches 4.7% of the wall
deep were not deleterious to the performance of the pipe; they were
not the focus for failure. Scratches that were 9% deep initiated
failure and reduced performance. But the reduction in fatigue
lifetime was due to reducing the effective wall thickness, rather
than the scratches acting as stress raisers. Care should be exercised
with this conclusion, since different scratches will have different
root radii to influence strength in different ways (1).

Studies (13) with buried pipes have revealed that some plastics
pipes can deform due to surface loading. Christie and Phelps (14)
examined how external compression influenced fatigue lifetime, when
the UPVC pipe under test was internally notched. Small external
compressions reduced fatigue strength markedly; at 5% compression the
fatigue lifetime was down to one tenth of the lifetime of pipe not
subjected to compression. This appears to be an area worthy of further
study using un-notched pipe or pipe notched on the outside wall.

Finally, it is noted that fracture mechanics studies on
fatigue crack growth have explored the influence of polymer molecular
weight and temperature. High molecular weights retard crack growth
(1)(15), which should extend fatigue lifetimes of pipes made from
higher molecular weight resins. Second, for UPVC plates it has been
shown fatigue lifetimes decrease with increasing temperature (16).
Since to date data has been obtained at temperatures between 20 and
24°C, lower fluid temperatures should enhance fatigue strength. For
higher fluid temperatures (> 25°C) down-rating is needed, the level
of which has not been clearly identified.

UPVC Fittings and Joints

Jacobi (17) has examined the fatigue strength of equal tee
fittings designed for 90mm outside diameter pipe, having a 7.5mm pipe
wall. The fittings were solvent jointed to pipe and the system
subjected to sinusoidal pressure amplitudes superimposed on a constant
line pressure. Tests were conducted at 20°C and a frequency of 1.33
Hz. Figure 8 presents the data from Jacobi as log (pipe hoop stress
range) against log N_f. Two curves are included, one for the original
design of fitting and a curve for the re-designed fitting. Tested
under constant pressure no failures were seen in the fittings. Under
fatigue, the original design of the fitting failed in a brittle
manner, the cracks initiating at the through section/branch arm
intersection. Re-design of the fitting enhanced the fatigue strength.
However, for both the original design and the re-design, the fitting
was weaker than the pipe. Menges and Roberg (18) also identified
fittings as a point of weakness under fatigue loading, with the
fitting failing before the pipe.

For the strength of UPVC joints under fatigue, Onishi et al (16)
have shown that hot plate joints have lower fatigue strengths.
However, UPVC pipes are more usually joined by solvent jointing or the
use of rubber 'O' rings. No published information was found on the

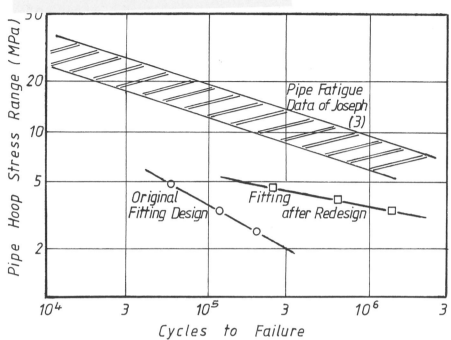

Figure 8. The fatigue response of a UPVC equal tee. Note first that the design was important in determining response, and that the fitting failed before the pipe.

strength of solvent joints under dynamic fatigue, although a recent paper showed creep loading could induce joint failure (19). This appears an area worthy of further study.

FATIGUE BEHAVIOUR OF MDPE/HDPE PIPE SYSTEMS

MDPE and HDPE Pipes

Lörtsch (20) explored the fatigue strength of HDPE pipes at 20°C, using high values for σ_{max} so as to induce failure in reasonable times. Under these conditions there was no evidence of a fatigue weakness; rather there was evidence "that pulsation increases the strength of the pipe". Barker et al (2), using older, low stress crack resistant HDPE pipes (Lifetimes at 80°C and 4.5 MPa of about 60 hours under constant pressures) showed again that 80°C fatigue induced failures were controlled by the static stress lifetime. These pipes therefore did not exhibit a fatigue weakness, even when tested at 80°C; there failure was due to the accumulated creep damage as defined in equation 2. This is illustrated in Figure 2 where τ_{FAT} is largely constant for varying values of t $_{max}$.

Modern MDPE pipes have improved stress crack resistance; under constant internal pressure loading, lifetimes in excess of 10,000 hours at 80°C and σ_H = 4.5 MPa are recorded. For these extra-tough MDPE pipes, recent evidence shows fatigue can induce brittle failures

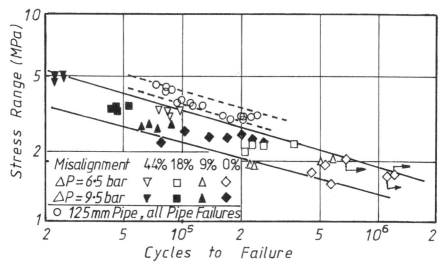

Figure 9. The 80° fatigue response of 125mm SDR11 MDPE pipes (21) failing in the pipe (o) and the behaviour of 90mm MDPE pipe systems containing a misaligned butt joint (24)(25). For misaligned pipes the failures were at the joint and the stress range used is the amplified butt joint stress (25). It is not applicable to correlate the pipe data with the misaligned butt joint lifetimes.

at 80°C in reasonably short times. Tests with un-notched 125mm SDR11 pipes at 80°C, with R = 0 and $\Delta\sigma_H$ = 4.5 MPa, showed fatigue induced brittle failures after about 250 hours of testing when using a trapezoidal loading profile with a frequency of about 0.083 Hz (21). These were fatigue induced failures. Work on externally notched MDPE gas pipes like-wise found that 80°C fatigue induced brittle failures in short test times (22).

The influence of pipe hoop stress range on the 80°C fatigue lifetime of un-notched MDPE pipes is shown in Figure 9. All failures were brittle (23), the substantial damage was from the fatigue element of loading (21)(23), and the influence of log $\Delta\sigma_H$ on log N_f was approximately as described by equation 3. Thus the 80°C data shows that the fatigue life of an MDPE pipe can be extended by reducing the pressure change, so reducing the chance of a "brittle-like" failure. Estimates of the 20°C strength are covered later in this paper.

MDPE/HDPE Butt Fusion Joints

Parmar and Bowman (24)(25) have examined the 80°C fatigue strength of butt fusion joints in 63, 90 and 125mm SDR11 blue water grade MDPE pipe. The focus of the work was to explore the influence of axial misalignment at the butt joint. It was shown that axial misalignments of 10% or greater (of the pipe wall thickness) induced failure at the joint (24), with increasing misalignment progressively reducing the 80°C fatigue strength (25), see figure 9. Work on aligned butt joints showed failures at (24) and remote (21)(23)(24) from the join. Thus the work on the fatigue strength of butt joined polyethylene pipes supports the 80°C constant stress studies (26)(27) that well made butt joints are as strong as the pipe itself.

The publications referred to above cite the fatigue response of complete butt joints. Other studies have compared the   fatigue strength of base resins to butt joints using notched tensile samples. It was observed that the polyethylene at the butt joint was weaker than the base resin (28). However, as indicated above, correctly made butt joints are as strong, or stronger, than the pipe when the fatigue tests evaluate the total joint. Care should therefore be excercised with data obtained on notched samples taken from butt joints.

MDPE and HDPE Pipe Fittings

Published literature shows 80°C fatigue stressing can induce HDPE (4)(6) and MDPE (21)(29) pipe fittings to fail, with the fittings failing prior to the pipe, see Figures 10 (a) and (b). The failures were brittle (4)(21), the accumulated creep damage small (4)(21)(29) and the plots of $\log \Delta \sigma_H$ against $\log N_f$ gave reasonable straight line plots over the substantial range of $\Delta \sigma_H$ values investigated, see Figure 10. These observations infer that these HDPE and MDPE fittings are failing by a fatigue mechanism, with the consequence that reductions in $\Delta \sigma_H$ substantially extended the fatigue lifetime. An additional point should be observed from Figure 10, and that is there is some evidence for a fatigue limit at low values of $\Delta \sigma_H$. This is an important point, and it needs resolving as to whether there is a fatigue limit per se, or whether the material ages during testing and in so doing become stronger.

In studies on electrofusion couplers (29), socket couplers (21)(23) notched pipes (22) and notched tensile samples (28) it is well documented that different grades of polyethylene, and different batches or lots within a single grade, give different fatigue strengths. For the elecrofusion couplers the value of N_f could vary by a factor of ten (21). This demonstrates clearly that for MDPE or HDPE pipe grade resins there is no well defined fatigue strength, unlike the case of UPVC pipes where the fatigue strength is well defined despite the pipes being sourced from companies on different sides of the Atlantic ocean. Thus the fatigue strength of MDPE and HDPE pipes and fittings is grade specific with some having significantly better fatigue strength. This variation in fatigue strength with the grade of resin means no single equation can describe the fatigue strength of polyethylene pipe systems.

The Influence of Notches, Temperature and Combined Loading

For a single batch/lot of 125mm SDR11 MDPE pipe, the influence of the depth of external notches on the 80°C fatigue strength of pipes has been assessed (23). Notches of a depth of 4.1% of the pipe wall thickness did not act as focus for failure, while notches 9.3% deep and above did focus the failure. At about 19% deep the fatigue lifetime was down to one tenth of the pipe alone. Interestingly these results are similar to those cited above, from Vinson (10), for UPVC pipes, where 4.7% notches had little influence while deeper notches did focus failure and reduce the fatigue strength of UPVC pipes.

The substantial work undertaken to date on the fatigue response of polyethylene pipe systems has been at 80°C. For most applications it is necessary to estimate to 20 or 23°C fatigue strength. For a single

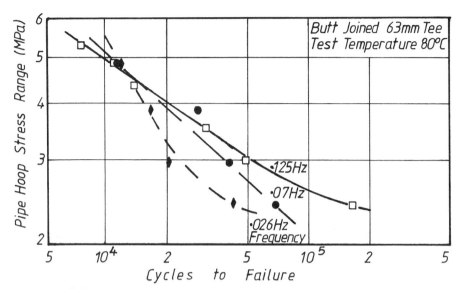

Figure 10 (a). The fatigue response of Continental European produced HDPE equal tee fittings. Note that at thehighest frequency a straight line holds over most of the $\Delta\bar{\sigma}_H$ range. Low frequencies reduce the fatigue strength. Data from Barker et al (4).

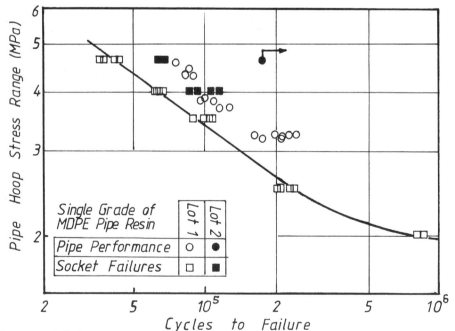

Figure 10 (b). Fatigue strength of 125mm SDR11 MDPE pipe systems. Two lots of the same resin were trialed. In both cases the socket coupler failed prior to the pipe (21). Note the evidence for a fatigue limit.

grade of MDPE, estimates of the 20°C fatigue strength of butt, socket
and electrofusion jointed 125mm SDR11 systems have been made, based on
data obtained at 80, 70 and 60°C (21)(23). The relationship, for a
fixed $\Delta\sigma_H$ value, between the fatigue life (N_f)   and   the   test
temperature (T), measured in degrees Kelvin, was proposed to be

$$\log N_f = A - B.T^{-1} \qquad\qquad 6$$

where A and B are experimentally determined material constants.
Equation 6 implies a log N_f versus T $^{-1}$ plot should give a straight
line, and this was observed (21) for the socket coupler failures, as
shown in Figure 11. Using the log N_f versus T $^{-1}$ plots, and equation
6, estimates of the 20°C fatigue strength for the various failure
sites were calculated (21)(23), and are presented in Table I.

TABLE I. THE MEASURED 80°C AND ESTIMATED 20°C FATIGUE STRENGTH OF
BUTT, SOCKET AND ELECTROFUSION JOINED MDPE PIPE SYSTEMS

	JOINTING METHOD		
	BUTT	SOCKET	ELECTROFUSED
Failure Site	in the pipe	socket fitting	electrofusion fitting
Nominal (N) or Actual (A) $\Delta\sigma_H$	4.38 MPa(A)	4.65 MPa (N)	4.65 MPa (N)
Mean Measured 80°C Fatigue Life	82,300 Cycles	39,300 Cycles	39,800 Cycles
Estimated 20°C Fatigue Life	271 x 10⁶ Cycles	21.1 x 10⁶ Cycles	19.0 x 10⁶ Cycles

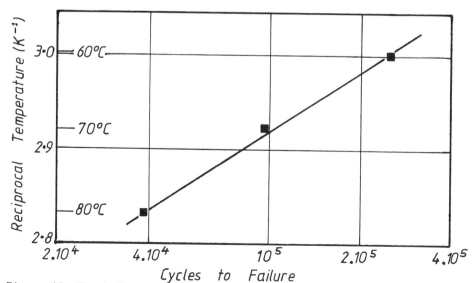

Figure 11. The influence of test temperature on the measured mean fatigue
lifetime of 125mm MDPE socket couplers. Data from Bowman (21).

The observed good 80°C fatigue strength of butt joined pipe is preserved at 20°C. Other studies on butt joined MDPE pipe (30) also indicate good fatigue strengths at lower and the more realistic service temperatures.

Potable water, sewerage and other pipelines can, in service, be subjected to mixtures of constant and fluctuating pressure loadings (31). To simulate this form of loading, creep loadings have been interspersed between discrete packets of fatigue. The fatigue packets have been of either 10 or 100 cycles, and the creep intervention times have varied between 0.1 and 11.6 hours. The maximum fatigue pressure (8 bar gage) has been equal to the creep pressure loading, with the frequency of the fatigue cycles at about 0.068 Hz for a trapezoidal loading profile with R = 0. The samples tested were 125mm MDPE socket couplers (21)(23), tested at 80°C till failure. The concepts of fractional damage, described by Bowman and Barker (6) and referred to earlier, have been applied to interpret the data. The total number of loading cycles to induce failure, together with the accumulated creep damage (from the creep intervention periods only) and the combined damage (Equation 4) are recorded below on Table II.

TABLE II   80°C PERFORMANCE OF 125mm SDR 11 SOCKET JOINED MDPE PIPE SYSTEMS TESTED UNDER ALTERNATING CREEP AND FATIGUE LOADING

Creep Inter- vention Time(h)	Total Mean Fatigue Cycles $(\times 10^{-4})$	Fractional Fatigue Damage	Total Mean Creep Loading (h)	Fractional Creep Damage	Combined Fractional Damage (Equation 4)	Total Mean Test Time (h)
	N_f	N_f/N_f^o	τ_{FAT}	τ_{FAT}/τ_{SR}	$N_f/N_f+\tau_{FAT}/\tau_{SR}$	
Pure Fatigue	9.97	1.0	0	0.0	1.0	407.0
0.1	8.48	0.850	84.8	0.027	0.877	421.5
0.3	5.38	0.540	161.5	0.052	0.592	383.5
1.0	4.93	0.495	493.1	0.159	0.654	696.8
11.6	1.43	0.144	1657.8	0.533	0.678	1715.5
Pure Creep	0	0	3100.0	1.0	1.0	3100.0

Note 100 loading cycles in each fatigue packet.

Table II shows that the introduction of periods of creep loading between packets of fatigue reduced the number of cycles of fatigue required to induce the socket coupler to failure. Or viewed the other way, the more frequent application of fatigue progressively reduced the total creep intervention time. But it should be noted, first that the combined damage was less than one when both creep and fatigue were present, and second that the total test time to failure was not markedly reduced by combining the two loading modes.

In addition to the data reported above, Bowman (21)(23) also investigated varying the number of fatigue cycles per packet, reducing it from 100 to 10. This was more deleterious, reducing N_f to 2.16×10^4 for a creep intervention period of 18 minutes. It can thus be concluded that for some polyethylene pipe systems the mixing of creep and fatigue loading is deleterious, reducing by a significant margin the total fatigue damage a system can sustain without precipitating failure.

Having examined the available literature on the fatigue response of UPVC and MDPE/HDPE pipes and fittings, it is now relevant to explore the fatigue loading seen in service. The in-service fatigue stressing is then compared to the capability of the UPVC and MDPE/HDPE pipe systems to sustain those loadings. Recommendations then follow on preferred materials for pipe systems likely to experience fatigue loading in service.

THE FATIGUE STRENGTH OF UPVC AND MDPE/HDPE PIPE SYSTEM

Some Sources of Fatigue Loading

Fluctuations in the pressure of fluids within buried plastic pipe systems can arise from a number of sources, but two are considered to be widespread in their action and therefore of interest. First, in potable water pipelines demand variations can induce, in a typical day, two to five low pressure ($\leq$ 4 bar) changes. These demand induced pressure changes (diurnal fatigue) see the line pressure decrease below the maximum value (which is usually found at night). For a 50 year design life the demand induced fatigue loadings would number between 3.7×10^4 and 9.0×10^4 cycles. In addition to these fatigue cycles there would be creep (constant pressure) loadings.

The second major source of internal pressure fluctuations arises from the operation of pumps and valves. These have been observed to induce pressure changes of 6 bar and above (31), and calculated to be of the order of 7.2 bar for HDPE and about 9 bar for UPVC (32), the difference being due to the different wall thicknesses and elastic modulii of pipes and materials respectively. These larger pressure fluctuations have the line pressure rising above the normal line pressure, with the fatigue loadings occurring in small packets which may number up to 6 packets per hour (31). Joseph, (11), for UPVC pipelines, analysed these discreet packets of fatigue, and concluded that a single decaying pressure packet was cumulatively equal to two large pressure pulses. Thus, in a 50 year life a pumped pipeline could experience up to 5×10^6 cycles of high pressure, surge fatigue loading. In addition to the fluctuating pressue there will be, between the packets of fatigue loading, periods of constant pressure.

Estimated 20°C Fatigue Strength of UPVC and MDPE Pipe Systems

Estimates of the mean 20°C fatigue lifetime of UPVC pipes of differing SDR values are contained in Table III. The fatigue strength lifetime is based on the work of Leevers and Joseph (3). Also included in Table III is an estimate of the likely fatigue strength of oriented UPVC pipe, based on the data of Dukes (12). For UPVC pipe systems no allowance has been made for the strength of the fitting, despite

these appearing, in some studies, to be weaker than pipe under fatigue
(17)(18). A review of the failures in UPVC pipe systems did not
identify fittings as a particular problem (33), hence they are not
considered here.

TABLE III 20° C FATIGUE PERFORMANCE AND EXPECTED SERVICE LIFE OF UPVC
PIPES AND MDPE PIPE SYSTEMS FOR FATIGUE INDUCED FAILURES

UPVC PIPES			125mm SDR 11 MDPE SYSTEMS		
4" SDR 15	4" SDR 19	ORIENTED PIPE (6")	BUTT FUSED	SOCKET FUSED	ELECTROFUSED

IV(a) DIURNAL LOADING, Pressure change 4 bar, 5 cycles per day

Equivalent $\Delta\sigma_H$ (MPa)	2.0	3.6	7.3	2.0	2.0	2.0
N_f for Continuous Loading (x 10^{-6})	27.7	14.8	>15.0	2175	242	528
N_f for Alternating Creep and Fatigue (x 10^{-6})	–	–	–	543	48	104
Expected Service Life (yrs)	>10^4	>10^3	>10^3	>10^5	>10^4	>10^5

IV (b) SURGE FATIGUE, Pressure change 8 bar, 12 cycles per hour

Equivalent $\Delta\sigma_H$ (MPa)	5.9	7.2	14.6	4.0	4.0	4.0
N_f for Continuous Loading x 10^{-6}	2.7	1.47	7.0	354	32.6	34.3
N_f for Alternating Creep and Fatigue (x 10^{-6})	–	–	–	71	6.5	6.9
Expected Service Life (yrs)	26	14	66	673	62	65

For the assessment of the fatigue strength of MDPE/HDPE pipe systems, separate consideration is given to the behaviour of butt, socket and electrofusion joined systems. Bowman (21) has taken 80°C data and, using the Arrhenius relationship (equation 6), estimated the 20°C fatigue lifetime.this data is contained in Table III where it can be seen that the fatigue strength of the MDPE pipe systems has been down-rated due to the possibility of creep and fatigue loading being present together. The data in Table III clearly shows a different response for the different methods of joining MDPE pipe systems.

Resistance of UPVC and MDPE Pipe Systems to Diurnal Fatigue Loading

Estimates of the likely fatigue loadings arising from diurnal fatigue have been made for a pipe design lifetime of 50 years, and range from 3.7×10^4 to 9×10^4 cycles of low pressure fatigue. This fatigue loading is compared with the expected fatigue strength of UPVC pipes and MDPE pipe systems, as annotated in Table III.

The available evidence shows diurnal fatigue, that is pressure changes resulting from demand variations in buried potable water lines, will not cause UPVC pipes nor MDPE pipe systems to fail for the 50 (or 100 year design) lifetimes. However, fatigue loading has been shown to reduce the creep strength when the two are combined, see Table III. But the residual creep strength, as measured at 80°C, is considered more than sufficient so as not to cause these MDPE pipe systems to fail due to the accumulated creep loadings. Therefore, UPVC pipes and MDPE pipe systems are highly tolerant of damage arising from pressure fluctuations that result from demand variations; diurnal fatigue poses no problems.

Resistance of UPVC and MDPE Pipe Systems to Surge Fatigue

The pressure changes associated with surge fatigue are both more frequent and of a larger stress range. Yet at the same time, it is more difficult to estimate the likely fatigue damage. In one of the preceeding sections an estimate of 5×10^6 cycles in a 50 year life was given, and it is noted that in some cases this may be an overestimate, in others an underestimate. This caluclated fatigue damage is compared to estimates of the fatigue strength of UPVC pipes and MDPE pipe Systems contained in Table III.

For conventional UPVC pipe, it appears that surge fatigue loading is capable of inducing premature failure. This is in-line with conclusions of Joseph (11), although it should be noted that this study has used an estimate of the mean fatigue life of pipe, while Joseph used the minimum life (see equation 5). Oriented UPVC pipes, appear to have an improved fatigue strength such that they are capable of withstanding the estimated surge fatigue damage in a 50 year life, see Table III. Further work does, however need to be undertaken on complete pipes to confirm the initial data obtained on pipe rings.

For MDPE pipelines the evidence is that all three methods of fusion jointing give systems capable of withstanding the damage induced by surge fatigue loading, with butt jointed systems the best. This statement is made with an allowance made for creep-fatigue interactions (6), see Table II. It should also be noted that the fatigue strength of MDPE and HDPE pipe resins differ grade to grade.

Some of the most modern polyethylene pipe grade resins have fatigue strengths significantly in excess of those used for the study by Bowman (21)(23).

Concluding, it should be noted that the number of buried pipelines seeing significant surge fatigue loading is small. Furthermore, it is possible to suppress the amplitude of the pressure pulses associated with the operation of pumps and valves. But for those pipelines where surge fatigue is likely to be present the evidence is that MDPE pipe systems offer good fatigue strength. And in particular, pipe systems made with good fatigue resistant MDPE and HDPE resins, and jointed using good butt fusion equipment (34), will give low pressure pipe systems tollerant of the damage arising from surge fatigue.

REFERENCES

1. Hertzberg, R.W. and Manson, J.A., Fatigue of Engineering Plastics Academic Press Inc, New York, 1980.
2. Barker, M.B., Bowman, J and Bevis, M., Journal of Materials Science 18 (1983) 1095.
3. Joseph, S.H. and Leevers, P.S. Ibid 20 (1985) 237.
4. Barker, M.B., Bentley, S.R., Bevis, M. and Bowman, J., Kunststoffe 74 (1984) 506.
5. Kraus, H. Creep Analysis, Wiley-Interscience Publications, New York, 1980.
6. Bowman, J. and Barker, M.B., Polymer Engineering and Science, 26 (1986) 1582.
7. Leevers, P.S. and Joseph, S.H., Private Communication, 1982.
8. Hucks, R.T., Journal AWWA, Water Technology/Distribution, July 1972, 443.
9. Radon, J.C., Journal Micromolecular Science - Physics, B14 (1977) 511.
10. Vinson, H.W., Proc. Int. Conf. on Underground Plastic Pipe, New Orleans, 1981.
11. Joseph, S.H., Plastics and Rubber Processing and Applications, 4 (1984) 325.
12. Dukes, B.W., paper 17 Proc. 6th Int. Conf. on Plastics Pipes, York England, March 1985.
13. Trott, J.J. and Gaunt, J., Proc. 3rd Inst. Plastics Pipes Symposium, Southampton England, September 1974.
14. Christie M.A. and Phelps, B., paper 17 of Proc. Int. Conf. on Fatigue of Polymers, London England, June 1983.
15. Hahn, M.T., Hertzberg, R.W., Manson, J.A. and Rimnac, C.M., Journal of Materials Science 17,(1982), 1533.
16. Onishi, I., Kimuza, H., and Uematie,M., Journal of the Japanese Welding Society, 35 (1966) 50.
17. Jacobi, H.R., Kunststoffe 55,(1965) 39.
18. Menges, G. and Roberg, P., Plastverarbeiter 21, 1970, VI - I.
19. Yue, C.Y. and Cherry, B.W., Plastics and Rubber Processing and Applications 12,(1989),105.
20. Lörtsch, W., Kunststoffe 55,(1965),460.
21. Bowman, J., paper 11 of the Proc. of the Int. Conf. on Plastics Pipes VII. Bath England, September 1988
22. Greig, J.M. and Lawrence, C.C. paper 14 idem.

23. Bowman, J., "Fatigue Performance of Blue MDPE Pipes and Fittings" Report for WRC Swindon, Swindon England, March 1988.
24. Parmar, R. and Bowman, J., Polymer Engineering and Science 29, (1989),1396.
25. Bowman, J. and Parmar, R., Ibid 29,(1989),1406.
26. Deidrich, G. and Gaube, E., Kunststoffe 60,(1970), 74.
27. Wolters, M. and Venema, B., Paper 26 of Proc. V Int. Conf. on Plastics Pipes, York England, September 1982.
28. Nishimura, H., Shibaro, H and Kitao, K. Proc. of 10th Plastic Fuel Gas Pipe Symposium, New Orleans USA, October 1987, 117
29. Bowman, J., Plastics and Rubber Processing and Applications 9, (1988), 147.
30. Parmar, R., Ph.D thesis, Brunel University, 1986.
31. Kirby, P.C., Paper 26 of Proc. 4th Int. Conf. on Plastics Pipes Brighton England, March 1979.
32. Janson, L-E., "Plastic Pipes for Water Supply and Sewage Disposal", published by Neste Chemicals, Stockholme, 1989.
33. Kirby, P.C., Proc. Int. Conf. on Underground Plastic Pipes, ASCE New Orleans USA, March 1981.
34. Bridgstock, E. Proc. XI Plastic Fuel Gas Pipe Symposium, San Francisco USA, October 1989, 127.

Design

Amster K. Howard

LOAD-DEFLECTION FIELD TEST OF 27-INCH (675-mm) PVC (POLYVINYL CHLORIDE) PIPE

REFERENCE: Howard, A. K., "Load-Deflection Field Test of 27-inch (675-mm) PVC (Polyvinyl Chloride) Pipe," Buried Plastic Pipe Technology, ASTM STP 1093, George S. Buczala and Michael J. Cassady, Eds., American Society for Testing and Materials, Philadelphia, 1990.

ABSTRACT: The Bureau of Reclamation constructed a special test section of 27-inch (675-mm) diameter PVC (polyvinyl chloride) pipe in November 1987 near Elba, Nebraska. Measurements were made of pipe deflections, pipe invert elevations, soil properties, and in-place unit weights. Pipe deflections will be continually monitored for several years. This paper describes the installation and pipe deflection measurements through the 2-year reading.

KEYWORDS: pipe, PVC pipe, flexible pipe, deflection, test section, soil mechanics, soil tests, time factors, soil-structure interaction

INTRODUCTION

This paper reports the results of a test section of buried 27-inch (675-mm) PVC (polyvinyl chloride) pipe installed near Elba, Nebraska, during November 1987. The Bureau of Reclamation constructed the test section at a special site to evaluate the short- and long-term behavior of PVC pipe installed with three different bedding conditions. Measurements made during installation of the test section and through 2 years following installation are reported. Measurements are to be made for 5 years following installation.

The test section was not made part of a functioning distribution system in order to gain access to take pipe diameter measurements whenever required. Results include measurements of pipe diameters as the pipe deflects, pipe invert elevations, and unit weights and physical properties of the soils used in construction.

BEDDING CONDITIONS

"Bedding" refers to placement of soil beneath and beside the pipe up to a height of 0.7 of the outside diameter of the pipe or up to the top of the pipe. "Backfill" refers to placement of soil over the pipe, and "cover" is the vertical distance from the top of the pipe to the top of the backfill.

Mr. Howard is a Research Civil Engineer with the U.S. Bureau of Reclamation, PO Box 25007, Denver CO 80225.

The three pipe bedding conditions examined for this study are illustrated on figure 1. The three conditions will be referred to as "dumped," "95 percent," and "85 percent" sections, and are described as follows:

- **Dumped section.** - Native soil from the trench excavation was dumped into the trench beside the pipe without any compaction. The backfill over the pipe to the ground surface was also dumped native material.

- **95 percent section.** - Native soil from the trench excavation was placed in 8- to 9-inch loose lifts beside the pipe and compacted to at least 95 percent compaction. These lifts were placed until the compacted bedding was up to at least 0.7 of the outside diameter of the pipe. The remainder of the backfill up to the ground surface was dumped native material.

- **85 percent section.** - Native soil from the trench excavation was placed in loose lifts and compacted to about 85 percent compaction for the whole trench section, that is, from the trench bottom to the ground surface.

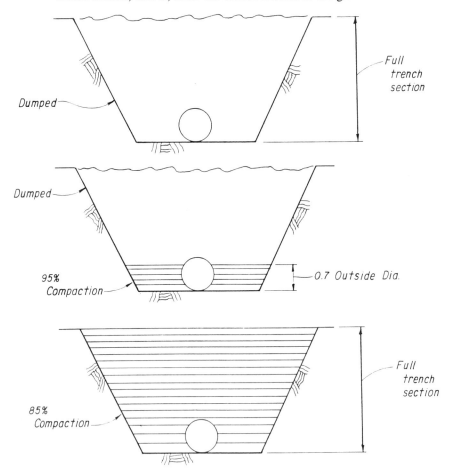

FIG. 1 -- Bedding and backfill conditions.

TEST SITE

The typical trench section is shown on figure 2 and pipe diameter measurement locations are shown on figure 3. The original trench section was to have about 18 inches (450 mm) of clearance on each side of the pipe, or a total bottom width of 5 feet 4 inches (1600 mm). The total depth was to be 18 feet (5.5 m) so there would be 15 feet (4.5 m) of cover over the pipe. At a depth of about 13 feet (4 m), a layer of clean, fine sand was encountered. As the sand dried, it began to slough creating vertical walls in the sand. Since the sloughing would undercut the overlying clay material, the excavation was terminated at a depth of about 15 feet 6 inches (4.7 m). Another result of the sloughing was that the trench width at the spring line of the pipe was 11 to 13 feet (3.4 to 4 m). This trench width is about 5 pipe diameters, which means the pipe was installed in a nontypical condition. In order to obtain as much cover (load) as possible over this pipe, the soil was mounded over the trench to create a final cover over the pipe of 15 feet (4.5 m).

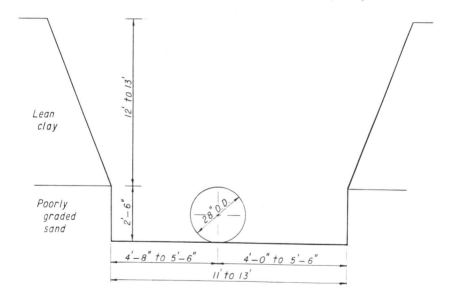

FIG. 2 -- Typical trench section for test section.

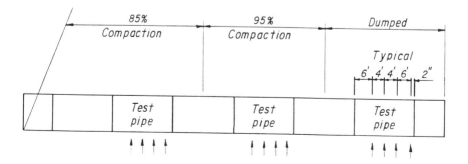

FIG. 3 -- Diameter measurement locations.

PIPE

The PVC pipe was 27-inch (675-mm) nominal inside diameter, SDR-51, pressure rated to 80 lb/in², and the pieces were 20 feet (6.1 m) long. The pipe was made with an integral bell to utilize a gasket for sealing, meeting the specifications defined in ASTM F 477.

The pipe is described in a catalog as "Agricultural PVC Pipe" having the following properties:

Outside diameter   = 27.953 inches
Inside diameter   = 26.857 inches
Wall thickness   = 0.548 inch
Modulus of elasticity = 400,000 lb/in²

Several measurements of the pipe wall thickness were made at the cut end of the outlet pipe using a vernier caliper. The measurements ranged from 0.595 to 0.629 inch (15.11 to 15.98 mm) with an average of 0.617 inch (15.67 mm).

The pipe stiffness factor for use in the equation for predicting the pipe deflection under load is expressed as:

$$\text{pipe stiffness factor, lb/in}^2 = \frac{EI}{r^3} = 0.149\ PS$$

where:
- E = modulus of elasticity, lb/in²
- I = moment of inertia of section of pipe wall, in⁴/in
- r = pipe radius, inches
- PS = pipe stiffness (term most commonly used)

The moment of inertia for a straight wall pipe is equal to $t^3/12$ where "t" is the pipe wall thickness. Using the following nominal values, the pipe stiffness factor, EI/r^3, was calculated to be 2.2 lb/in² (15.2 kPa).

E = 400,000 lb/in² (2760 kPa)
t = 0.55 inch (13.97 mm)
D = 27.0 inches (686 mm)

If the measured wall thickness of 0.62 inch (15.67 mm) was used, the pipe stiffness factor would be 3.2 lb/in² (22.1 kPa), or about 50 percent higher. However, measurements were made on only one pipe at one section and may or may not be representative of the entire test section. In addition, because predictions of pipe deflection are generally based on nominal values, the nominal pipe stiffness factor is used in this study for comparison purposes.

SOIL PROPERTIES

Foundation and Trench Walls

The soil in the foundation and in the trench walls from the trench bottom up to about the top of the pipe was classified as a POORLY GRADED SAND. Four in-place densities were measured in this material. Relative densities ranged from 61 to 88 percent with an average of 72 percent. Trench wall conditions would be considered trench type I as used in Reclamation [1].

Native Soil

The native soil excavated from the trench above the POORLY GRADED SAND was classified as LEAN CLAY. The soil classifications are in accordance with ASTM D-2487.

CONSTRUCTION SEQUENCE OF TEST SECTION

Dumped Section

The soil was placed in two loose lifts beside the pipe, one lift from trench bottom to pipe spring line and the other from spring line to the top of the pipe. For each lift, loose soil was leveled using garden rakes and shovels.

The backfill over the pipe was placed in 3-foot (1-m) thick loose lifts up to a final cover height of 15 feet (4.5 m). These lifts were leveled using the hydraulic excavator bucket.

85 Percent Section

From the trench bottom to the top of the pipe, the soil was placed in 8-inch (200-mm) loose lifts and compacted with one pass of a mechanical compactor to about 6 inches (150 mm). This was continued over the pipe up to a cover height of 3 feet (1 m). Then progressively thicker lifts were used, and these were compacted using wheel traffic from a front-end loader.

95 Percent Section

From trench bottom up to 0.7 of the outside pipe diameter, the soil was placed in about 8-inch (200-mm) loose lifts and compacted with several passes of a mechanical compactor to a compacted height of about 6 inches (150 mm). The required number of passes was monitored by measuring the in-place density using a sand cone device. After having placed compacted soil to a height of 0.7 of the outside diameter of the pipe, loose soil was placed and leveled up to the top of the pipe. The backfill sequence of placing soil over the pipe was the same as described for the dumped section.

UNIT WEIGHT OF BACKFILL OVER PIPE

Dumped and 95 Percent Sections

Five in-place unit weight tests were performed in the uncompacted backfill soil over the dumped section and the 95-percent section. Two of the tests were performed in the backfill over the 95-percent section and three were performed in the backfill over the dumped section. However, test results were so similar that unit weight of the uncompacted backfill is discussed without regard to location.

The wet unit weight of the uncompacted backfill ranged from 78.7 to 84.2 lbf/ft^3 (12.4 to 13.2 kN/m^3) with an average of 81.3 lbf/ft^3 (12.7 kN/m^3). For calculation of the predicted pipe deflection, a unit weight of 81 lbf/ft^3 (12.7 kN/m^3) was used.

Percent compaction of the uncompacted backfill ranged from 66.8 to 74.3 percent with an average of 70.7 percent.

85 Percent Section

Five in-place unit weight tests were performed in the compacted backfill soil over the 85-percent section.

Wet unit weights of soil compacted over the top of the pipe ranged from 90.0 to 100.6 lbf/ft³ (14.1 to 15.8 kN/m³) with an average of 96.6 lbf/ft³ (15.2 kN/m³). For calculation of the predicted pipe deflection, a unit weight of 97 lbf/ft³ (15.2 kN/m³) was used.

Percent compaction of the compacted backfill ranged from 81.0 to 89.8 percent with an average of 86.4 percent.

DEGREE OF COMPACTION OF BEDDING SOIL

To determine the degree of compaction of the bedding soil (soil placed beside the pipe), percent compaction was determined for each test reach. The degree of compaction is required in order to determine E', Modulus of Soil Reaction, used in calculating predicted pipe deflection [2, 3]. The degrees of compaction used are *dumped, slight, moderate,* and *high.*

Dumped Section

The dumped section had no compaction except for occasional foot traffic associated with spreading the soil in level increments at spring line and at the top of the pipe. The unit weight and percent compaction of the bedding were assumed to be the same as those discussed under the preceding "Unit Weight of Backfill Over Pipe" section. The degree of compaction would be *dumped.*

85 Percent Section

Two in-place unit weight tests were performed when the bedding soil was at spring line and two tests when the bedding was at 0.7 outside diameter (o.d.). Percent compaction ranged from 85.3 to 91.0 percent with an average of 88.5 percent. The degree of compaction would be *moderate.*

95 Percent Section

Two in-place unit weight tests were performed with the bedding at spring line and two tests when the bedding was at 0.7 o.d. Percent compaction ranged from 94.3 to 96.7 percent with an average of 95.7 percent. The degree of compaction would be *high.*

PIPE DIAMETER MEASUREMENTS

Measurement points for vertical diameters were established by locating and marking the invert of the pipe using steel balls and then marking the top of the pipe using a plumb bob. A special device was placed on the vertical diameter marks, and the ends were used to locate the horizontal diameter. Care was taken that the device was perpendicular to the axis of the pipe. A screw was inserted into the pipe wall at the marked locations of the vertical and horizontal diameters.

The diameters were measured with an inside micrometer that could be read to 0.001 inch (0.025 mm). These measurements were made with the ends of the inside micrometer on the screw heads.

The readings are accurate to about plus or minus 0.010 inch (0.25 mm) because of the variation in the pressure used to tighten the micrometer in the final reading position. The readings through final backfilling were all made by the same person.

All elongations and deflections discussed are the *vertical* elongations and deflections of the pipe unless otherwise described. Elongation is defined as an increase in the vertical diameter of the pipe due to bedding soil being placed beside the pipe and compacted. Deflection is defined as a decrease in the vertical diameter of the pipe due to backfill soil being placed above the top of the pipe.

The percent vertical deflection or elongation (ΔY) is defined as:

$$\Delta Y\ (\%) = \frac{\text{change in diameter}}{\text{original diameter}} \times 100$$

For elongation, "change in diameter" is the diameter measured at some stage in the bedding process minus the diameter of the pipe when the pipe was in place on the trench bottom before any bedding operation was begun. For deflection, "change in diameter" is the diameter measured when bedding was completed up to the top of the pipe minus the diameter measured during or after the backfilling process. The "original diameter" used for both elongation and deflection calculations was the nominal inside diameter of the pipe, 27 inches (675 mm).

Elongation is shown as a negative value and deflection as a positive value.

PIPE ELONGATION DURING BEDDING

Flexible pipe can elongate (increase in vertical diameter and decrease in horizontal diameter) due to compaction of the bedding soil alongside the pipe. The diameters (horizontal and vertical) of the pipe were measured with the pipe resting in place on the trench bottom before any bedding soil was placed. Diameter measurements were again made after each lift of soil was placed and compacted. The dumped bedding was placed in two lifts, and diameter measurements were made after each placement.

The horizontal diameter change was larger than vertical diameter change as summarized in the following table:

TABLE 1 -- Elongation

Test reach	Percent average elongation with soil at top of pipe	
	Vertical	Horizontal
Dumped	-0.2	-0.3
85 percent compaction	-1.6	-1.6
95 percent compaction	-3.0	-3.1

The amount of elongation was directly related to the compactive effort applied to the bedding soil. The measurements show that just dumping soil beside a pipe can result in elongation. Compacting the bedding soil to over 95 percent compaction can elongate the pipe about 3 percent.

The percent vertical elongation values appear to be typical based on other reported measured values [4].

PIPE DEFLECTION DURING BACKFILLING

Flexible pipe deflects (decreases in vertical diameter and increases in horizontal diameter) due to backfill load on the pipe. The initial diameter (or zero) reading for calculating deflection was the pipe diameter measured when bedding soil was at the top of the pipe. From this zero point, any changes in pipe diameters are due to backfill placed over the pipe.

The following table summarizes the average deflection:

TABLE 2 -- Deflection

Test reach	Percent average vertical deflection at 15 feet (4.5 m) of cover
Dumped	9.4
85 percent compaction	1.0
95 percent compaction	0.9

Percent vertical deflection versus cover height is plotted for each test reach as shown on figure 4.

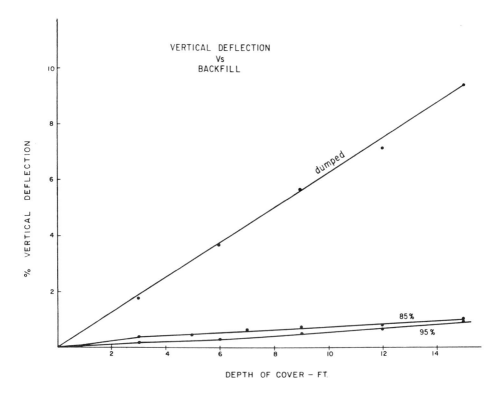

FIG. 4 -- Pipe deflection versus cover height.

Vertical Versus Horizontal Diameter Changes

Horizontal deflections were smaller than vertical deflections as summarized in the following table:

TABLE 3 -- ΔX/ΔY Ratio

Test reach	Vertical ΔY	Horizontal ΔX	Ratio ΔX/ΔY
Dumped	9.4	8.4	0.89
85 percent compaction	1.0	0.8	0.80
95 percent compaction	0.9	0.5	0.56

Average percent deflection at 15 feet (4.5 m) of cover

For pipe that deflects elliptically, the ratio of the horizontal to vertical deflections should be about 0.91 [3].

Net Change in Pipe Diameter

The net change in pipe diameter from measurements made when the pipe was in place on the trench bottom and after backfilling was completed is shown on the following table:

TABLE 4 -- Net Diameter Change

Test reach	Elongation	Deflection	Net change
Dumped	-0.2	9.4	9.2
85 percent compaction	-1.6	1.0	-0.6
95 percent compaction	-3.0	0.9	-2.1

Percent vertical change

On the day the 15 feet (4.5 m) of cover was completed, the pipes with compacted beddings had not returned to their original diameter.

Theoretical Versus Actual Deflections

Theoretical deflections at 15 feet (4.5 m) of cover for each bedding condition were calculated using the following equation [3]:

$$\Delta Y\ (\%) = T_f \frac{0.07\gamma\ h}{EI/r^3 + 0.061\ E'}$$

where:

$\Delta Y(\%)$ = theoretical vertical deflection in percent
T_f = timelag factor, 1.0
γ = backfill soil unit weight in lbf/ft^3 = 81 lbf/ft^3 for dumped and 95 percent sections, or 97 lbf/ft^3 for 85 percent section
h = cover height in feet over pipe = 15 feet
EI/r^3 = pipe stiffness factor in lb/in^2 = 2.2 lb/in^2
E' = modulus of soil reaction in lb/in^2, varies with compaction and soil type

This equation is a commonly used variation of the Iowa Formula [5, 6]. A timelag factor of 1.0 was used to calculate the initial (day backfilling completed) deflections.

The soil type used would be "fine-grained soil with less than 25 percent coarse-grained particles." For the three bedding conditions, E' values would be as follows [3]:

TABLE 5 -- E' Values

Test reach	Degree of compaction	Modulus of soil reaction E' in	
		lb/in²	(kPa)
Dumped	Dump	50	(345)
85 percent compaction	Moderate	400	(2 760)
95 percent compaction	High	1,500	(10 350)

Pipe in the dumped section deflected about half the predicted value. The E' value was backcalculated to be 111 as compared to the recommended value of 50.

Pipe in the 85-percent section deflected about one-fourth the predicted value. The E' value was backcalculated to be 1,634 as compared to the recommended value of 400.

Pipe in the 95-percent section deflected within the anticipated deflection range. The E' value was backcalculated to be 1,513 as compared to the recommended value of 1,500.

TIMELAG OF PIPE DEFLECTIONS

A flexible pipe continues to deflect over time for two reasons [3]:

1. Increase in the soil load on the pipe

2. Compression and consolidation of the soil at the sides of the pipe.

Diameter measurements were made at the following time periods following completion of backfilling: 1, 3, 7, and 14 days; 1, 2, 3, and 6 months; 1 and 2 years. Future readings will be made at 3, 4, and 5 years.

Timelag is defined as the ratio of the deflection measured at some time period following completion of backfill to the deflection measured at completion of backfill.

The following table gives timelag factors for vertical deflections measured at 1 and 6 months and 2 years.

TABLE 6 -- Timelag Factors

Test reach	Percent vertical deflection				Timelag factor		
	0 day	1 mo	6 mo	2 yr	1 mo	6 mo	2 yr
Dumped	9.5	10.8	11.9	13.3	1.1	1.3	1.4
85 percent compaction	1.0	1.5	1.8	2.1	1.5	1.8	2.0
95 percent compaction	0.9	1.3	1.6	1.7	1.4	1.8	1.9

The anticipated timelag factors, over several years, are 1.5 for the dumped section and 2.5 for the 85-percent and 95-percent sections [3]. About 75 percent of the timelag factor should be reached in 3 to 6 months. Figures 5 through 7 show the percent vertical deflection versus time for the three test reaches. As shown in these figures, most of the increase in deflection with time has occurred within the 3- to 6-month period.

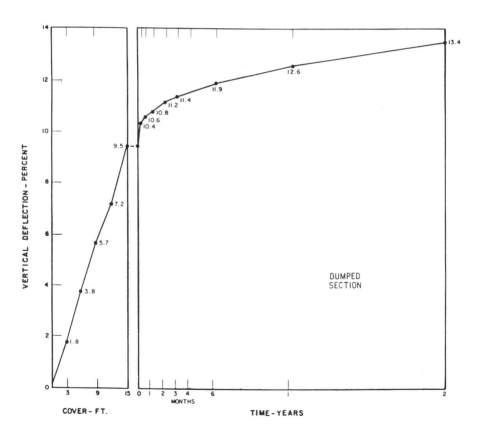

FIG. 5 -- Deflection with time - dumped section.

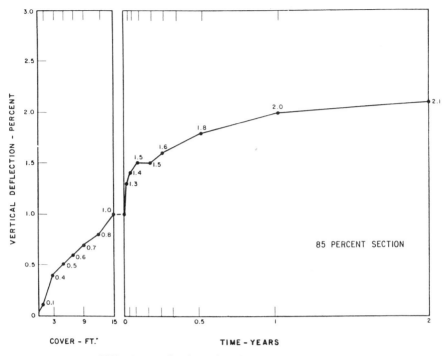

FIG. 6 -- Deflection with time - 85 percent section.

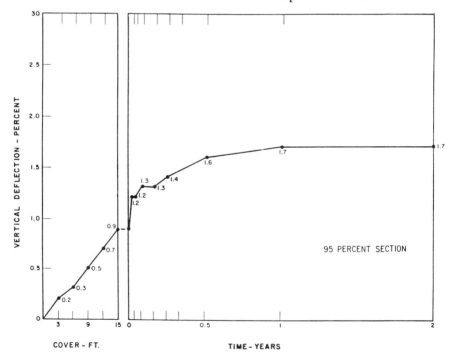

FIG. 7 -- Deflection with time - 95 percent section.

ELONGATION AND DEFLECTIONS OF PIPE JOINTS

Diameter measurements of pipe joints were made at the spigot side of the joint at the upstream end of each test pipe. These measurements were made about 2 inches (50 mm) from the end of the pipe.

The joint is stiffer than the barrel of the pipe, and smaller elongation and deflection values were recorded at the joints.

Elongation

Horizontal diameter change was larger than vertical diameter change as summarized in the following table:

TABLE 7 -- Elongation of Joint

Test reach	Percent elongation of joint with soil at top of pipe	
	Vertical	Horizontal
Dumped	-0.1	-0.1
85 percent compaction	-0.8	-0.9
95 percent compaction	-1.9	-2.0

The amount of elongation was directly related to the compactive effort applied to the bedding soil. The measurements show that just dumping soil beside the pipe can result in joint elongation. Compacting the bedding soil to over 95 percent compaction can elongate the joint about 2 percent.

Deflection

Deflection of joints due to backfilling over the pipe is shown on the following table along with the ratio of horizontal to vertical diameter:

TABLE 8 -- Deflection of Joint

Test reach	Percent deflection of joints at 15 feet (4.5 m) of cover		
	Vertical ΔY	Horizontal ΔX	Ratio $\Delta X / \Delta Y$
Dumped	8.0	7.2	0.90
85 percent compaction	0.7	0.7	1.00
95 percent compaction	0.5	0.5	1.00

The ratio of horizontal to vertical deflection of the joints is 0.9 or more.

Net Change in Pipe Diameter

The net change in pipe diameter at the joints from measurements made when the pipe was in place on the trench bottom and after backfilling was completed is shown on the following table:

TABLE 9 -- Net Diameter Change of Joint

Test reach	Percent vertical change of joint		
	Elongation	Deflection	Net change
Dumped	-0.1	8.0	7.9
85 percent compaction	-0.8	0.7	-0.1
95 percent compaction	-1.9	0.5	-1.4

As with net change in the barrel of the pipe, on the day the 15 feet (4.5 m) of cover was completed, the pipe with compacted beddings had not returned to its original diameter.

Timelag

The following table gives timelag factors of the joints for the vertical deflections measured:

TABLE 10 -- Timelag Factors of Joint

Test reach	Joint percent vertical deflection				Joint timelag factor		
	0 day	1 mo	6 mo	2 yr	1 mo	6 mo	2 yr
Dumped	8.0	9.4	10.5	12.5	1.2	1.3	1.6
85 percent compaction	0.7	1.0	1.4	1.6	1.5	2.0	2.4
95 percent compaction	0.5	0.7	1.1	1.2	1.5	2.3	2.4

Comparison of Joint and Barrel of Pipe

Relative stiffness of the joint is illustrated in the following table comparing elongation and initial deflection of this joint with average values for the barrel of the pipe:

TABLE 11 -- Barrel-Joint Comparison

Test reach	Percent vertical change			
	Elongation		Initial deflection	
	Barrel	Joint	Barrel	Joint
Dumped	-0.2	-0.1	9.4	8.0
85 percent compaction	-1.6	-0.8	1.0	0.7
95 percent compaction	-3.0	-1.9	0.9	0.5

Change in the joint compared to change in the barrel of the pipe ranges from about 50 to 85 percent.

PIPE INVERT ELEVATIONS

The elevation of the pipe invert was monitored during installation using surveying equipment to measure the elevation of the top of the screw heads in the pipe invert.

Of particular interest was any raising of the pipe due to compaction of bedding below the spring line of the pipe. For lightweight pipe, compactive effort in the haunch area of the pipe can raise the pipe. To prevent any significant raising, sandbags were placed on top of the pipe in the 95-percent section.

Placement and compaction of soil in the 95-percent section up to the spring line of the pipe raised the pipe about 0.04 foot (12 mm). Continuation of compacted bedding up to 0.7 o.d. raised the pipe another 0.01 foot (3 mm). The 85-percent section did not have sandbags on top of the pipe, and placement and compaction of soil up to spring line and then to 0.7 o.d. did not affect invert elevation significantly.

In all three sections, loading the pipe by placement of the backfill over the pipe showed a general trend of the pipe settling only about 0.01 to 0.02 foot (3 to 6 mm).

Elevation readings made 2 weeks following completion of backfilling indicated further settlement of about 0.01 foot (3 mm). The 1-year readings show that the pipe has settled about 0.1 foot (30 mm).

Compared to the amount of elongation and deflection that occurred, movement of the pipe invert was relatively small.

REPORT

A complete tabulation of all measurements and calculated deflections in addition to a more detailed description of the test installation is presented in a Reclamation report [7]. The test section was initiated and constructed by personnel from the Reclamation Nebraska-Kansas Project Office in Grand Island, Nebraska. Their work is especially acknowledged, particularly Mike Kube, Chief of the Office Engineering Branch, and Larry Cast, Project Geologist.

SUMMARY AND CONCLUSIONS

A special test section of 27-inch (675-mm) diameter PVC pipe was constructed in November 1987 near Elba, Nebraska. Pipe deflections, pipe invert elevations, soil physical properties, and in-place unit weights were measured. Pipe deflections are to be monitored periodically to evaluate time-deflection behavior of the pipe. Measurements from the test section through the 2-year readings gave the following results:

- Pipe deflections in the dumped and 85-percent sections are much less than predicted.

- Pipe deflection in the 95-percent section is within the range of predicted values.

- Pipe elongation (increase in vertical diameter) created during placement of bedding soil beside the pipe was typical based on other reported values.

- Pipe joints deflections ranged from about 50 to 85 percent of the deflection measured in the pipe barrel.

REFERENCES

[1]   Bureau of Reclamation, "Pipe Bedding and Backfill," Geotechnical Branch
       Training Manual No. 7, Denver, Colorado, June 1981.
[2]   Howard, A. K., "Modulus of Soil Reaction Values for Buried Flexible Pipe,"
       Journal of the Geotechnical Engineering Division, ASCE, vol. 103, No. GT1,
       Proc. Paper 12700, January 1977.
[3]   Howard, A. K., "The USBR Equation for Predicting Flexible Pipe Deflection,"
       Proceedings of the International Conference on Underground Plastic Pipe,
       ASCE, New Orleans, Louisiana, March 1981.
[4]   Howard, A. K., "Diametral Elongation of Buried Flexible Pipe," Proceedings of
       the International Conference on Underground Plastic Pipe, ASCE, New
       Orleans, Louisiana, March 1981.
[5]   Spangler, M. G., "The Structural Design of Flexible Pipe Culverts," Iowa
       Engineering Experiment Station Bulletin No. 153, Ames, Iowa, 1941.
[6]   Watkins, R. K., and M. G. Spangler, "Some Characteristics of the Modulus of
       Passive Resistance of Soil: A Study of Similitude," Highway Research Board
       Proceedings, vol. 37, Washington, D.C., pp. 576-583, 1958.
[7]   Howard, A. K., Kube, M., and Cast, L., "Fullerton PVC Pipe Test Section,"
       Report No. R-89-07, Bureau of Reclamation, Denver, Colorado, August 1989.

Ernest T. Selig

SOIL PROPERTIES FOR PLASTIC PIPE INSTALLATIONS

REFERENCE: Selig, E. T., "Soil Properties for Plastic Pipe Installations," Buried Plastic Pipe Technology, ASTM STP 1093, George S. Buczala and Michael J. Cassady, Eds., American Society for Testing and Materials, Philadelphia, 1990.

ABSTRACT: Soil property requirements for the basic trench and embankment installation conditions are discussed. Characteristics of compacted soils are described and representative stress-strain parameters given. Preliminary values of existing ground stiffness properties are suggested. The applications of these properties for analyzing pipe deflection, wall thrust and buckling strength are indicated.

KEYWORDS: soil properties, stress-strain behavior, strength, compaction, flexible pipe, plastic pipe, Young's modulus, Poisson's ratio, bulk modulus, constrained modulus, deflection, buckling, wall thrust.

INTRODUCTION

The installed shape of a buried plastic (flexible) pipe is strongly influenced by the soil placement process and the resulting soil stiffness properties. The long-term pipe deflections are controlled by soil deformation subsequent to installation in addition to the time-dependent pipe response. This soil deformation results from soil consolidation, creep, moisture changes, and erosion, as well as from loading changes. Pipe buckling stability is highly dependent on the value of soil stiffness. The pipe wall stresses and strains induced by earth and live loading are dependent on the relative stiffness of the soil and pipe. The type of soil and level of compaction are the fundamental factors determining these characteristics for placed soils. The soil type, in situ state, and stress history are the corresponding factors determining the relevant characteristics for undisturbed ground. To help illustrate these principles the relationships between soil type, amount of compaction and compaction effort will be discussed and their influence on resulting soil properties will be shown. The role of these soil properties in analyzing plastic pipe deflection, wall thrust, and buckling stability will be indicated.

Dr. Ernest T. Selig is Professor of Civil Engineering at the University of Massachusetts, Amherst, MA 01003.

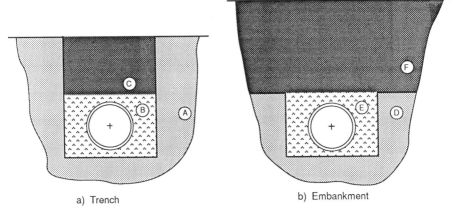

a) Trench   b) Embankment

Fig. 1 -- Pipe installation type.

INSTALLATION TYPE

The two basic plastic pipe installation type are shown in Fig. 1. The trench case (Fig. 1a) represents a situation in which the existing ground (zone A) is excavated to the depth required for pipe installation. The resulting trench is backfilled with two zones of compacted soil. Zone B is the zone immediately surrounding the pipe which requires certain restrictions on the placement and compaction to avoid distressing the pipe, and restrictions on the type of soil to provide needed stiffness and stability. The remainder of the trench (zone C) is usually filled with the excavated soil appropriately placed and compacted. The specific trench dimensions as well as the dividing line between zones B and C depend on the requirements of the installation.

The embankment case (Fig. 1b) shows the pipe installed in a shallow trench excavated in the existing ground (zone D) and backfilled with zone E material meeting requirements similar to those of zone B. An earth embankment (zone F) is then constructed on top of the existing ground. This configuration is known as a negative projecting embankment pipe installation [1]. The pipe may also be installed above the existing ground, in which case zone E is laterally supported by embankment soil in zone F rather than by existing ground.

The soil property requirements for plastic pipe design are different in various ways for each zone in Fig. 1.

SOIL REQUIREMENTS

Existing Ground

In the case of existing ground in zone A the stress level remains essentially unchanged by the pipe installation. The main requirement is stability of the trench walls and bottom during construction. This is provided as needed by bracing and dewatering. Unless the existing ground is unsuitable, as may be the case with peats and organic deposits, the existing soil properties are accepted and the design and construction are carried out considering these properties. For analyzing the soil-pipe interaction, soil strength and stiffness during filling of the trench are the primary parameters required for zone A soil.

The requirements are different for existing ground in zone D because the stresses are significantly increased by construction of the embankment. It is necessary to insure that the ground is stable under the weight of the embankment and that excessive immediate and consolidation settlement will not occur. If the soil in zone D is not saturated then volumetric compression will occur under the embankment load. Whether or not the soil is saturated, shear strains will occur under the embankment load. Both of these characteristics result in immediate settlement. If the soil is saturated or becomes so because of compression under increased load, then consolidation will take place over a period of time after construction as the excess pore water pressure is dissipated. Thus for zone D soil knowledge of the strength and consolidation characteristics is required as well as the nonlinear stress-strain properties during construction.

Soil Envelope

Zones B and E which immediately surround the pipe will be termed the soil envelope. This envelope includes the bedding, the side fill, and the top fill (Fig. 2). The haunch zone is included within the bedding and side fill as shown in Fig. 2. Zones B and E will be considered together because their required properties are essentially the same.

The stability of flexible plastic pipe is substantially controlled by the properties of the material in the soil envelope. The following are the requirements of this envelope:

1. Constructability - ability to be placed and compacted to the desired properties without distorting the pipe.

2. Provide the stiffness needed for limiting the pipe deformations (the particularly important areas are those shown by arrows A and B in Fig. 3a).

3. Provide the stiffness needed to achieve adequate pipe buckling strength.

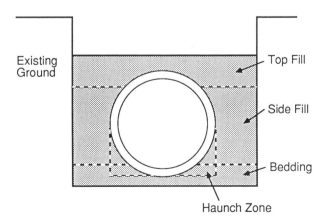

Fig. 2 -- Soil envelope.

4. Be stable under long-term moisture changes.

5. Exhibit little creep and consolidation deformation.

6. Provide drainage of excess pore water pressure.

7. Reduce the earth and live load carried by the pipe wall.

8. Prevent erosion or piping of surrounding fine soil as a result of pipe leaks or ground water movement.

These soil envelope requirements dictate the use of compacted coarse-grained soils (mainly sand and gravel components) in most cases. The material in the envelop thus may be referred to as structural backfill.

Trench Backfill

Zone C represents the trench backfill remaining above the structural backfill zone B. If a pavement or a structure requiring limited settlement is to be placed on the surface above the trench, then zone C soil must provide firm support (arrow C in Fig. 3a). Suitable material adequately compacted for zones B and C will be needed to prevent settlement as shown in Fig. 3b. The main mechanisms of settlement in zone C are: 1) volume reduction and shear strain from the surface load, particularly from repeated wheel loading, and 2) shrinkage from cycles of moisture change. These problems diminish with increased level of compaction, but even so soils whose behavior is controlled by fine-grained (silt and clay) components generally will not perform satisfactorily in this application. Thus coarse-grained soils (sand and gravel components) are most appropriate.

When surface settlement is not a concern, then zone C may be backfilled with the excavated soil using appropriate compaction. This is the most economical solution.

Embankment

In a negative projecting installation (Fig. 1b), the embankment, zone F, acts primarily as dead load. However some arching of the embankment load will occur which results in the pressure applied to

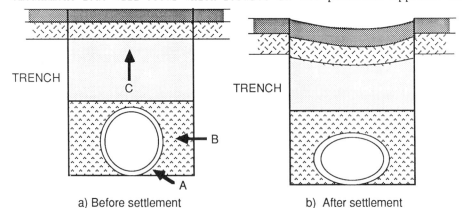

a) Before settlement   b) After settlement

Fig. 3 -- Settlement with too little compaction.

the top of zone E being either more or less than the average pressure at the base of the embankment. The unit weight of embankment fill is thus its most important property. Also important is the soil stiffness in the lower part of the embankment, i.e., within 3 trench widths of the top of zone E.

If the pipe were installed in either a positive projecting or imperfect trench condition [1], then the embankment stiffness properties would become much more important.

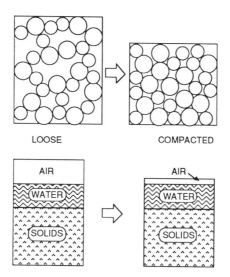

LOOSE   COMPACTED

Fig. 4 -- Illustration of compaction.

CHARACTERISTICS OF COMPACTED SOILS

Compaction Reference Test

Compaction is immediate densification of soil by mechanical means. The water content remains constant and the void air space is reduced (Fig. 4). Consolidation, in contrast, is gradual squeezing out of water from saturated soils (no air in voids) which results in some densification.

Compaction is performed to achieve suitable properties of soil being placed. Increasing the amount of compaction increases strength, decreases compressibility, decreases permeability, reduces collapse potential, and reduces swelling and shrinking with moisture change. The magnitude of these effects depends on the soil type.

Standardized tests by ASTM and AASHTO are used to determine the amount of compaction that can be achieved for each soil with specified standard compaction efforts. For cohesionless, free-draining material (clean sands and gravels) the soil is vibrated vertically in a rigid mold with a surcharge weight placed on the soil surface (ASTM Test for Maximum Index Density of Soils Using a Vibratory Table D4253) as illustrated in Fig. 5a. The maximum density achieved is used as a reference for field compaction.

For other soils, compaction is achieved in a mold by a falling weight impacting the soil (Fig. 5b). The standardized impact tests are known as standard compaction effort (ASTM Test for Moisture-Density Relations of Soils and Soil-Aggregate Mixtures Using 5.5-lb (2.49-kg) Rammer and 12-in. (305-mm) Drop D698; same as AASHTO T-99) or modified compaction effort (ASTM Test for Moisture-Density Relations of Soils and Soil-Aggregate Mixtures Using 10-lb (4.54-kg) Rammer and 18-in. (457-mm) Drop D1557; same as AASHTO T-180). The modified test applies 4 to 5 times greater compaction effort to the soil than the standard test.

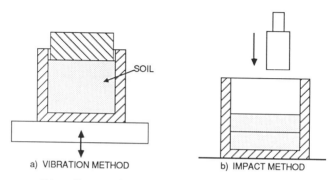

a) VIBRATION METHOD   b) IMPACT METHOD

Fig. 5 -- Laboratory compaction test.

In the impact tests soil is compacted at different water contents with the same effort and the resulting compaction is represented by the calculated dry unit weight. The characteristic compaction curves for the two efforts are illustrated in Fig. 6. The moisture content corresponding to the maximum dry density (MDD) in each case is known as optimum moisture content because the soil is easiest to densify at this moisture content. Figure 6 shows that as the compaction effort increases the maximum dry density increases and the optimum moisture content decreases.

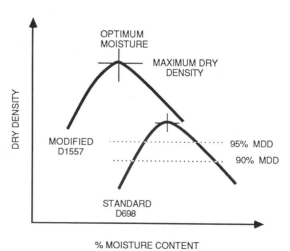

Fig. 6 -- Compaction test results.

Maximum density is not actually the highest that can be achieved for a given soil, but rather the maximum for a constant effort. Field compaction methods usually produce less than the maximum density in the standard test (ASTM D698). The density achieved in the field divided by the reference density and expressed as a percent is termed percent compaction. Values of 90 and 95% are shown in Fig. 6.

The characteristic curves shown in Fig. 6 apply to most soils. However the numerical values of the parameters vary with the soil type as illustrated in Fig. 7. Even within a given soil type the values of optimum moisture content and maximum dry density vary with changes in such characteristics particle gradation and plasticity. For this reason the ASTM compaction test needs to be repeated frequently during field construction to account for normal soil variability in order to be able to accurately check percent compaction.

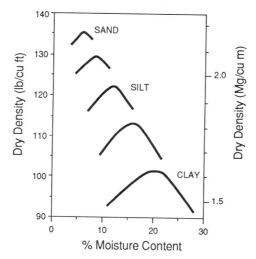

Fig. 7 -- Effect of type of soil
on compaction.

Compactability

For the same effort, the percent compaction achieved varies significantly with the soil type (Fig. 8a). Granular soils are much easier to compact than silty soils, which are easier to compact than clayey soils. In this example the 100% effort represents the ASTM D698 test effort. This is calculated as the product of hammer weight times drop height times number of drops (impacts) divided by the volume of compacted soil, i.e., total hammer potential energy per unit volume of soil. For the D698 test this value is about 12000 ft-lb/cu ft (580 kN-m/m^3); for the D1557 test this value is about 56000 ft-lb/cu ft (2700 kN-m/m^3). To achieve the curves in Fig. 8a the standard test was repeated numerous times but with the number of hammer drops and height of drop reduced to provide a range of compactive efforts.

Figure 8a shows that considerably higher compaction effort is required to obtain a specified percent compaction for clay than for sand. What is not universally recognized is that even when the same percent compaction is achieved, the resulting stiffness and strength properties are not the same for all soils. This results in a dramatic difference in stiffness among soils when related to compaction effort as illustrated in Fig. 8b. Quantitative examples of these comparisons may be found in Refs. [2-4]. These characteristics are not generally considered in compaction specifications because the same percent compaction is commonly specified regardless of the backfill soil type.

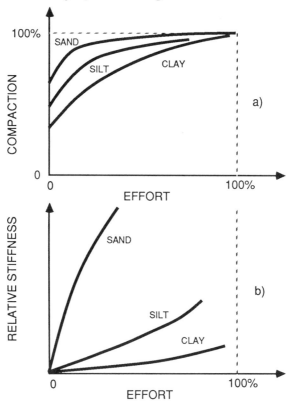

Fig. 8 -- Effect of soil type on variation
of percent compaction and soil
stiffness with compaction effort.

The relative compactability illustrated in Fig. 8 is very important in flexible pipe installation, because, for a given soil stiffness required to support the pipe, the less the required compaction effort the less the pipe distortion during placement of the soil envelope. This is one of the reasons for using coarse-grained soils for the envelope.

Changes After Compaction

When soils are subject to wetting and drying cycles after compaction, they will decrease in volume over time from the effects of the water. With increasing compaction the magnitude of this effect diminishes. The magnitude of volume change is much more significant for clays than for silts, and for silts it is much more significant than for sands.

The strength and stiffness of any soil will be higher when compacted at water contents less than optimum, than at optimum, but clay soils will swell more if the water content should increase later. This will cause a reduction in strength and stiffness. Conversely strength and stiffness will be lower when compacted at water contents higher than optimum, but fine-grained soils, especially clays, will shrink more upon drying. Compaction of soils that are too wet should be avoided because low strength and stiffness will result.

When soils are placed loosely around buried pipe they are subject to substantial volume reduction if they should become saturated. This phenomenon, known as collapse, will result in pipe deflection after construction. The reason for this behavior is that loosely placed soils are unsaturated and develop their resistance to deformation from effective stress induced by capillary water tension. When these soils become saturated the capillary tension is lost, causing the soil particles to settle into a denser packing.

The collapse characteristic is illustrated in Fig. 9 from tests on a silty sand. To perform the test the soil first was lightly compacted at around optimum moisture content in an oedometer. For one test (dashed curve) the soil was loaded in steps and then unloaded with the moisture content remaining at around optimum. In the other test (solid curve) the sample was loaded at optimum moisture content to 3.5 psi (24 kPa) and then allowed to saturate. As water entered the sample a sudden large strain occurred under constant load. Further loading while saturated gradually produced additional strain as in the moist sample case.

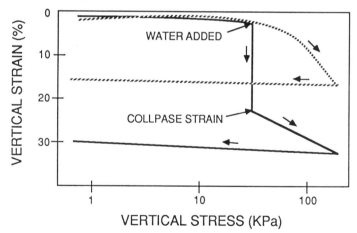

Fig. 9 -- Collapse of lightly compacted
silty sand from soaking.

Tests on a variety of specimens showed that the magnitude of the collapse strain decreased as the amount of compaction increased, and diminished to an insignificant amount when the percent compaction reached about 85 to 90% D698 or about 85% D1557 maximum dry density.

Another cause of pipe deformation after construction is migration of fine soil particles from the trench walls into the soil envelope.

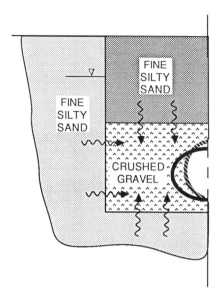

Fig. 10 -- Migration of fine soil
into coarse soil envelope.

This occurs as a result of groundwater movement when the soil envelope gradation is much coarser than that of the existing ground.   An example is given in Fig. 10.   This problem should not occur if the traditional filter criteria [5] are used in selecting the soil envelope gradation or if a geotextile (filter fabric) is used to provide proper separation.

PROPERTIES OF COMPACTED SOILS

Finite element analysis has been shown to be a good approach for evaluating soil-pipe interaction.   Duncan [6] has proposed the use of a hyperbolic model for representing the non-linear, stress-dependent soil behavior.   Modifications to this model were proposed by Boscardin, Selig, Lin and Yang [3] and design parameters determined for a variety of soils and compaction levels from laboratory tests [2].   These parameters were modified for flexible pipe by Haggag [4]. The values were then extended to other soil type and compaction levels by the writer and incorporated into CANDE [7].   The values are given in Table 1.   These parameters may be used to calculate tangent Young's modulus and bulk modulus as a function of stress state using the appropriate equations in the literature [2,3].

The linear-elastic model is a special case of the hyperbolic model in which the parameters are constant, independent of stress state.   This is the simplest model for representing soil behavior in soil-pipe interaction analysis.   Two independent elastic constants are needed.   The choice is normally from among Young's modulus (E_s), bulk modulus (B), Poisson's ratio (ν_s), and shear modulus (G).   Values of Young's modulus were estimated from the hyperbolic model for various values of maximum principal stress (σ_1) with the minimum principal stress (σ_3) equal to one-half to one times the maximum principal

Table 1 -- Recommended hyperbolic parameters for compacted soils.

Tested Soil	USCS	AASHTO	% T-99	% T-180	Soil No.	Wet Density (lb/ft³)	(Mg/m³)	K	n	R_f	c (psi)	c (kPa)	ϕ_0 (deg)	$\Delta\phi$ (deg)	B_i/P_a	ε_u
Gravelly Sand (SW)	SW, SP, GW, GP	A1, A3	100	95	27	148	2.37	1300	0.90	0.65	0	0	54	15	272.0	0.007
			95	90	21	141	2.25	950	0.60	0.70	0	0	48	8	187.0	0.014
			90	85	1	134	2.14	640	0.43	0.75	0	0	42	4	102.0	0.036
			85	80	22	126	2.02	450	0.35	0.80	0	0	38	2	31.8	0.057
			80	75	2	119	1.90	320	0.35	0.83	0	0	36	1	15.3	0.078
			61	59	3	91	1.46	54	0.85	0.90	0	0	29	0	4.3	0.163
Sandy Silt (ML)	GM, SM, ML; Also GC, SC with < 20% passing #200 seive.	A2, A4	100	95	28	134	2.14	800	0.54	1.02	5.5	38	36	0	197.5	0.021
			95	90	23	127	2.03	440	0.40	0.95	4	28	34	0	120.8	0.043
			90	85	4	120	1.92	200	0.26	0.89	3.5	24	32	0	46.0	0.071
			85	80	24	114	1.82	110	0.25	0.85	3	21	30	0	23.8	0.100
			80	75	5	107	1.71	75	0.25	0.80	2.5	17	28	0	12.8	0.134
			49	46	6	66	1.06	16	0.95	0.55	0	0	23	0	3.3	0.305
Silty Clay (CL)	CL, MH, GC, SC	A5, A6	100	90	29	125	2.00	170	0.37	1.07	11	76	12	0	81.3	0.064
			95	85	25	119	1.90	120	0.45	1.00	9	62	15	4	53.0	0.092
			90	80	7	112	1.79	75	0.54	0.94	7	48	17	7	25.5	0.121
			85	75	26	106	1.69	50	0.60	0.90	6	41	18	8	13.0	0.149
			80	70	8	100	1.60	35	0.66	0.87	5	34	19	8.5	8.8	0.178
			45	40	9	56	0.90	16	0.95	0.75	0	0	23	11	1.8	0.391
CH		A7	100	90	7	112	1.79	75	0.54	0.94	7	48	17	7	25.5	0.121
			95	85	26	106	1.69	50	0.60	0.90	6	41	18	8	13.0	0.149
			90	80	8	100	1.60	35	0.66	0.87	5	34	19	8.5	8.8	0.178
			45	40	9	56	0.89	16	0.95	0.75	0	0	23	11	1.8	0.391

stress. The hyperbolic parameters used were those in Table 1. Values of bulk modulus were estimated in the same manner. Then Poisson's ratio, ν_s, was derived from the relationship

$$\nu_s = 0.5 \left(1 - \frac{E_s}{3B}\right) \quad .$$

The resulting parameter values are given in Table 2.

Table 2 -- Elastic soil parameters.

Soil Type: SW, SP, GW, GP					
Stress level	95% D698			85% D698		
psi (kPa)	E_s	B	ν_s	E_s	B	ν_s
1 (7)	1600 (11)	2800 (19)	0.40	1300 (9)	900 (6)	0.26
5 (34)	4100 (28)	3300 (23)	0.29	2100 (14)	1200 (8)	0.21
10 (70)	6000 (41)	3900 (27)	0.24	2600 (18)	1400 (10)	0.19
20 (140)	8600 (59)	5300 (37)	0.23	3300 (23)	1800 (12)	0.19
40 (280)	13000 (90)	8700 (60)	0.25	4100 (28)	2500 (17)	0.23
60 (410)	16000 (110)	13000 (90)	0.29	4700 (32)	3500 (24)	0.28

Soil Type: GM, SM, ML, and GC, SC with < 20% fines					
Stress level	95% D698			85% D698		
psi (kPa)	E_s	B	ν_s	E_s	B	ν_s
1 (7)	1800 (12)	1900 (13)	0.34	600 (4)	400 (3)	0.25
5 (34)	2500 (17)	2000 (14)	0.29	700 (5)	450 (3)	0.24
10 (70)	2900 (20)	2100 (14)	0.27	800 (6)	500 (3)	0.23
20 (140)	3200 (22)	2500 (17)	0.29	850 (6)	700 (5)	0.30
40 (280)	3700 (25)	3400 (23)	0.32	900 (6)	1200 (8)	0.38
60 (410)	4100 (28)	4500 (31)	0.35	1000 (7)	1800 (12)	0.41

Soil Type: CL, MH, GC, SC					
Stress level	95% D698			85% D698		
psi (kPa)	E_s	B	ν_s	E_s	B	ν_s
1 (7)	400 (3)	800 (6)	0.42	100 (1)	100 (1)	0.33
5 (34)	800 (6)	900 (6)	0.35	250 (2)	200 (1)	0.29
10 (70)	1100 (8)	1000 (7)	0.32	400 (3)	300 (2)	0.28
20 (140)	1300 (9)	1100 (8)	0.30	600 (4)	400 (3)	0.25
40 (280)	1400 (10)	1600 (11)	0.35	700 (5)	800 (6)	0.35
60 (410)	1500 (10)	2100 (14)	0.38	800 (6)	1300 (9)	0.40

Note: Units of E_s and B are psi (MPa).

Deflections of buried flexible pipe are commonly calculated using the Iowa formula [1] which uses the modulus of soil reaction (E') as the parameter representing soil stiffness. Since E' is not a directly measureable soil parameter, but must be determined by back-calculation using observed pipe deflections, studies have been carried out to seek a correlation between E' and soil stiffness parameters such as Young's modulus (E_s) and constrained modulus (M_s), where E_s and M_s are related through Poisson's ratio (ν_s) by

$$M_s = \frac{E_s (1 - \nu_s)}{(1 + \nu_s)(1 - 2\nu_s)} \quad . \tag{1}$$

These studies [8-10] and analysis by the writer indicate that for

$$E' = k \, M_s \quad , \tag{2}$$

the value of k may vary from 0.7 to 2.3, with k = 1.5 as a representative value. For ν_s = 0.3, combining Eqs. 1 and 2 gives

$$E' = 2E_s \quad , \tag{3}$$

although the factor k could easily be higher than a value of 2.

The E' values developed by Howard [11] based on back-calculation from field observations may be converted to E_s values for comparison with the values in Table 2 for σ_1 = 5 to 10 psi (34 to 69 kPa). The comparison is as follows for compaction levels of 85 to 95% D698:

$$E_s \text{ (psi/MPa)}$$

Soil Type	Howard	Table 2
CL	200/1.4	250-1100/1.7-7.6
ML	500/3.5	700-2900/4.8-20
SW	1000/7	2100-6000/14-41

PROPERTIES OF EXISTING GROUND

A thorough review of the characteristics of existing ground is beyond the scope of this paper, and indeed encompasses most of the field of soil behavior. The complexity of soil behavior is part of the problem in defining the required soil properties for analysis. Equally critical is the spatial variability of natural soils combined with the practical necessity to estimate the properties from a very limited amount of sampling and testing.

Time-dependent stress-strain response, characterized by consolidation and creep, is often an important consideration for existing ground. However, the present state-of-the-art does not provide means for incorporating this response in pipe design except with very rough approximations.

A typical static-triaxial test stress-strain curve with unloading and reloading is illustrated in Fig. 11. This figure shows that unloading and reloading behavior is considerably more linear than the primary loading curve. This observation together with the recognized complexities of existing ground already discussed has resulted in approximating existing ground in zone A (Fig. 1) by constant modulus values representing linear elastic behavior. This approach is not as satisfactory for zone D (Fig. 1) because the stress-strain relationships may be very non-linear. If the embankment loading

produces stresses well above those previously experienced by
thenatural ground (considering stress history) then nonlinear modeling
such as used for compacted soil may be desired.

Fig. 11 -- Static triaxial test
results: increase in
axial strain with
increase in
axial stress.

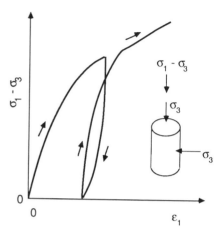

The existing ground parameters proposed for concrete pipe design
[12] are listed in Table 3. These are preliminary estimates which
need considerable refining by more study.

Table 3 -- Existing ground properties.

Material	Wet density (pcf)	Wet density (Mg/m^3)	E_s (psi)	E_s (Mpa)	ν_s
1. Coarse-grained					
A. Dense	145	2.32	10000	69	0.49
B. Medium	130	2.08	6000	41	0.35
C. Loose	115	1.84	2000	14	0.20
2. Fine-grained					
A. Stiff	125	2.00	6000	41	0.3
B. Firm	118	1.89	3500	24	0.4
C. Soft	110	1.76	1000	7	0.49
3. Concrete	150	2.40	3×10^6	21×10^3	0.17
4. Rock					
A. Weak	145	2.32	0.1×10^6	700	0.2
B. Competent	160	2.56	5×10^6	34×10^3	0.3

APPLICATIONS OF SOIL PROPERTIES

There are three common calculations in pipe design using soil
properties: 1) deflection, 2) wall thrust, and 3) buckling strength.
Examples of each will be given to illustrate the use of soil
properties.

Deflection

The use of the Iowa formula to calculate pipe deflection has already been mentioned. The deflection given is the horizontal diameter change produced by earth load placed above the crown of the pipe. Deflection caused by placing the soil envelope around the pipe is not included in the Iowa formula. The earth load needs to consider arching action caused by the installation conditions, for example the difference between trench and embankment as shown in Fig. 1. The required soil parameter (really a soil-structure interaction parameter) is E'. Design values of E' may be estimated from the Howard table [11], or from experience with similar installations.

An alternative approach which uses the conventional soil properties E_s and ν_s is the elasticity solution by Burns and Richard [13]. As for the Iowa formula, the deflection is just for earth load above the crown, which also needs to be adjusted for arching because the solution is based on a pipe deeply buried in a homogeneous soil and subjected to uniform surface pressure. The Burns and Richard solution not only provides horizontal pipe deflection, but also pipe deflection, wall thrust, bending moment and radial pressure at any point on the circumference for both no-slip and frictionless conditions at the soil-pipe interface. Soil properties may be estimated by: 1) using values in Table 2, 2) conducting field or lab tests on representative soil, or 3) back calculation with the elasticity solution for similar installations. The Burns and Richard solution is available as part of the CANDE computer program [7,14].

In critical or unusual cases more precise deflection analysis may be performed using finite element methods such as in CANDE. The soil may be represented by properties in Tables 1 and 3, unless data are available from tests on the specific soils involved. In most installations at least two zones of soil surround the pipe such as shown in Fig. 1. Only the finite element method is capable of determining the composite effect of these separate zones from a knowledge of properties of the individual zones.

Wall Thrust

Wall thrust can be estimated by the Burns and Richard solution or by the finite element methods described for the deflection analysis. The Marston-Spangler method may also be applicable [1].

Buckling Strength

Buckling strength is an important consideration in the design of buried flexible pipe. Buckling strength is normally determined for plastic pipe using equations based on some form of elastic spring soil model (Fig. 12) such as derived by Luscher [15]. The soil properties represented by the spring constant suffer the same limitation as E' in that they can not be directly measured, although approximate correlations with M_s and E_s have been proposed. Empirical corrections for depth of cover have also been suggested.

The approach representing soil as an elastic continuum (Fig. 13) is recommended as more suitable because it gives a more realistic representation of the soil-pipe interaction, it used directly measureable soil parameters, E_s and ν_s, and it provides a means of accounting for such factors as pipe shape, shallow cover and nonhomogeneous soil conditions [16,17]. The solution is presented in the form of critical hoop (wall) thrust, N_c, which is compared with actual wall thrust to determine the factor of safety against buckling. The critical hoop thrust is given by

$$N_c = 0.55 \ N_{ch} \ R_h \quad , \quad (4)$$

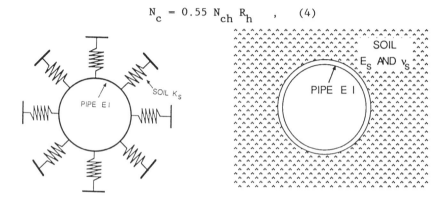

Fig. 12 -- Soil spring model   Fig. 13 -- Soil continuum model
       for buckling.   for buckling.

where R_h is a correction factor for shallow burial and nonhomogeneous soil (see examples in [4,17]), and N_{ch} is the critical thrust for a circular pipe deeply embedded in a homogeneous soil. For a smooth soil-pipe interface (conservative assumption) and for $EI/E_s^* < 0.01$, then

$$N_{ch} = 1.2 \ (EI)^{1/3} \ (E_s^*)^{2/3} \quad , \quad (5)$$

where
    E = pipe Young's modulus,
    I = pipe wall moment of inertia,
    $E_s^* = E_s/(1 - \nu_s^2)$,
    E_s = soil Young's modulus,
    ν_s = soil Poisson's ratio.

For deep burial in homogeneous soil then Eq. 4 becomes

$$N_c = 0.7 \ (EI)^{1/3} \ [E_s/(1 - \nu_s^2)]^{2/3} \quad . \quad (6)$$

The soil properties, E_s and ν_s, may be estimated from Table 3.

SUMMARY

The main requirements for the different soil zones encountered in buried plastic pipe installations were discussed. Characteristics of compacted soils were described, including the relative ease of compaction and the changes after compaction. Representative values of stress-strain properties were provided for compacted soils and for existing ground. Applications of these properties in analysis of pipe deflections, wall thrust and buckling stability were described.

REFERENCES

[1]   Spangler, M. G., and Handy, R. L., Soil Engineering, Harper and Row Publishers, NY, 1984.

[2]   Selig, E. T., "Soil Parameters for Design of Buried Pipelines," Proceedings, Pipeline Infrastructure Conference, ASCE, Boston, MA, 1988, pp. 99-116.

[3]   Boscardin, M. D., Selig, E. T., Lin, R. S., and Yang, G. R., "Hyperbolic Parameters for Compacted Soils," Journal of Geotechnical Engineering, ASCE, Vol. 116, No. 1, January, 1990, pp. 88-104.

[4]   Haggag, A. A., "Structural Backfill Design for Corrugated-Metal Buried Structures", Doctoral Dissertation, Department of Civil Engineering, University of Massachusetts, Amherst, MA, May, 1989.

[5]   Cedergren, H. R., Drainage of Highway and Airfield Pavements, Wiley, 1974.

[6]   Duncan, J. M., Byrne, P., Wong, K. S., and Mabry, P., "Strength, Stress-Strain and Bulk Modulus Parameters for Finite Element Analysis of Stresses and Movements in Soil Masses", Report No. UCB/GT/80-01, Department of Civil Engineering, University of California, Berkeley, CA, August, 1980.

[7]   Musser, S. C., Katona, M. G., Selig, E. T., CANDE-89 User Manual, Federal Highway Administration, Turner-Fairbank Highway Research Center, McLean, VA, 1989 (in publication).

[8]   Neilson, F. D., "Modulus of Soil Reaction as Determined from Triaxial Shear Test", Highway Research Record No. 185, Washington, D.C., 1967, pp. 80-90.

[9]   Allgood, J. F., Takahashi, H., "Balanced Design and Finite Element Analysis of Culverts", Highway Research Record No. 413, Washington, D.C., 1972, pp. 44-56.

[10]   Hartley, J. P. and Duncan, J. M., "E' and its Variation with Depth", Journal of Transportation Engineering, Vol. 113, No. 5, September 1987, pp. 538-553.

[11]   Howard, A. K., "Modulus of Soil Reaction (E') Values for Buried Flexible Pipe", Engineering and Research Center, Bureau of Reclamation, Denver, Colorado, 1976.

[12] Heger, F.J., Liepins, A. A., and Selig, E. T., "SPIDA: An Analysis and Design System for Buried Concrete Pipe," Advances in Underground Pipeline Engineering, Proceedings of ASCE, August 1985, pp. 143-154.

[13] Burns, J. Q. and Richard, R. M., "Attenuation of Stresses for Buried Cylinders", Proceedings of the Symposium on Soil Structure Interaction, University of Arizona, Tucson, Arizona, September 1964, pp. 378-392.

[14] Katona, M. G., et al., "CANDE: Engineering Manual-A Modern Approach for the Structural Design of Buried Culverts", Report No. FHWA/RD-77, NCEL, Port Hueneme, CA, October 1976.

[15] Luscher, U., "Buckling of Soil-Surrounded Tubes", J. Soil Mech. Found. Div., Proc. Am. Soc. Civ. Engrs., Vol. 92, No. SM6, Nov. 1966, pp. 211-228, (discussed in Vol. 93 (1967): No. SM2, p. 163; No. SM3, pp. 179-183, No. SM5, pp. 337-340, Author's closure in Vol. 94, No. SM4, 1968, pp. 1037-1038.

[16] Moore, I. D., Selig, E. T., and Haggag, A., "Elastic Buckling Strength of Buried Flexible Culverts," Transportation Research Board 1191, Culverts and Tiebacks, 1988.

[17] Moore, I. D., and Selig, E. T., "Use of Continuum Buckling Theory for Evaluation of Buried Plastic Pipe Stability", Buried Plastic Pipe Technology, ASTM STP 1093, George S. Buczala and Michael J. Cassady, Eds., American Society for Testing and Materials, Philadelphia, 1990.

A. P. Moser, O. K. Shupe, and R. R. Bishop

IS PVC PIPE STRAIN LIMITED AFTER ALL THESE YEARS?

REFERENCE: Moser, A. P., Shupe O. K., and Bishop, R.R., "Is PVC Pipe Strain Limited After All These Years," Buried Plastic Pipe Technology, ASTM STP 1093, George S. Buczala and Michael J. Cassady, Eds., American Society for Testing Materials, Philadelphia, 1990.

ABSTRACT: PVC (Polyvinyl chloride) sewer pipes have seen wide use in the United States and this has prompted concern for an appropriate material property design limit. It had been proposed that the imposition of a strain limit derived from long-term creep testing would also be appropriate for buried gravity flow pipes subjected to constant strain. Laboratory tests of pipe ring samples exposed to various strains and temperatures have been conducted for the past 13 years on filled and unfilled PVC compound formulations. Samples of pipe, from a test installation of buried pipe, have been excavated after 14 years and a post evaluation has been conducted. These test results are used to draw some conclusions concerning the applicability of a material strain limit for constant strain design conditions.

KEYWORDS: buried pipes, PVC (polyvinyl chloride) Pipes, stress-relaxation, strain, filled PVC

INTRODUCTION

The use of PVC (polyvinyl chloride) pipe as sewer pipe in the United States began in the early to mid 1960's as early manufacturers of PVC resin looked for potentially high volume applications for their resin. Throughout the sixty's, PVC pipe of various types were provided for gravity sewer applications. Formal Standards [ASTM D3033 "Type PSP Poly(Vinyl Chloride) (PVC) Sewer Pipe and Fittings," and D3034 "Type PSM Poly(Vinyl Chloride) (PVC) Sewer Pipe and Fittings"] were adopted in 1972 launching a virtual explosion of PVC sewer pipe use. Today, 90 percent of all sewer pipes in sizes 4 - 15 inches used in the United States, are made of PVC. (Note: ASTM D3033 was dropped as a formal standard in 1989.)

The first issue of ASTM D3034 and D3033 contained material requirements for a single PVC cell class of 12454B as described in ASTM D1784 "Rigid Poly(Vinyl Chloride) (PVC) Compounds and Chlorinated Poly(Vinyl Chloride) (CPVC)." The second issue published in 1973 contained a 13364B cell class as a second option.

Dr. Moser is the Head of the Mechanical Engineering Department, and Dr. Shupe is Professor of Mechanical Engineering at Utah State University, Logan, Utah 84321-4130. Mr. Bishop is Director of Technical Services at Carlon, 25701 Science Park Drive, Beachwood, Ohio 44122.

This option was prompted by the Arab oil embargo era and incorporated the use of fillers which increased the modulus of elasticity from 400,000 to over 500,000 psi. Using filled compounds also decreased tensile strength and tensile elongation while finished pipe met the same finished product requirements established in the original D3034-72. Sewer pipes of both compounds have found wide use in the past 17 - 18 years.

Two fundamental questions which arose in the early 1970's are expressed as follows: 1) What particular PVC compounds are suitable as sewer pipe? and 2) What material property limits should be used for structural design purposes? At least partial answers to these questions have been published in the literature over the years. An early approach suggested by Chambers and Heger [1] to limit the strain to 50 percent of an assumed ultimate strain of one percent has been shown to be very conservative by Moser [2], Janson [3], and Molin [4].

Tests to help fully answer these questions were established in 1975 and 1977 at Utah State University. Partial reportings of results of these tests were published by Moser [1] and by Bishop [5] in 1981. These tests have continued. Constant strain tests conducted on bar and pipe ring samples have been under test for 13 years. Data from these tests are now reported herein. Also, buried pipes which have been installed for 14 years have been excavated and a post evaluation of the pipe samples has been conducted and is reported.

TIME DEPENDENT PERFORMANCE OF BURIED PVC PIPE

In September of 1975, an embankment installation reaching a depth of cover of 22 feet was constructed over four test pipe sections that extended radially from a single access manhole. The test site became known as 'the mole hole' and has provided an excellent opportunity to easily monitor buried performance of PVC pipes for the past 14 years. In the fall of 1989, the test pipes were excavated for a post test examination. The test site is part of a gravel pit where the insitu soil is a fine blow sand with 18 percent silt. The soil is moisture sensitive and is subject to soil collapse when saturated. The site itself experiences seasonal ground water level changes which place the pipe below the water table in the spring months and above the water table in summer and most winter months.

Pipes were made of two different PVC compounds. Two samples were 12364B cell class per ASTM D1784. They have a calcium carbonate filler content of 40 parts to each 100 parts of resin by weight. Two other samples were foamed PVC with a specific gravity of 1.2. Table 1 provides basic dimensions and property data for these two pipe compounds. Typical properties for unfilled, unfoamed PVC cell class 12454B are also given in Table 1 for comparison purposes.

Long Term Deflection Data

In-ground vertical deflection data have been taken for 14 years and are plotted in Figure 1. A stable deflection period was reached at 40 days (960 hrs) after installation, and was constant until the first instance of the ground water table reaching the level of the pipe zone bedding. During the first spring season at about 150 days (3600 hrs) following installation, the ground water table rose above the level of the pipe. This groundwater condition caused the soil to consolidate and the load to increase. This produced a somewhat rapid increase in deflection for all pipe samples during this period. A new stable or equilibrium deflection level was reached at about 400 days (9600 hrs). The water table has continued to fluctuate on an annual basis for the 14 year test period. These subsequent water table movements have

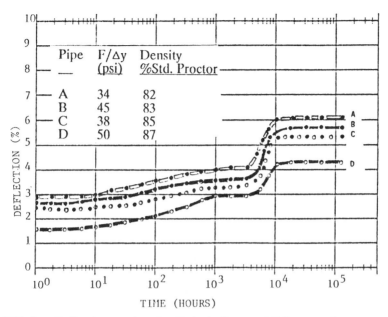

FIG. 1 -- Deflection vs. time for 10-inch diameter PVC sewer pipe.
(22 foot deep embankment installed Sept. 1975)

TABLE 1 -- Basic properties of pipe samples

Compound	Cell Class[1]	Pipe Stiff (psi)	Thickness (in)	E (psi)	Sp. Gravity
Filled	12364B	45-50	.327-.331[3]	630,000	1.62
Foamed	Exp.[2]	32-36	.381-.417	218,000	1.2
Unfilled/ Unfoamed	12454B	46 min.	0.300	400,000	1.4

[1] Per ASTM D1784
[2] Experimental (Not classified)

influenced deflection readings only slightly since the initial saturation of the pipe zone in 1976.

Again, the soil around these pipes was a silty fine sand. For this soil, over 92 percent standard Proctor density is necessary to insure a void ratio less than the critical value. The installed densities were less than 92 percent, resulting in void ratios greater than critical. Thus, when the water table rose into the pipe zone, soil consolidation took place and caused pipe deflections to increase. This indicates that for pipe installation below the groundwater table, additional deflection control can be obtained if the density is such that the void ratio is below the critical value.

The test site area has also been subjected to small earthquake tremors during the test period. Any effects are included in the deflection results but cannot be isolated.

Post Evaluation of Buried Samples

Pipe samples excavated from the site were examined visually and no signs of cracking, crazing or other polymer damage were evident. Specific gravity, pipe stiffness and wall thickness measurements were taken for each sample and are given in Table 2. Notably, the pipe stiffness for the foamed samples varied from 34-38 psi initially and now range from 36-40 psi. The filled pipe samples varied from 45-50 psi initially and now measure 44-49 psi after 14 years of buried service. This small variation is probably within the expected experimental error and no change in the pipe's capacity to resist deflection has occurred over this time period.

These pipes were each subjected to a 60 percent deflection test, to determine ductility. These tests were as prescribed in ASTM Standards D3034, F789 "Type PS-46 Poly(Vinyl Chloride (PVC) Plastic Gravity Flow Sewer Pipe and Fittings," F679 "Poly(Vinyl Chloride (PVC) Large-Diameter Plastic Gravity Flow Sewer Pipe and Fittings," and F794 "Poly(Vinyl Chloride (PVC) Large-Diameter Ribbed Gravity Sewer Pipe and Fittings Based on Controlled Inside Diameter." Each sample sustained that deflection level without cracking (see Fig 2).

TABLE 2 -- Post Excavation Properties of Embankment Pipe Samples

Pipe Sample Designation	Compound Type	Thickness Average (in)	Specific Gravity	Pipe[1] Stiff. (psi)	60% Flattening
A	Foamed	0.381	1.2	36.8	no cracking
B	Filled	0.327	1.6	44.0	no cracking
C	Foamed	0.417	1.2	40.5	no cracking
D	Filled	0.331	1.6	49.0	no cracking

[1]Post evaluation pipe stiffness per ASTM D2412

STRESS RELAXATION TESTS

Stress relaxation tests were performed on ring sections cut from PVC pipe (see Figure 3). These test specimens were each diametrically deformed to a specified deflection. The load necessary to hold this deformation constant was determined as a function of time. This series of tests has been in progress for over thirteen years.

FIG. 2 -- Sixty percent deflection test

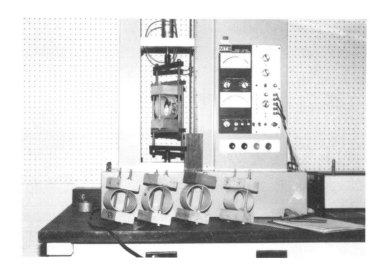

FIG. 3 -- Pipe ring specimens undergoing stress relaxation and testing

Each specimen was maintained at one of three temperatures: ambient (70°F)(21.1°C), 40°F (4.4°C), and 0°F(-17.8°C). The ambient temperature was held to ± 5°F(±2.8°C). A refrigerator was used to maintain the 40°F temperature and was found to fluctuate between 38°F and 41°F. The 0°F specimens were placed in a freezer and the temperature varied between a -5°F and 0°F. The purpose of the lower temperature test was to slow down the stress relaxation which would amplify any tendency toward brittle fracture. (For dimensions of test specimen, see Table 3).

TABLE 3 -- Pipe ring properties used in stress relaxation tests
Pipe rings were cut from 4-inch diameter PVC pipe

Material PVC	Wall Thickness (inch)	Length (inch)	Average Flexure Modulus (psi)	Average Pipe Stiff. (psi)
Filled	0.132 ± 0.05	2.0	540,000	87
Unfilled	0.153 ± 0.04	2.0	470,000	117

Three PVC compounds were tested: two filled compounds and an unfilled compound. One filled compound contained forty parts calcium carbonate by weight and is designated as ASTM cell class 12364B. The other filled compound contained thirty parts of calcium carbonate by weight and is designated ASTM cell class 13364B. The unfilled compound is designated as ASTM cell class 12454B. Results from the two filled compounds could not be differentiated. Thus, for discussion purposes, they have been combined into one group called "filled".

Some of the pipe ring test specimens were notched to produce stress risers. The pipe rings were compressed (deflected) vertically. The notches were placed along the length in four places corresponding to the locations of the highest tensile stresses -- twelve and six o'clock positions on the inside surface and the three and nine o'clock positions on the outside surface. These longitudinal notches were cut to a depth of 0.012 ± 0.006 inches. The purpose of these notches was to produce stress risers which would amplify any tendency of brittle fracture. In all, there are 91 specimens being tested in the study which started January 1977 (see Table 4 for details).

Figures 4 through 9 contain stress relaxation curves. As of January 1990, after 13 years, none of the test specimens had failed. The stress relaxation data can be represented by straight lines on log-log plots. The data are similar for either filled or unfilled PVC pipe when tested at the same temperature. The slopes of the stress relaxation curves show that the relaxation rate is less for lower temperatures in both the filled and unfilled PVC. The addition of the filler material, doesn't cause brittle fracture to occur with time if the pipe section is not loaded to its failure point initially. The difference in the stress relaxation curves for filled and unfilled PVC is that more force was required to deflect the unfilled specimens to their initial desired percent deflection. This is because of the thicker wall and the resulting higher pipe stiffness for the unfilled PVC specimens. However, the addition of the calcium carbonate filler does increase the elastic modulus.

In comparing the stress relaxation curves for the notched and unnotched specimens, within the filled and unfilled groups respectively, no significant difference could be observed. The increased strain at the base of the notches had no apparent effect on the stress relaxation characteristics of either filled of unfilled PVC. Therefore, it was concluded that PVC is not notch sensitive when it is deformed diametrically in a constant deflection test.

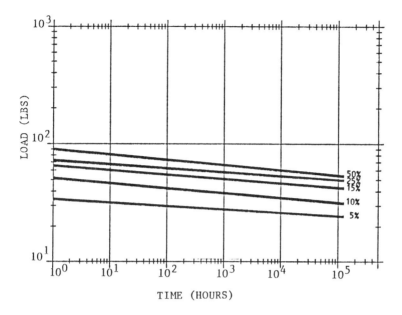

FIG. 4 -- Relaxation curves for filled, unnotched PVC specimens at specified deflections and a temperature of 40°F.

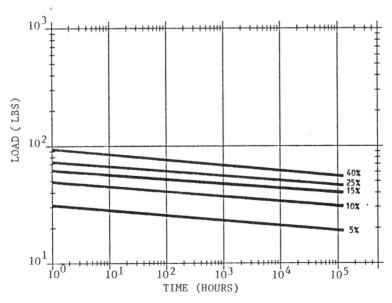

FIG. 5 -- Relaxation curves for filled, notched PVC specimens at specified deflections and a temperature of 40°F.

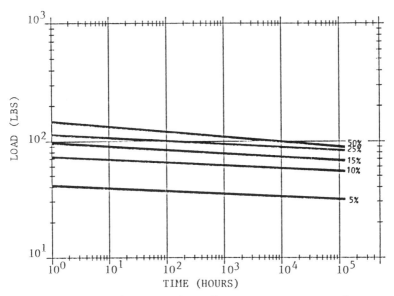

FIG. 6 -- Relaxation curves for unfilled unnotched PVC specimens at specified
deflections and a temperature of 40°F.

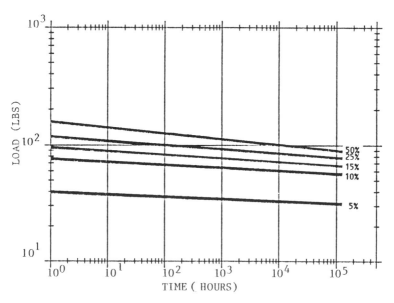

FIG. 7 -- Relaxation curves for unfilled, notched PVC specimens at specified
deflections and a temperature of 40°F.

Table 4 -- Grouping of the 91 pipe specimens in the stress relaxation tests.

Groups	Sets	Deflections of Specimens					
		1	2	3	4	5	6
Group 1		(in percent)					
Specimens were filled and unnotched	Set 1, Ambient	5	10	15	25	50	
	Set 2, 40° F	5	10	15	25	50	
	Set 3, 0° F	5	10	15	25	50	
Group 2							
Specimens were filled and unnotched	Set 1, Ambient	5	10	15	25	50	
	Set 2, 40° F	5	10	15	25	50	
	Set 3, 0° F	5	10	15	25	50	
Group 3							
Specimens were filled and notched	Set 1, Ambient	5	10	15	25	40	
	Set 2, 40° F	5	10	15	25	35	
	Set 3, 0° F	5	10	15	25	35	
Group 4							
Specimens were filled and notched	Set 1, Ambient	5	10	15	25	40	35
	Set 2, 40° F	5	10	15	25	40	
	Set 3, 0°	5	10	15	25	40	
Group 5							
Specimens were unfilled and unnotched	Set 1, Ambient	5	10	15	25	50	
	Set 2, 40° F	5	10	15	25	50	
	Set 3, 0° F	5	10	15	25	50	
Group 6							
Specimens were unfilled and notched	Set 1, Ambient	5	10	15	25	50	
	Set 2, 40° F	5	10	15	25	50	
	Set 3, 0° F	5	10	15	25	50	

It is interesting to note the relaxation that has taken place in the thirteen year period is significantly less than has been supposed by some. The total stress relaxation associated with the five percent initial deflection is small for the ambient temperature and is negligible for the 40°F and 0°F temperatures. The slightly higher relaxation rate takes place for a higher imposed constant deflection (or the initial load) the greater the relaxation rate. This is evident because the slope of the relaxation line is steeper for specimens which have the greatest imposed deflection or initial load.

Creep information may be obtained from these curves. For example, from Fig. 8, it can be seen that a sample placed with a constant load such that it deflects five percent at one hour will creep to over 10 percent at 10,000 hours. This is for the case of an ambient temperature. Fig. 9 shows that the creep at 10,000 hours is very small for a five percent initial deflection. However, this same figure shows that the sample

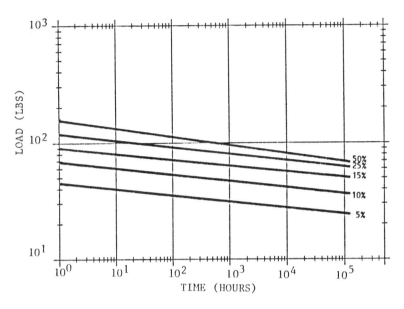

FIG. 8 -- Relaxation curves for unfilled, unnotched PVC specimens at specified deflections and a temperature of 70°F.

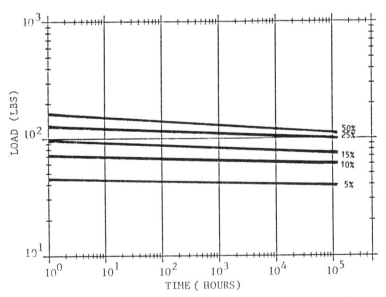

FIG. 9 -- Relaxation curves for unfilled, unnotched PVC specimens at specified deflection and a temperature of 0°F.

will creep to about 50 percent deflection from initial deflection of 25 percent in a 10,000 hour period.

Stiffness data for the stress relaxation specimens are given in Table 5. The initial stiffnesses were determined using the one-hour relaxation loads. That is, these stiffnesses are the one-hour load per length divided by the imposed deflection. Stiffness measured at the end of the 13 year period are incremental stiffnesses. Each specimen was deflected an additional five percent from its preset value. The stiffnesses were then calculated by dividing the incremental load per length by the five percent incremental deflection. These long term values are the instantaneous stiffnesses and are the stiffnesses that resist any additional deflection. These data show that pipe stiffnesses and modulus for PVC pipe do not decrease with time.

TABLE 5 -- Pipe stiffness of constant strain ring samples.

Sample Description		Temperature[1]	Pipe Stiffness (psi)			
			Initial[2]		13 years[3]	
Filled	Notched	(°F)	5%	25%	5%	25%
yes	no	0	71	39	69	63
yes	yes	40	76	38	74	65
yes	yes	0	75	41	69	63
no	no	40	101	60	89	91
no	no	0	102	65	91	110
no	yes	40	101	63	96	87

[1] Constant temperature during 13 year test. Sample conditioned to 73° F for stiffness testing.

[2] Pipe stiffness determined by secant method after being held at the specified deflection for one hour.

[3] 13 year stiffness determined by applying an additional five percent deflection increment to the specified deflection.

CONCLUSIONS

1. Stress relaxation in filled and unfilled PVC can be approximated by a straight line on log-log paper and the relaxation rate is temperature dependent. The rate of relaxation decreased with a decrease in temperature.

2. Filled or unfilled PVC pipes do not appear to be notch sensitive when loaded under constant deformation.

3. Buried pipe and soil systems stabilize to an equilibrium condition which typifies a fixed deflection or fixed strain condition.

4. Under conditions of fixed strain, buried PVC pipes maintain the same capacity to resist additional deflection increments as when initially installed.

5. Filled PVC compounds of cell class 13364B and 12364B along with unfilled cell class 12454B can sustain deflections of 40-50 percent without loss of stiffness or ductility for periods exceeding 13-14 years.

6. Apparent or creep modulus is an inappropriate property to predict long term deflection of buried PVC gravity sewer pipe. Pipes continue to respond to additional deflection increments by resisting movement at the same stiffness as newly made pipe.

REFERENCES

[1]   Chambers, R. E., and Heger, F. J., "Buried Plastic Pipe for Drainage of Transportation Facilities," Simpson Gumpertz and Heger, Inc., Cambridge, Massachusetts, 1975.

[2]   Moser, A. P., "Strain as a Design Basis for PVC Pipes?" Proceedings of the International Conference on Underground Plastic Pipe, American Society for Civil Engineering Conference, New York, 1981, pp. 89-102.

[3]   Janson, L-E., "Plastic Gravity Sewer Pipes Subjected to Constant Strain by Deflection," Proceedings of the International Conference on Underground Plastic Pipe, American Society of Civil Engineers. New York, 1981, pp. 194-116.

[4]   Molin, J., "Long Term Deflection of Buried Plastic Sewer Pipes," Advances in Underground Pipeline Engineering, American Society of Civil Engineers, New York, 1985, pp. 263-277.

[5]   Bishop, R. R., "Time Dependent Performance of Buried PVC Pipe," Proceedings of the International Conference on Underground Plastic Pipe, American Society Civil Engineering Conference, New York, 1981, pp 202-212.

Peter G Chapman

FIELD EXPERIENCE, PERFORMANCE TESTING AND DESIGN OF VERY
FLEXIBLE THERMOPLASTIC PIPE SYSTEMS

REFERENCE: Chapman, P. G., "Field Experience,
Performance Testing and Design of Very Flexible
Thermoplastic Pipe Systems", Buried Plastic Pipe
Technology, ASTM STP 1093, George S. Buczala and
Michael J. Cassady, Eds., American Society for
Testing and Materials, Philadephia, 1990.

ABSTRACT: As pipe stiffness is reduced, prediction of deflection
and deformation of buried pipes becomes less reliable. However,
for thermoplastic pipe materials operating under predominantly
constant strain conditions, the level of strain is not a critical failure
parameter, and prediction of deformations and strains is likewise
less relevant. Buckling and wall compression are the other mechanisms for system failure, but are unlikely to be critical factors except
at very low or very high soil covers. Cost savings can therefore be
achieved through the use of very low stiffness pipes, where the
application and field conditions permit. Experimental work is described involving controlled loading tests on buried pipes, monitoring of field installations, and a large scale installation at the World
Expo site in Brisbane using very low stiffness polyvinyl chloride
(PVC) storm drains.

KEYWORDS: Flexible pipes, thermoplastic, stiffness, installation,
structural profiled wall pipe.

INTRODUCTION

Large scale use of Thermoplastic Pipes, in particular polyvinyl chloride

---

Mr Chapman is Technical Manager at Vinidex Tubemakers Pty Limited, 15
Merriwa Street, Gordon, New South Wales, Australia 2072.

(PVC), for buried gravity sewer and storm drainage applications began in Australia in the early 1970's. The evolution of development paralleled that in other countries, through the smaller diameter range 100 - 150 mm to larger diameters as the material economics and product acceptance improved.

Due to the higher costs of thermoplastic resins in Australia, by comparison with world prices, the competitive status of thermoplastic pipes against other materials has traditionally been limited. Plain walled pipes are today manufactured to 400 mm diameter in PVC, 1000 mm in high density polyethylene (HDPE). The advent of spirally wound structural profiled wall pipes shifted the economic balance, and pipes of this form can now be manufactured to 2000 mm diameter.

One such profiled wall PVC pipe, under the trade name Rib Loc™*, was introduced to the Australian market in 1981. It is manufactured by spiral winding an extruded strip with T-ribs and interlocking edges as shown in Fig 1. The strip is wound cold using relatively simple machinery which is adjustable in diameter. Thus the winding operation may be conducted remote from the extrusion site, which has some advantages in terms of stock holding and transport economics. The seam may or may not be solvent welded depending on the application.

A range of profiles is manufactured to suit the range of applications and diameters targeted. The strip stiffness, EI (material modulus x cross-sectional moment of inertia) for each strip is fixed, but the facility to wind strips to any diameter provides the ready capability of producing pipes with ring stiffness EI/D^3 over a very wide range.

A similar product is manufactured in the United States of America but this is hot formed to produce pipes of much higher stiffness.

Economic pressures on both manufacturers and users naturally raised the question as to the minimum pipe stiffness that could usefully be employed. The criteria governing minimum stiffness were thus the subject of considerable discussion and experimentation during the development of this product, and this work is the primary subject of this paper.

Fig 1 -- A typical Rib Loc strip profile

AUSTRALIAN STANDARDS FOR FLEXIBLE PIPE DESIGN

The classical Spangler Marston principles were embodied in an Australian Code of Practice, "Plastics Pipelaying Design", AS 2566 - 1982, for thermoplastic

* Trademark of Rib Loc Australia Pty Ltd

pipes as early as 1972, following generally the same load computational procedures as normally applied at that time for rigid pipes, but using the modified Iowa formula for prediction of deflection.

$$\delta = K\ W/(8S + .061E')$$

where
    W = the vertical load per unit length of pipe,
    K = a constant related to installation factors
    S = the ring stiffness of the pipe (EI/D^3),
    E' = modulus of passive soil resistance

Because of the doubts concerning the extrapolation of this theory to small diameter plastic pipes, experimental work was undertaken by Standards Australia to provide supportive data[1] for the recommended soil modulus values for use in design.

The observation was made in the course of this work that a correlation existed between the apparent soil modulus and the dimension ratio of the pipe, viz. pipes of lower wall thickness registered a lower soil modulus. This empirical relationship is embodied in the design method, such that the soil is characterised by a soil constant Y, and the soil modulus is derived from this value through the linear relationship:

$$E' = Y/(D_m/\ t)$$

where
    D_m/t = mean dimension ratio
          = pipe diameter at the neutral axis of the wall/wall thickness

The soil constant Y is recommended in the Code for five soil types and three levels of compaction. In the test work, three dimension ratios were used, 63, 43, and 17.

It is noted that the relationship is fundamentally illogical, since the soil response cannot be affected by a geometric property of the pipe. The soil can only respond physically to a mechanical property of the pipe, logically its lateral stiffness. The observed correlation is preserved if the relationship is expressed:

$$E' = Y_1\ S^{1/3}$$

where
    Y_1 = a new soil constant scaled appropriately to account for other constants.

[1] Burn, L S, "Deformation of Buried Plastic Pipes", Commonwealth Scientific Industrial Research Organisation, Division of Building Research, internal paper 86/31, unpublished.

This approach enables the code to be applied to profiled wall pipes specified by stiffness S rather than dimension ratio D_m/t as in a plain wall pipe.

Although the relationship is supported by somewhat scant data, it is seen to be conservative since the above method results in a lower estimate of soil modulus for lower pipe stiffness, with a corresponding increase in predicted deflection.

A further point worthy of note concerning this code is that the pipe stiffness term in the Iowa formula uses a very conservative long term (50 year) material creep modulus, for prediction of deflections due to continuous loadings. Although this is still the subject of some discussion, current thinking [1] generally accepts that the short term modulus is more applicable for the largely constant strain conditions involved in buried flexible pipe deformations.

Since the soil modulus recommendations were back-calculated using the same conservative stiffness values, the correct predictions would nevertheless be obtained, at least within the range of test conditions covered. However, the low values of stiffness have the effect of reducing the significance of this term in the Iowa equation, so that the effect of pipe stiffness might be suppressed when extrapolating outside of the test conditions.

In spite of these short-comings the Code of Practice has provided valuable guidance to designers in predicting the general response of buried thermoplastic pipes.

USE OF THE IOWA EQUATION FOR VERY LOW STIFFNESS PIPES

Because the pipe stiffness term in the Iowa equation has decreasing significance as the stiffness is reduced, some authors have questioned the validity of the theory for such conditions. Certainly the Iowa equation suggests that deflection becomes insensitive to pipe stiffness, and proposes a limiting value of deflection at zero pipe stiffness, but this simply implies that it is possible to dig a self-supporting tunnel, and this in fact is correct by common knowledge, the only proviso being that the skin of the tunnel remains intact under loading and deformation.

There is therefore no problem with the logic of the Iowa equation, nor its prediction that very low stiffness pipes can be used without excessive deflections.

However, the usefulness of Iowa predictions for very low stiffness pipes is necessarily limited to the broad level, since their accuracy is clearly limited by the accuracy of prediction of the soil properties, which can only be approximately estimated and controlled.

The Iowa formula does not predict shape deformation. Other techniques, such as finite element modelling, are seen to offer an improved approach from this point of view, but their accuracy is still limited by the input data.

In the final analysis, the requirements for accuracy must be considered in the light of how relevant such information may be in achieving a successful installation. As discussed following, for thermoplastic pipes, the application of more sophisticated arithmetic is not likely to improve the result.

FLEXURAL STRESS AND STRAIN LIMITATIONS

The main purpose of deformation prediction is to enable determination of deflection and stresses and strains due to ring bending.

Extensive experience with plain wall thermoplastic pipe since 1950 produced no evidence that stress/strain conditions likely to produce material rupture arise in normal buried pipe applications.

However, during the 1970's, lack of data concerning very long term effects, and some disastrous experience with reinforced plastics with relatively low strain limits, prompted the application of conservative limits also to thermoplastics.

Since structural profile wall sections increase substantially the stress and strain levels present in a pipe wall, these conservative limits retarded the development and use of such pipes. This was unfortunate, and probably unnecessary, since structural profiled thermoplastic pipes had in fact been in use in the form of corrugated land drainage pipes for many years with no evidence of failure due to material flexural strain.

Likewise in the case of Rib Loc, no evidence of material rupture due to long term strain levels has been registered in field or laboratory studies, and it may be noted that initial strains of up to 2% are generated in the outer rib fibres during the cold winding process, which are additional to the flexural strains induced in service.

It is now generally conceded [2,3,4] that the acceptable flexural strain levels in the thermoplastic pipes under constant strain conditions are much higher than originally supposed. Buried flexible pipes are not strictly subjected to completely constant strain conditions, but it is probable that, provided the rate of stress relaxation exceeds the rate of loading, no strain limit can be found for a visco-elastic material.

Whether or not strain limits exist for thermoplastic materials seems to be academic; they would not be approached under service conditions since other criteria for satisfactory performance of a pipe will be reached at lower levels of deformation.

COMPRESSIVE STRESSES AND BUCKLING

Compressive stresses developed in the side walls of the pipe may constitute a limiting factor. For plain walled pipes it can readily be shown that these stresses are not likely to be critical under any practical situation. However, structural profiled wall pipes can theoretically be designed with very efficient flexural cross-sections, and low cross-sectional areas, so that the compressive stresses developed are much higher.

The limit on the efficiency of a structural section is usually related to its stability against local buckling of the webs of the section under flexure. This is a matter for design of the profile, and criteria need to be established to ensure satisfactory performance in this respect. An arbitrary but effective criterion is that the load/ deflection curve under parallel plate loading shall show no maximum at deflections up to 30% of diameter. Whilst 30% is well beyond the functional range of deflections acceptable in practice, conservatism is warranted since a radius of curvature equivalent to 30% elliptical deflection might be produced under practical non-elliptical deformation conditions. Criteria similar to this are being adopted in United States, European and Australian product standards.

Provided the design of the profile is satisfactory from this point of view, sidewall compression is found to be a limiting factor only for deeper burial, or at very low cover and/or high wheel loads.

Whilst side wall compression may be the limiting factor for light weight structural profiled pipes, this does not imply a lower limit on pipe stiffness.

For low cover and high wheel load conditions, a further factor may dominate design, viz. shear failure of the soil arch above the pipe. Under these circumstances pipe stiffness becomes irrelevant, since even very stiff flexible pipes cannot directly support the loads. Research into this area is currently being conducted in Australia[2].

Buckling collapse of the pipe section may also constitute a failure mechanism for very flexible pipes. The degree of support offered by the soil is a major factor here. The methods of analysis variously proposed in the literature generally assume that the pipe is surrounded by a fluid medium at a pressure equal to the soil pressure imposed on the top of the pipe, plus vacuum pressures that may be developed internally, and compare this to a buckling collapse pressure calculated from the geometric mean of the pipe's unsupported critical collapse pressure and the Spangler soil modulus. Some compensation is usually applied for reduction of collapse pressure due to deformation. The empirical deficiencies in the analysis are covered by a large factor of safety.

Whilst more rigorous approaches have been proposed, the stumbling block is as usual the inadequate knowledge and control of the soil properties. And since this failure mechanism is not observed under any normal practical condition, a greater

[2] Rib Loc Australia Pty Ltd, private communication.

degree of precision is not warranted. However, in considering the use of very low stiffness pipes, the question needs re-examination.

From experience in Australia, there is one set of circumstances where buckling collapse may become critical, viz. where vacuum conditions may be developed in a pipe with low soil cover. Partial vacuum may be developed in storm drains under flood conditions. The assumption of soil support against vacuum collapse is only valid if support is provided around the full circumference of the pipe. This is not so on the top of the pipe at low covers. One design method assumes full soil support at a cover of three pipe diameters, with the buckling pressure decreasing linearly to the unsupported critical buckling pressure of the pipe at zero cover. However, more test work is needed in this area.

CRITERIA FOR MINIMUM STIFFNESS

From the above discussion, it can be seen that, except in special cases, none of the classical direct methods of analysis for deformation, material or structural failure mechanisms will provide sensible limits to pipe stiffness for thermoplastic pipes. There are, however, practical factors that influence minimum stiffness selection. These are unquantifiable from theory, and we must rely on practical field experience for guidance towards satisfactory specifications.

Whilst it may be possible to dig a self-supporting tunnel, it is not possible to dig a trench and replace the soil around a hole in the bottom! Re-compaction of the soil is necessary, and the pipe must have sufficient lateral stiffness to withstand the compaction procedures, without undue distortion. As observed by many other field researchers, installation deflection is by far the largest component of the total deflection in real systems.

What constitutes undue distortion may depend on many considerations. Clearly we wish to limit deflection and distortion for a number of practical reasons:

1. Integrity of joints
2. Passage of cleaning equipment
3. Reduction in hydraulic capacity
4. Aesthetics (for want of better terminology)

Integrity of joints is a matter for pipe design, and performance tests can easily be established if required to demonstrate the capability of joints in pipes and fittings under adverse conditions of deflection and differential loadings. They are incorporated in most sewer pipe standards. For storm drain specifications, joint integrity is regarded as much less critical, and in fact systems with unsealed joints are frequently used.

Cleaning and inspection is essential for sewers, but less so for storm drains. Most equipment can tolerate reductions in diameter of 10%, and there would be no great problem in designing equipment for greater tolerance if necessary.

Hydraulic capacity is not impaired within design tolerances up to deflections of 10%. Again for storm drain work a greater tolerance could be accepted.

Aesthetics may well be the most demanding criterion of all. Deflections up to 5% are barely noticeable, but they certainly are at 10%. A 15% deflection looks positively embarrassing, and it is very difficult to convince people that it was designed to do that!

Generally there is no great argument on deflection limits. The world seems to have settled on 5-6% as a target, with some tolerance for actual construction of say 2-2$^{1}/_{2}$%. More latitude is usually given for storm drains. Scandinavia seems to accept that even 15% long term deflection will not seriously affect functionality, provided joint capability is demonstrated.

Whatever the reasons and whatever the number, the problem is to determine what level of pipe stiffness will provide reasonable assurance that deflections and distortions will be held within the desired limits, given the typical accuracy of data concerning soil properties and the degree of control that can be exercised over installation.

This is the function of two main factors:

A: The type of system, sewer or storm drain. As noted above there are functional reasons why storm drains are less critical, and there are also other differences: sewers tend to be buried to greater depths, and are more frequently required to transit through difficult ground conditions. The consequences of failure of a sewer is obviously more serious, so a lower risk element is required.

B: The size of pipe. We are concerned with relative deflection, δ/D, rather than absolute deflection. Construction loadings tend to be similar for all pipes regardless of diameter. Resistance to deflection by a given load is not a function of stiffness, but rather of the parameter EI/D^2, which is a measure of its direct load capacity. A man standing on a 100 mm pipe of ring stiffness 1 kN/m^2 would squash it severely, whereas a 600 mm pipe of the same stiffness would suffer no appreciable deflection. Likewise the effect of a machine compactor on the two pipes is vastly different. Ring stiffness is only relevant for loads which are more or less proportional to diameter, such as soil loads. For this reason smaller diameter pipes need to be stiffer than larger diameter.

Simple economics points to the same conclusion: the ratio of pipe cost to installation cost changes radically with diameter. The relative cost of a stiffer pipe in small diameters is small. Conversely, substantial savings are available through the use of lower stiffness pipes at large diameter. For larger diameter installations, it is economically possible to improve site investigation, design, backfill materials, construction techniques and supervision to enable more flexible pipes to be used.

Australian Standards for PVC sewer pipes were originally set with two classes of mean dimension ratio D_m/t 50 and 37, providing stiffness EI/D^3 of approximately 1.8 and 4.5 kN/m² (ASTM D2412 F/δ = 14 and 35 psi respectively). Experience over the 1970's, mainly with 150 mm pipes showed that the lighter class of pipe was not sufficiently robust for general sewer reticulation work and it was eventually removed from the Standard. A heavier class was introduced for diameters up to 300 mm of dimension ratio 29, stiffness approximately 9.0 kN/m² (70 psi).

At the same time PVC pipes in diameters 90 to 300 mm of dimension ratio 50 have been found satisfactory for general storm drain applications. There are of course some situations where the stiffer sewer class pipes are preferred. Experience with larger diameter lower stiffness pipes in storm drains has been good, and the Australian Standard for PVC Stormwater pipes will shortly be revised to include pipes of reducing stiffness above 300 mm diameter.

Although the same principles apply to thermoplastic pipes generally, it is not necessarily appropriate to extrapolate experience in one material to another in terms of actual stiffness levels. For example, polyethylene behaves somewhat differently to PVC in two respects. Firstly, within the installation time framework when stiffness is important, the initial stiffness as specified declines considerably more rapidly for polyethylene than for PVC. During installation, constant strain conditions do not apply, and a short term creep modulus might be more appropriately used in determining comparable stiffnesses. Secondly, although no quantitative data is available, it seems likely from experience that skin friction plays some part in soil/pipe interaction, particularly in the installation phase. Polyethylene has a lower frictional coefficient than PVC. From these considerations, it might be expected that higher initial stiffness values could be required for polyethylene pipes.

VERY FLEXIBLE PIPE DRAINAGE SYSTEM AT EXPO 88

World Expo 88 was held in Brisbane from April to October 1988. The site of the event was a 30 hectare area of flat land on the banks of the Brisbane River. Construction of site services commenced in 1986. Brisbane's climate is sub-tropical, and storm drainage systems must cope with a high rainfall intensity. A high capacity drainage system for the site was essential. Further the system would be subjected to significant live loading during the construction phase.

In keeping with the modern technology theme which is always part of such occasions, and with due attention to cost savings, consultants Gutteridge Haskins and Davey, acting for the project contractors Thiess Watkins (Constructions) Ltd, studied the possibility of using very flexible pipes for the drainage system. However, for such a prestigous event, the consequences of flooding due to system failure could not be contemplated, and it was deemed circumspect to carry out site trials before proceeding, in particular to determine construction specifications and whether the locally available river sand would be satisfactory for pipe surround.

Trials of Rib Loc spirally wound structural profiled wall pipe were conducted in August 1986. Controlled test installations of four sizes of pipe were conducted, with two classes of surround material, and cover heights ranging from 0.4 to 1.5 metres. Size and stiffness of pipes are given in Table 1.

Table 1 -- Expo site drainage pipe details

Internal dia		Ring stiffness $S = EI/D^3$	Pipe Stiffness[b] ASTM D2412	Length installed
mm	(in)	(kN/m^2)	(kN/m^2(psi))	metres
225	(9)	2.8	148 (21.0)	990
300	(12)	1.2	66 (9.6)	1690
375[a]	(15)	1.0	54 (7.8)	370
450	(18)	0.6	31 (4.5)	500
525[a]	(21)	1.5	79 (11.5)	190
600	(24)	1.0	54 (7.8)	125

[a] These sizes not included in preliminary trials
[b] Pipe stiffness from parallel plate loading tests, ASTM D2412, is $F/\Delta Y$, at a specified deflection. The theoretical relationship at zero defletion is $PS = 53.7\ EI/D^3$.

The construction procedure used was as follows:

1. 75 mm bedding, levelled but not compacted.
2. Pipes placed and side filled to 2/3 pipe diameter.
3. Hand tamped to compact haunching and side support.
4. Surround material placed to 150 mm overlay.
5. Mechanical compaction (Wacker Packer).
6. Backfilled with excavated material and machine compacted.

Note that this procedure has some significant differences from that adopted in most installation specifications for pipes. It was established from earlier trials and observations of behaviour of very flexible pipes, and is quite important in obtaining correct response. The 75 mm cushion of uncompacted bedding is an important factor (however, loosened material below this bedding should be compacted and stabilised if necessary). Compaction in incremental layers is very difficult to achieve with very flexible pipes, and is more likely to do more harm than good. The pipe must have some support up to a level above the springline before compaction begins. Then only hand compaction should be used. Mechanical compaction is not applied until the pipe is completely covered.

It is obvious therefore that surround material must be such that low compactive effort is required. The more flexible the pipe, the lower the compactive effort that can be applied. The best possible material is single size crushed stone, or specification graded material.

The installations were tested under wheel loading using the front wheels of a 5 tonne front-end loader carrying a full scoop, as representative of the construction conditions that would be encountered. Deflection was tested by pulling through circular plate gauges set at 2% and 5% of internal diameter. All pipes passed, although it was noted that the 450 mm pipes suffered some shape distortion.

On the basis of these tests, the Rib Loc pipes were accepted and the site drainage system was installed using the same pipe specifications.

In February 1990, an inspection was carried out at one of the drainage systems to determine the longer term response of the very flexible pipe system.

Although access was somewhat limited, it was possible to remotely photo-graph a large number of lines via the inspection shafts. The spiral configuration of the wall of the pipes provided an excellent means of qualitatively assessing shape distortion in three dimensions, and even enabling approximate quantitative evalu-ation in many cases. A representative selection of photographs with comments are shown in Figures 2 - 7. It should be noted that apparent flattening at the invert in some pipes carrying water is an illusion.

The results and conclusions formed were as follows:

1. The majority of lines in the installation performed very well, with deflections and distortion well within functional limits. Some lines were observed where deflection and shape distortion would normally be considered unacceptable. In the majority of these cases, the problem was due to inadequate pipe stiffness to handle the machine compaction techniques used.

2. There was no evidence of material failure due to bending strains, even with observed curvatures producing estimated strains up to 4.4%. See Fig 7.

3. The stiffness of the 600 mm and 525 mm pipes was adequate. Some evidence of shape deformation with 600 mm pipes indicates that 1.0 kN/m^2 (ASTM D2412 7.8 psi) is about the lower limit of stiffness for pipes of this diameter.

4. The 450 mm pipes experienced some problems, and a stiffness of 0.6 kN/m^2 (ASTM D2412 4.6 psi) is insufficient for general purpose drainage speci-fications at this diameter.

5. The 300 mm pipes, in spite of their higher stiffness, performed poorly in some instances. Partly this may have been due to their lower cover, and the consequent effect of construction traffic loads. However, smaller diameter pipes are more frequently used at lower cover heights, and their stiffness should accordingly be adequate to compensate. A minimum stiffness of 2.0 kN/m^2

Fig 2 -- 600 mm S = 1.0 kN/m²
2.8 m cover
Vertical deflection 0.4 - 1.6%
Uniform elliptical deformation

Fig 3 -- 600 mm S = 1.0 kN/m²
2.1 m cover
Vertical deflection 3.1 - 3.6%
Slight shape distortion

Fig 4 -- 450 mm S = 0.6 kN/m²
1.8 m cover
Vertical deflection 3.8 - 4.3%
Uniform elliptical deformation

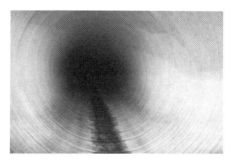

Fig 5 -- 450 mm S = 0.6 kN/m²
2.2 m cover
45° deflection 4.2 - 4.7%
Eccentric compaction

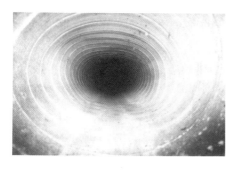

Fig 6 -- 450 mm S = 0.6 kN/m²
0.9 m cover
Vertical deflection 15-16%
Squaring, est. max. strain 4.4%

Fig 7 -- 300 mm S = 1.3 kN/m²
1.0 m cover
Vertical deflection 9-10%
Crown flattening

(ASTM /D2412 16 psi) could be appropriate. This would match the criteria suggested of constant EI/D^2 or relative stiffness.

SUMMARY AND CONCLUSIONS

For thermoplastic pipes, classical analyses for deflections, material strains, or structural stability provide little guidance (except in special cases) as to lower limits to pipe stiffness which in reality are controlled by practical considerations largely unquantifiable from theory. As such we are reliant on field testing and experience with working installations to determine appropriate specifications.

Basic logic suggests that lower stiffness thermoplastic pipes in the "very flexible" category can and should be considered at larger diameters. There is an economic balance between pipe stiffness and construction costs. Reducing stiffness requires improved construction techniques and materials. Caution needs to be exercised at low cover heights with traffic loads or potential vacuum conditions.

Judgements concerning minimum stiffness depend on the application and material. Sewer mains are more demanding than storm drains. Differences in material characteristics can cause misleading comparisons of stiffness when considering pipe response in the installation time framework.

Satisfactory results can be achieved in general storm drainage applications with PVC pipes of stiffness 2.0 kN/m^2 for diameters to 300 mm, reducing for larger sizes according to relative stiffness EI/D^2 of 0.6 kN/m.

ACKNOWLEDGEMENTS

The author thanks Vinidex Tubemakers Pty Limited for permission to publish this paper, and acknowledges the assistance of his colleagues, in particular Mr Michael Skinner for his contribution. Special thanks to Joan Krahe for her patience and perseverence, and congratulations on conquering the line spacing idiosyncrasies of the character δ.

REFERENCES

[1]   Janson, L E, "Investigation of the Long-term Creep modulus for Buried Polyethylene Pipes Subjected to Constant Deflection", Proceedings of the International Conference on Advances in Underground Pipeline Engineering, ASCE, August 1985.

[2]   Moser, A P, "Strain as a Design Basis for PVC Pipes", Proceedings of the International Conference on Underground Plastic Pipe, ASCE, April 1981.

[3]   Janson, L E, "Plastic Gravity Sewer Pipes Subjected to Constant Strain by

Deflection", Proceedings of the International Conference on Underground Plastic Pipe, ASCE, April 1981.

[4]   Janson, L E, "The Relative Strain as a Design Criterion for Buried PVC Gravity Sewer Pipes", Proceedings of the International Conference on Advances in Underground Pipeline Engineering, ASCE, August 1985.

Mark E. Greenwood and Dennis C. Lang

VERTICAL DEFLECTION OF BURIED FLEXIBLE PIPES

REFERENCE: Greenwood, Mark E. and Lang, Dennis C. "Vertical Deflection of Buried Flexible Pipes", Buried Plastic Pipe Technology, ASTM STP 1093, George S. Buczala and Michael J. Cassady, Eds., American Society for Testing and Materials, Philadelphia, 1990.

ABSTRACT: For over 50 years, studies of soil-pipe interaction have provided methods to predict vertical deflection of buried flexible pipes. Several of these methods are widely used in standards. However, these methods do not account for many of the recognized parameters that affect buried pipe behavior. Based on recent research findings, empirically-based modifications are introduced to the original Spangler approach to obtain a new calculation method for estimating vertical flexible pipe deflection. Development, applications and limitations of this method are presented. Comparison studies of the new method and that presented in ANSI/AWWA C950-88 "AWWA Standard for Fiberglass Pressure Pipe" (which is based on the modified Iowa formula) versus actual field measurement data are included. These studies present correlation of predicted versus measured values as well as similarities and differences between the methods.

KEYWORDS: Fiberglass Reinforced Plastics, pipe, buried pipe, deflection, long-term, AWWA C950

INTRODUCTION

The behavior of the pipe-soil system requires determination of the interaction which occurs between the pipe, embedment soil and native in situ soil. Each of these elements acts together to determine total system behavior. The response can be measured by pipe deformation.

Mark E. Greenwood is a Research Associate, Reinforcements and Resins Laboratory, Owens-Corning Fiberglas® Corp., Technical Center, 2790 Columbus Road, Granville, OH 43023-1200; Dennis C. Lang is a Market Development Engineer, High Performance Fibers, Allied-Signal, Inc., P.O. Box 31, Petersburg, VA 23804.

For over 50 years, studies of soil-pipe interaction have provided a substantial amount of information which has enhanced our understanding of the problem. The classical works of Marston [1] and Spangler [2] mark the beginning of these investigations and provided a fundamental understanding of applied earth loads and buried pipe response. Nearly 20 years later, in 1958, Watkins and Spangler [3] published a calculation procedure to determine pipe deflection. This equation has become to be known as the modified Iowa formula.

Research on buried pipe response has continued since Spangler's original work. Such interest is indicative of continued acceptance and use of flexible conduits in a variety of buried applications. Recent efforts include development of various analytical models such as elastic solutions [4,5] and Finite Element Methods [6,7]. In the 1970's, Leonhardt [8,9] introduced a calculation system which formed the basis of a German design document, ATV [10]. Howard [11] had determined design values for the modulus of soil reaction, E', which expanded the range of application of the modified Iowa formula. Although additional investigations have been conducted since Spangler's early work, the modified Iowa formula exerts significant influence on current design procedures and has withstood the test of time. [12] Today this method has been adopted by a variety of flexible pipe product standards such as ANSI/AWWA C950-88 "AWWA Standard for Fiberglass Pressure Pipe" for fiberglass pipe design.

Realizing that deviations between predicted and measured data can occur and recognizing the importance of deflection prediction to pipe performance, Owens-Corning Fiberglas® (OCF) conducted its own comprehensive research in soil-pipe interaction. The OCF installation research program had two objectives. The first objective was to define and quantify buried flexible pipe behavior by examining field measurement data, laboratory test results and finite element modeling. Results of this portion of the program have been published by several research participants [13,14,15,16,17,18].

The second objective was to develop a design methodology which correlates to the experimental results. Vertical pipe deflection is an important consideration in buried pipe design. Buried pipes are not allowed to exceed certain limiting deflection values depending on their application. This allowable value is determined by applying a factor of safety to the pipe's extrapolated 50-year performance in a particular application.

Vertical deflection is also important in terms of installation quality control. A common practice is to relate installation quality to an initial deflection limit. Measured initial deflections taken after completion of backfill to grade should not exceed the allowable initial value.

Based on recent research, a new understanding of the parameters influencing buried pipe response has been obtained. This information provides the basis for modifications applied to the original Spangler approach to obtain an improved deflection prediction method.

OBJECTIVE AND APPROACH

The objective of this paper is to present a method to predict vertical deflections of buried flexible pipe that is significantly better than the method described in ANSI/AWWA C950-88. The proposed method must consider the significant factors that govern buried pipe behavior, including initial installation and should be relatively easy to use. The application of the method should not require complex analysis such as finite element analysis.

The approach taken to develop a vertical deflection prediction method was to emperically upgrade the modified Iowa formula which predicts horizontal deflection by using results of controlled laboratory pipe tests. A comprehensive deflection prediction equation that is theoretically based is not being proposed. Validity of the empirically upgraded equation was evaluated through comparison with field measurement data.

EQUATION DEVELOPMENT

There are many parameters which may effect pipe deflection. Several of these parameters are significant and govern long-term response. Listed below are governing parameters which collectively determine pipe response. They were determined by research findings presented in the current literature and results of OCF installation studies.

1. Pipe stiffness
2. Soil stiffness (soil type, density, modulus and moisture content)
3. Applied loads (vertical and lateral pressure loads due to overburden and applied surface loads)
4. Trench configuration (trench geometry, native in situ soil condition and embedment type)
5. Haunch support (degree of uniform support at the bottom quarter points of the pipe)
6. Non-elliptical deformation (deviation from truly elliptical deformation
7. Initial ovalization (vertical elongation due to placement of embedment during construction)
8. Time (pipe and soil properties as well as applied loads change, each as a function to time)
9. Variability (construction variability due to excavation, soil placement and compaction)

Each parameter will be discussed independently. The predictive method resulted from assembling these parameters into one equation.

Deflection Equation Components

Pipe deflection consists of three distinct elements. Symbolically, these can be expressed as percent fractions of pipe diameter as follows:

$$\delta_{VP} = \delta_{VL} - \delta_{VO} + \delta_{VA} \tag{1}$$

Respectively, the upper limit of predicted vertical deflection (δ_{VP}) equals the deflection due to load (δ_{VL}) minus deflection due to construction (initial ovalization; δ_{VO}) plus a deflection term to reflect field installation variability (δ_{VA}).

Vertical deflection due to load is a function of three elements and represented as

$$\delta_{VL} = W/(S_p + S_S) \qquad (2)$$

Equation 2 states that deflection caused by load is proportional to the total applied external load (W) and inversely proportional to the sum of the pipe stiffness (S_p) and soil stiffness (S_S).

The components of Equations 1 and 2 will be discussed separately and then reassembled to form the complete calculation method.

Load Induced Deflection

Deflection resulting from applied earth and surface loads is the first component to be developed. Refer to Figure 1 for the location of various physical parameters and geometry of a typical buried pipe installation.

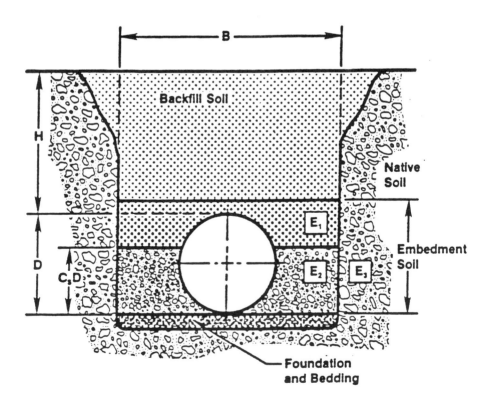

FIGURE 1—Typical buried pipe installation

Magnitude and distribution of applied loads: Both the magnitude and distribution of loads vary with time.   The effect of time on the magnitude of load is likely to be more significant than the change in load distribution.   This observation forms the basis of an important assumption: the influence of time on applied loads on buried pipe is a result of the degradation of arching and not a redistribution of loads.   With this practical assumption, the definition of load can be easily obtained.

Spangler [2] represented the distribution of loads as illustrated in Figure 2.   Vertical load at the top of the pipe is uniformly distributed over the pipe diameter.   The vertical reaction load acting at the bottom of the pipe is uniformly distributed over a bedding angle.   These vertical loads are symmetric about the vertical axis of the pipe.

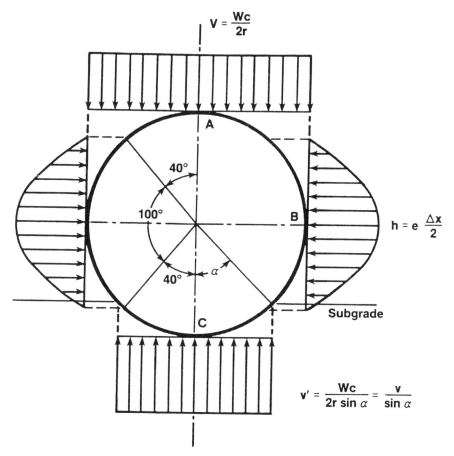

$$V = \frac{Wc}{2r}$$

$$h = e\,\frac{\Delta x}{2}$$

$$v' = \frac{Wc}{2r\sin\alpha} = \frac{v}{\sin\alpha}$$

FIGURE 2—Spangler's assumed distribution of pressure around a flexible pipe under as earth fill

The horizontal loads are symmetric and passive. The passive loads increase with horizontal deflection and are distributed equally around the pipe springlines over a lateral distribution angle of 100°. This pressure distribution may not be accurate for all possible combinations of pipe and soil stiffness, but is is rational and widely accepted.

The effect of time is introduced into the load calculation by consideration of arching. The long-term, maximum soil load applied to the pipe is assumed to be the free field load caused by the prism of soil above the pipe. As prescribed in ANSI/AWWA C-950, all positive arching that may have been present is discounted for long-term deflection calculations. The short-term load is represented as a fraction of the prism load by application of a reduction coefficient, based on the Marston theory of arching [1,19]. Therefore, the soil load acting on a buried pipe is expressed as follows:

$$W_S = C_L \gamma H \tag{3}$$

$$C_L \quad \begin{cases} \dfrac{1 - e^{[-2K\mu'(H/B)]}}{2K\mu'(H/B)}; & \text{short term} \\[2ex] 1; & \text{long term} \end{cases} \tag{3A}$$
$$\tag{3B}$$

Where γ is the unit weight of the overburden soil; H and B are the depth and trench width dimensions as indicated in Figure 1 and $K\mu'$ is the product of Rankine's ratio and the coefficient of internal friction between the fill material and side of the trench. Typical values [19] of $K\mu'$ are given in Table 1. The soil arching factor, C_L, is a function of the type of soil from the top of the pipe to grade.

TABLE 1--Typical values of arching factor, K_μ'

Soil description	K_μ'
Non–cohesive	0.192
Sand, gravel	0.165
Saturated top soil	0.150
Ordinary clay	0.130
Saturated clay	0.110

If live loads (W_L) exist on the ground surface, the load on the pipe can be calculated by the projection of a loaded area as presented in most soil mechanics text books. AWWA C950-88 uses such a method. In the proposed equation, live loads are normalized with respect to pipe diameter. The live load is treated independent of time and added directly to the soil load to obtain the total applied load (W), represented by the following equation:

$$W = W_S + W_L \tag{4}$$

Pipe stiffness: Pipe stiffness is an independent variable which influences the three types of deflections defined by Equation 1. The dependency of initial ovalization and variability on pipe stiffness will be discussed later. In terms of deflection due to load, pipe stiffness can play a major role.

Pipe stiffness is indicative of the load bearing capacity of the pipe subjected to buried conditions. Field measurement data indicate that deflections increase for a period of time and then stabilize to a level that remains virtually unchanged. This data suggests that long-term soil-pipe system behavior is a combination of creep and relaxation with time.

In part, this behavior can be attributed to long-term stiffness of the pipe. The pipe stiffness term of Equation 2 can be expressed as

$$S_p = 8 \ C_{TP} EI/D^3 \tag{5}$$

Where E is the circumferential flexural modulus of the pipe material; I is the pipe wall cross-sectional moment of inertia; D is the pipe diameter and C_{TP} is the pipe stiffness retention factor. Conservative values of the retention factor are obtained by extrapolation of creep modulus data to 50 years. For short-term deflection calculations C_{TP} equals 1.

Introduction of the time-dependent stiffness of pipe specifically indicates the increase of deflection resulting from a change in the pipe's modulus of elasticity. This factor is based on actual data. It can account for differences in pipe wall construction and material.

Soil stiffness: Soil stiffness is the most important element in determining the deflection response of the pipe to applied loads. Soil stiffness is a measure of the soil's ability to help the pipe resist the applied loads. Considering all components, soil stiffness is the most complicated and difficult to define.

In terms of basic properties, soil stiffness is a function of the following parameters:

1. Soil type and density
2. Burial depth
3. Moisture content
4. Trench configurations

5. Lateral pressure distribution
6. Time

These items must be considered to properly describe buried pipe behavior.

Soil modulus: Several types of soil moduli are available for application to soil-structure problems. There are secant, tangent, and initial moduli determined by triaxial testing at specific confinement pressures or by flat plate bearing tests. Soil modulus can be determined by the flat plate dilatometer, pressuremeter and cone penetrometer.

Soil modulus can be defined at a variety of stress or strain levels and determined by various kinds of equipment. Unfortunately, definition and type of equipment both strongly influence the value obtained. To more completely describe a soil's ability to resist deformation, the following conditions were identified:

1. The soil modulus method of measurement must be sensitive to the parameters which govern pipe deflection.
2. Soil modulus must be a measurable parameter determined by a laboratory test method.
3. Determination of the soil modulus by testing should not require highly-specialized equipment or expertise.
4. Typical values can be tabulated which, in the absence of testing, can be used reliably as input to the soil stiffness expression.

The confined compression test using equipment as described in the consolidation test, ASTM D-2435 "Test Method for One-Dimensional Consolidation Properties of Soils", was chosen. The soils investigation portion of the OCF installation research included the confined compression testing of different soil types at various densities. From the stress-strain curves generated, secant moduli corresponding to equivalent depths of cover were determined. For each soil type and density tested, a confined compression modulus was obtained. This testing was conducted at both optimum and saturated moisture contents. The secant modulus is a function of soil pressure. This behavior indicates that the secant soil modulus is a function of burial depth. Hartley and Duncan [20] supports this conclusion and approach.

These raw data were tabulated and plotted as functions of depth and soil density for each soil type. Table 2 is offered as a means of identifying various soils into groups that behave similarly. These soil groupings are referenced in ASTM D2487-83 "Classification of Soils for Engineering Purposes". Rough transitions were smoothed and interpolated to obtain a final set of confined compression modulus values as listed in Table 3, which offers typical soil modulus values when actual test data are not available.

TABLE 2--Soil classification reference

Soil group	Range of fines %	Soil class
Clean gravel	< 5	GW, GP
	5-12	GW-GM, GW-GC GP-GM, GP-GC
Dirty gravel	12-50	GM, GC, GC-GM
Clean sand	<5%	SW, SP
	5-12	SW-SM, SW-SC SP-SM, SP-SC
Dirty sand	12-50	SM, SC, SC-SM
Inorganic clay and silt	>50	CL, ML, CL-ML CH, MH

Reference: ASTM D 2487-83

TABLE 3 -- Confined compression secant modulus (N/mm^2)

Soil type	Depth (M)	Installed standard Proctor density				
		95%	90%	85%	80%	70%
Clean gravel	2	17.2	15.2	13.1	11.0	6.9
	4	19.0	16.9	14.8	13.1	9.0
	6	20.7	18.6	16.6	14.8	10.7
	8	22.4	20.3	18.3	16.2	12.4
	10	24.1	22.1	20.0	17.9	13.8
Dirty gravel	2	16.2	14.1	10.0	7.6	4.5
	4	17.9	15.9	12.4	9.3	5.9
	6	20.0	17.9	14.5	10.7	6.9
	8	21.7	19.7	16.2	12.1	8.3
	10	23.8	21.7	18.3	13.8	9.3
Clean sand	2	15.2	13.1	6.9	4.1	2.1
	4	17.2	15.2	9.3	5.5	2.8
	6	19.3	17.2	11.7	6.9	3.4
	8	21.4	19.3	14.1	8.3	4.1
	10	23.4	21.4	16.6	9.7	4.8
Dirty sand	2	13.8	11.7	6.6	3.8	1.4
	4	15.9	13.8	9.0	5.2	2.1
	6	17.9	15.9	10.7	6.6	2.8
	8	20.0	17.9	12.4	7.6	3.1
	10	22.1	20.7	14.5	9.0	3.4
Inorganic clay and silt	2	8.6	7.9	5.9	2.4	0.7
	4	9.7	9.0	6.9	3.4	1.2
	6	11.0	10.3	7.9	4.5	1.7
	8	12.4	11.4	9.3	5.5	2.2
	10	13.8	12.8	10.3	6.6	2.8

Notes: 1 N/mm^2 = 145 psi
Table 2 defines soil groups

Time and moisture influence on soil modulus: Both increased time
and moisture content reduce soil modulus from an initial value. For
the method presented, these effects have been treated as though they
were uncoupled and independent. In the case of time, this reduction
was represented by creep due to viscoelastic, time-dependent
behavior. The effect of moisture was considered as a fully-saturated
state caused by high water table conditions above the pipe invert
resulting in lower soil modulus values. Treatment of each of these
effects on soil modulus can be described as follows:

Soil creep modulus can be represented by the following power law
expression [21,22].

$$E_S(t) = E_0 t^{-m} \tag{6}$$

Long-term soil modulus at some time, t, is related to initial modulus
via a power coefficient, t^{-m}. The value of m varies for different
soil types and is the straight line slope of a modulus versus log
time plot. Typical values of m are available for different
soils [21].

Equation 6 can be rewritten to define soil stiffness retention
factors as follows:

$$E_S(50) = C_T(50)E_0 \tag{7}$$

The power coefficient can be defined as a stiffness retention factor,
C_T, for a specific soil representing a 50-year, long-term value.
Typical values for the modulus retention factor were determined and
presented in Table 4.

TABLE 4 -- Soil creep, modulus retention factors

Soil group	Soil modulus retention factor, C_T
Clean gravel	1.0
Dirty gravel	0.9
Clean sand	0.8
Dirty sand	0.7
Inorganic clay and silt	0.4

The long-term creep modulus is equal to the product of the soil stiffness retention factor and the confined compression modulus for an unsaturated soil. Equation 7 represents how soil creep modulus is represented as a function of time. Initial soil modulus is represented by Equation 7 with the stiffness retention factor set equal to 1 for short-term deflection calculations.

Installations in which the water table rises above the pipe invert after installation are cases in which soil modulus must be reduced. Saturated confined compression testing provides values for soil modulus which may be representative of high water table conditions. A reasonable approximation for soil modulus when saturated conditions exist is the modulus of the soil with a 5% lower compaction for granular soils and 10% lower compaction for cohesive soils and granular soil with greater than 12% cohesive fines. Table 3 can be used for this estimate when actual data is not available.

Basic knowledge of soil density through in situ testing or experience and good judgement should provide an estimate of in situ soil compaction. Soil type, degree of compaction, water table location, and depth can be used to select an appropriate soil modulus value from Table 3 when native soil modulus testing results are not available.

In typical installations, the time dependent considerations need not be considered for the native soil trench. When the water table is above the pipe springline, the native soil modulus may be determined by reducing the standard Proctor density as previously described for backfill.

Split installations: When pipes are embedded in two different soil types or densities, a convenient approximation of embedment soil modulus can be calculated. Typically in a split installation, the pipe will have the bottom portion embedded in a higher modulus soil than the upper portion of the pipe. If C_s represents the percentage of the lower portion of pipe embedded in soil with a modulus E_2 and the upper portion of the pipe $(1-C_s)$ embedded in a soil with a modulus of E_1, a combined embedment modulus can be approximated as

$$E_s = C_s C_{T2} E_2 + (1-C_s) C_{T1} E_1 \qquad (8)$$

This approximation appears to be reasonable when values of C_s are at least 0.4. For conditions with C_s less than 0.4, use the lower soil modulus value as the embedment soil modulus.

Trench width: Trench width will affect soil stiffness depending on the in situ native soil conditions. If a narrow trench is used in soft native soil conditions, the effective soil stiffness around the pipe will be reduced. This reduction in soil stiffness is related to the difference between the embedment soil modulus and the native soil modulus. If a wider trench is used, the detrimental effect of the soft native soil is diminished. To consider the effect of trench width on soil stiffness, a relationship between the native soil modulus, embedment soil modulus and trench width must be obtained.

This relationship was developed by Leonhardt [8,9] and indicates the effect of trench width on soil stiffness. This factor is a function of the trench width to pipe diameter ratio and the embedment soil modulus to native soil modulus ratio. The "Leonhardt factor" is described by Equation 9.

$$\zeta = \frac{1.662 + .639(B/D-1)}{(B/D-1) + [1.662 - .361(B/D)-1)]E_S/E_3} \tag{9}$$

Equation 9 is applied as a multiplier to the embedment soil modulus. It acts like a reduction or intensification factor depending upon the conditions. Figure 3 illustrates the dependency of Equation 9 on trench width and soil modulus. The required input data to the trench width factor (Equation 9) are the trench width (B), pipe diameter (D), the embedment soil modulus (E_S), and the native soil modulus (E_3).

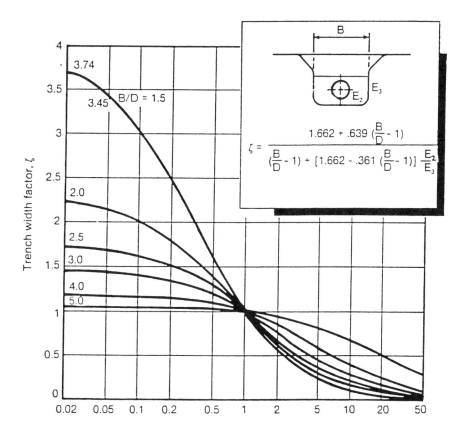

FIGURE 3—Trench width factor

Soil stiffness summary: The soil stiffness component of the deflection term caused by load can be fully described and is given by the following expression:

$$S_S = 0.6\zeta E_S \qquad (10)$$

Where E_S is obtained from Equation 8 and ζ from Equation 9. The 0.6 constant is used by Leonhardt to relate the stiffness of backfill in a trench condition versus a confined compression test.

Non-elliptical deformation: Depending on pipe stiffness and installation condition, a buried pipe may deform non-elliptically. The vertical deflection may not always equal the horizontal deflection.

Non-elliptical deformation is a result of nonuniform soil support around the pipe. The degree of nonuniformity is related to soil type and degree of compaction. Sensitivity to this nonuniformity is related to pipe stiffness.

Field and laboratory measurements [13,15,16] have shown that the ratio of vertical to horizontal deflection, $\Delta V/\Delta H$, is representative of the degree of non-elliptical deformation. The $\Delta V/\Delta H$ ratio has been found to be a function of soil type, degree of compaction, split embedment and pipe stiffness. Based on actual field and laboratory measurement data, typical values are summarized in Table 5.

TABLE 5 -- Deflection ratio , $\Delta V / \Delta H$

Soil group	Pipe STIS (N/M^2)	Standard Proctor density of backfill			
		> 95%	85%-95%	70%-84%	< 70%
Clean and dirty gravel	1,250	2.0(3.0)	1.5(2.5)	1.3(1.5)	1.2
	2,500	1.5(2.0)	1.3(1.8)	1.2	1.2
	5,000	1.3(1.5)	1.2	1.1	1.1
	10,000	1.1	1.1	1.1	1.0
	20,000	1.1	1.1	1.0	1.0
	>20,000	1.0	1.0	1.0	1.0
Clean sand	1,250	2.5(3.3)	2.0(3.1)	1.5(1.8)	1.2
	2,500	2.0(2.5)	1.5(2.0)	1.3(1.5)	1.2
	5,000	1.5(2.0)	1.3(1.5)	1.2	1.1
	10,000	1.2(1.5)	1.2	1.1	1.0
	20,000	1.1	1.1	1.0	1.0
	>20,000	1.0	1.0	1.0	1.0
Dirty sand	1,250	3.0(3.5)	2.5(3.3)	2.0(2.1)	1.2
	2,500	2.5(2.8)	2.0(2.5)	1.5(2.0)	1.2
	5,000	1.8(2.2)	1.5(1.7)	1.2	1.1
	10,000	1.4	1.3	1.1	1.0
	20,000	1.1	1.1	1.0	1.0
	>20,000	1.0	1.0	1.0	1.0
Inorganic clay and silt	1,250	3.5(3.8)	3.0(3.5)	2.3(2.5)	1.2
	2,500	2.5(3.0)	2.3(2.7)	1.7(2.3)	1.1
	5,000	2.0(2.5)	1.7(2.0)	1.3	1.1
	10,000	1.5	1.3	1.1	1.0
	20,000	1.1	1.1	1.0	1.0
	>20,000	1.0	1.0	1.0	1.0

NOTES: STIS = EI/D^3 (N/M^2) = 128.42 F/ ΔY (psi)
Table 4 defines density ranges
Numbers in parenthesis refer to a split
installation condition when E1<.5E2

The dependency of $\Delta V/\Delta H$ on pipe stiffness and installation condition is not commonly reflected in current design practice. Other investigations have identified this behavior [23,24], but sensitivity of $\Delta V/\Delta H$ on pipe and soil stiffness has not been previously quantified. Values of the $\Delta V/\Delta H$ ratio are in some cases quite large and significantly influence pipe behavior.

Nonuniform haunch support: Nonuniform haunch support occurs with a high degree of probability and must be considered in pipe design [13,15]. In terms of pipe deflection, poor haunch support results in higher deflection than if the soil support is uniform.

The degree of haunch support is related to the apparent bedding angle over which the vertical reaction is distributed at the bottom of the pipe. In Spangler's original work, a bedding factor, k_x, was introduced to indicate the distribution of load at the bottom of the pipe.

In the method presented, k_x is related to the likelihood of achieving uniform bottom support as related to the installation. Values for the factor are dependent on soil type and degree of compaction. Table 6 summarizes bedding factor values. Note that the use of k_x is different from that currently defined in AWWA C-950.

TABLE 6 -- Bedding factor values, k_x

Soil group	Range of fines %	Backfill standard Proctor density			
		> 95	85-95	70-84	<70
Clean gravel	< 5	0.083	0.083	0.083	0.083
	5-12	0.096	0.096	0.083	0.083
Dirty gravel	12-50	0.103	0.103	0.096	0.083
Clean sand	<5%	0.103	0.103	0.096	0.083
	5-12	0.103	0.103	0.096	0.083
Dirty sand	12-50	0.103	0.103	0.096	0.083
Inorganic clay and silt	>50	0.103	0.103	0.096	0.083

Deflection due to load, summary: Development of the basic
elements which together determine vertical deflection resulting from
load has been presented.   Equation 2 can be rewritten by including
all the elements discussed.

$$\delta_{VL} = \frac{k_x(\Delta V/\Delta H)(C_L\gamma H + W_L)}{8C_{TP}EI/D^3 + 0.061(0.6\zeta)E_S} \times 100 \qquad (11)$$

Pipe-soil interaction coefficient: When evaluating the behavior
of pipe deformation in various installation conditions, the influence
of soil stiffness and pipe stiffness did not appear to be totally
independent.   A pipe-soil interaction existed that was dependent on
the relative stiffnesses of soil and pipe.   This observation raised
the question that the simple soil pressure distribution model used by
Spangler may be insufficient to account for the interactions between
soil and pipe.   The simple models used by Leonhardt may be questioned
as well.   The magnitude and distribution of soil stress around a pipe
must be a function of the pipe stiffness, soil stiffness and soil
placement in addition to the response from pipe deflection.

Without an exhaustive theoretical model, an empirically defined
factor was added to the soil stiffness term of Equation 11 to reflect
the observed pipe behavior.   The pipe-soil interaction coefficient,
C_I, was derived by means of constrained optimization of Equation 12
versus measured soil box data, finite element analysis and
engineering judgement.

$$\delta_{VP} = \frac{k_x(\Delta V/\Delta H)(C_L\gamma H + W_L)}{8C_{TP}EI/D^3 + 0.061(0.6\zeta)C_I E_S} \times 100 \qquad (12)$$

An expression for C_I has been developed which is a function of pipe
stiffness and embedment soil degree of compaction and is given as
follows.

$$C_I = a(STIS/1250Pa)^b \qquad (13)$$

$$STIS(Pa) = EI/D^3 \qquad (13A)$$

Values for parameters a and b are provided in Table 7 and are a
function of degree of compaction.   Specific tangential initial
stiffness (STIS) included in Equation 13 is expressed in N/M^2 or
Pascals (1 psi = 6894.4 Pa).   The pipe-soil interaction coefficient
was found to be a function of pipe stiffness and was normalized with
respect to 1250 Pa (9 psi pipe stiffness) to provide dimensional
consistency.

TABLE 7 -- Values of parameters a and b for the
pipe-soil interaction coefficient

Backfill standard Proctor density	a	b
> 95	1.24	0.180
85 – 95	0.938	0.245
70 – 84	0.643	0.353
< 70	0.456	0.436

Relative to the ANSI/AWWA C 950-88 method for predicting deflection, Equation 12 includes parameters that more completely describe the prediction of vertical deflection caused by vertical loading. To further describe observed deflection of installed pipe, additional parameters must be considered. Deflection caused by compaction of backfill during construction and variability of installation are both critical parameters when evaluating pipe installations.

Deflection Due to Construction

Flexible pipes ovalize initially due to the soil placement and compaction process during installation [24,25]. Initial ovalization is defined as negative vertical deflection that occurs during installation resulting from embedment soil loads acting laterally on the pipe prior to backfill placement above the pipe crown. Lateral pressure results from backfill compaction. Initial ovalization is the response of the pipe to the installation process. The amount of ovalization is a function of soil type, degree of compaction and pipe stiffness. Higher compactive effort results in more initial ovalization of pipes. Installations with higher stiffness pipes result in lower initial ovalization.

Mathematical modeling of this process is unrealistic since it is very difficult to uncouple all effects analytically. Based on field and lab measurements, Table 8 has been developed to provide typical values which can be used for calculation purposes. This value has been introduced into Equation 1 as an input parameter δ_{VO}. The deflection resulting from initial ovalization and vertical loads is

$$\delta_{VP} = \frac{k_x(\Delta V/\Delta H)(C_L \gamma H + W_L)}{8C_{TP}EI/D^3 + 0.061(0.6)\zeta C_I E_S} \times 100 + \delta_{VO} \qquad (14)$$

TABLE 8 -- Initial ovalization, δ_{VO} (%)

Soil group	STIS (N/M^2)	Backfill standard Proctor density			
		> 95	85-95	70-84	<70
Clean and dirty gravel	1,250	1.0	0.5	0	0
	2,500	0.5	0	0	0
	5,000	0	0	0	0
	10,000	0	0	0	0
	>20,000	0	0	0	0
Clean sand	1,250	2.5	1.5	0.5	0
	2,500	1.5	1.0	0.3	0
	5,000	1.0	0	0	0
	10,000	0	0	0	0
	>20,000	0	0	0	0
Dirty sand	1,250	3.0	2.0	1.0	0
	2,500	2.5	1.5	0.5	0
	5,000	1.5	0.5	0	0
	10,000	0.5	0	0	0
	>20,000	0	0	0	0
Inorganic clay and silt	1,250	5.0	3.0	2.0	0
	2,500	4.5	2.5	1.0	0
	5,000	3.0	2.0	0.5	0
	10,000	1.5	1.0	0	0
	>20,000	0	0	0	0

Note: STIS (N/M^2) = 128.42 F/ ΔY (psi)

Deflection Due to Variability

Measured field deflections along pipelines are variable for a number of reasons. To estimate the upper limit of deflection, this variability must be considered when predicting deflection. The following list identifies the more significant causes of deflection variation along a pipeline.

1. Specified versus achieved density of compacted soil materials
2. Variability of split embedment level
3. Variation of native soil properties
4. Variability of trench width
5. Construction practice or contractor experience to achieve or adhere to a specified installation guideline
6. Deviation of actual from assumed conditions

Of these six probable causes of deflection variation, the first is likely to be the most significant. To better identify the magnitude of variation, a parametric study was conducted to determine the sensitivity of Equation 12 to variations in soil density. Soil densities were varied ±5 to 10 standard Proctor density (SPD) points with a wide range of pipe stiffness and installation types.

The deflection variability values, δ_{VA}, are summarized in Table 9. These values are a function of pipe stiffness and degree of compaction. These values have been found to correspond to actual long-term field data variation where multiple deflection measurements for the same conditions are available.

TABLE 9 -- Deflection resulting from variability, δ_{VA} (%)

STIS	Backfill standard Proctor density			
(N/M^2)	> 95	85–95	70–84	<70
1,250	2.0	1.75	1.5	1.0
2,500	2.0	1.75	1.5	1.0
5,000	1.75	1.5	1.25	1.0
10,000	1.5	1.25	1.0	1.0
20,000	1.0	1.0	1.0	1.0
>20,000	0.5	0.5	0.5	0.5

Note: STIS (N/M^2) = 128.42 F/ΔY (psi)

Equation 14 is representative of an expected average deflection. With variability included, the range of expected deflections would be described as follows:

$$\delta_{VP} = (\delta_{VL} - \delta_{VO}) \pm \delta_{VA} \qquad (15)$$

Deflection Equation Summary

The equation for predicting maximum vertical deflection, including all elements discussed, is as follows:

$$\delta_{VP} = \frac{k \left(\Delta V/\Delta H\right)(C_x \gamma^{H+W} L)}{8C_{TP}EI/D^3 + 0.061(0.6)\varsigma C_I E_S} \times 100 - \delta_{VO} + \delta_{VA} \qquad (16)$$

Average deflection can be predicted by not adding the variability term.

COMPARATIVE STUDIES

Correlation studies were conducted comparing predicted and measured field deflections using Equations 14, 16 and AWWA C950-88. The purpose was to evaluate the accuracy of the newly defined equations compared to ANSI/AWWA C950 in predicting average and maximum deflections.

Data Correlation

The following field sites form the base of measurement data against which the correlation was evaluated. These are summarized by site number in terms of physical descriptions.

Site 1: 2 Conditions. STIS 1800 Pa with variable, split (0.7D) embedments with dumped and compact sand; D=762mm; H=1.5M; B=2.1M. In Situ Native: Stable/dry, moderate degree of compaction (DOC). Published data - reference 16.

Site 2: 1 Condition. STIS 1450 Pa; compact sand embedment; D=2000mm; H=2.0M; B=3.0M. In Situ Native: Stable, moderate DOC.

Site 3: 1 Condition. STIS 1450 Pa; compact sand embedment; D=1800mm; H=1.5M; B=3.0M. In Situ Native: Stable, moderate DOC.

Site 4: 5 Conditions. Variable STIS 1215 - 25500 Pa with variable embedments including foot tamping of dumped sand, dumped gravel, and compact sand; D=800mm; H=3.5M; B=2.4M. In Situ Native: Unstable, slight DOC, high water table.

Site 5: 9 Conditions. Variable STIS 6400 - 23115 Pa with foot tamping of dumped gravel, clay, and native; D=609mm; H=3.6M; Variable B=1.2 - 2.7M. In Situ Native: Unstable at slight DOC, high water table.

Data from these sites represented 18 fiberglass pipe stiffness and installation combinations offering a wide range of conditions. These deflection data covered an elapsed time ranging between 1.5 and 12 years. Initial measured deflections were not available for all installations. The period of time that elapsed between installation and initial deflection measurement varied from a few hours to several days. For correlation, initial, average and maximum measured deflection data from each condition was compared to initial, average and maximum predicted deflections determined by Equations 14, 16 and the method in ANSI/AWWA C 950-88.

Table 10 summarizes the field data and predicted deflections using ANSI/AWWA C 950-88 and the newly-defined method. Deflection values are reported in percent of pipe diameter. All installations from site number 1 used a split installation with C_c = 0.7 and foot tamped native soil above the split. For site number 4, only one deflection per installation condition was reported. The data was used as an average deflection value.

TABLE 10 -- Correlation data of calculated and measured deflections
Vertical deflections in percent of diameter

SITE NO.	FIELD MEASUREMENTS			C-950 PREDICTIONS			PROPOSAL PREDICTIONS		
	INIT.	AVE.	MAX.	INIT.	AVE.	MAX.	INIT.	AVE.	MAX.
1-1	0.7	1.5	1.8	0.3	0.9	1.2	0.7	2.8	4.5
1-2	5.9	7.1	8.4	3.2	6.5	8.2	3.8	9.2	10.7
2	...	2.1	3.6	0.5	1.0	1.4	0.3	1.3	3.1
3	...	3.2	4.4	0.3	0.8	1.1	0.2	2.7	4.4
4-1	2.8	6.9	...	0.8	1.7	2.2	2.6	4.9	6.7
4-2	0.9	2.6	...	0.8	1.6	2.2	2.4	3.9	5.5
4-3	4.1	7.9	...	1.7	2.7	3.5	4.6	7.7	9.2
4-4	2.7	4.5	...	1.7	2.6	3.3	3.3	5.3	6.6
4-5	2.0	2.4	...	1.4	2.1	2.6	1.5	2.5	3.5
5-1	3.8	5.1	5.7	0.8	1.6	2.1	1.9	3.5	4.8
5-2	2.5	3.6	4.2	0.7	1.5	1.9	1.3	2.5	3.5
5-3	1.3	1.8	2.7	0.7	1.4	1.7	0.9	1.8	2.3
5-4	3.5	4.6	5.6	3.2	5.0	6.1	2.3	4.0	5.3
5-5	3.3	3.6	4.4	2.5	3.9	4.6	1.5	2.7	3.7
5-6	1.7	1.9	2.6	2.0	3.1	3.5	1.1	1.9	2.4
5-7	6.9	7.9	8.9	7.7	11.8	13.3	3.1	9.7	10.9
5-8	3.5	3.6	4.2	4.6	7.1	7.6	1.7	5.0	6.0
5-9	1.8	2.0	2.5	3.1	4.8	5.1	1.2	3.4	3.9

In developing the proposed deflection prediction method, a significant quantity of data has been generated for installations using granular backfill. The method was correlated to field installations in which granular backfill was used. Before the method is applied for installations using cohesive backfill soil, additional studies must be completed.

Figures 4-6 graphically demonstrate the correlation of the current ANSI-AWWA C 950-88 deflection prediction method with the measured deflection values. Correlation coefficient, r, and the slope of each least squares linear data fit is shown as well as the perfect correlation reference line. A perfect correlation is represented by an r = 1.0, slope = 1.0 and a curve fit that lies on the reference line.

Figures 7-9 graphically demonstrate the correlation of deflections predicted by equations 14 and 16 with the measured deflection values. Statistical data is provided.

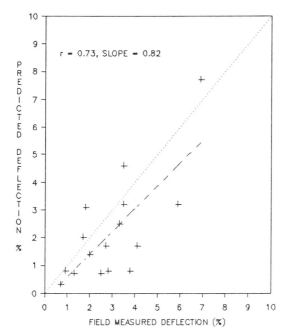

FIGURE 4--Initial deflection correlation
AWWA C950 vs. field measurements

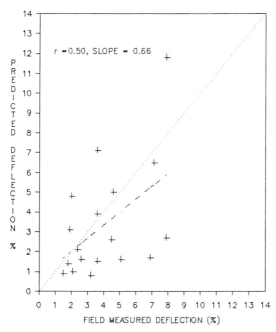

FIGURE 5--Average deflection correlation
AWWA C950 vs. field measurements

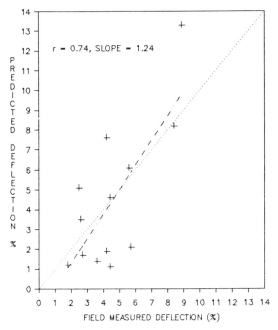

FIGURE 6--Maximum deflection correlation
AWWA C-950 vs. field measurements

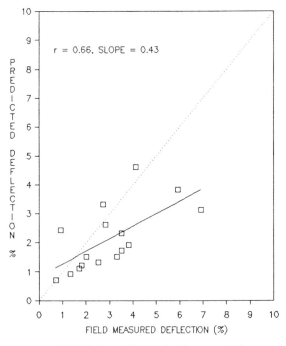

FIGURE 7--Initial deflection correlation
Proposed method vs. field measurements

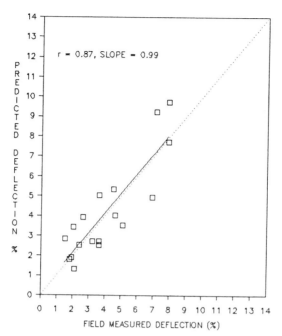

FIGURE 8——Average deflection correlation
Proposed method vs. field measurements

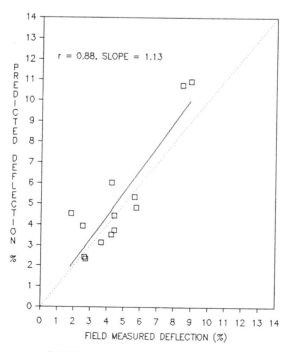

FIGURE 9——Maximum deflection correlation
Proposed method vs. field measurements

Discussion of Data

The correlation of measured initial deflections with predicted initial deflections is not encouraging for the proposed method. Probably the period of time from installation to initial deflection measurement significantly influenced the amount of soil load that was transferred to the pipes. This suspicion is justified by the somewhat better correlation of measured initial deflections and the AWWA equation which does not assume an initial reduction in soil load resulting from arching. The slope of 0.43 indicated the measured deflection tended to be higher than that predicted. In the majority of the cases, measured and predicted initial deflections correlated well when using the proposed procedure. More data correlation will yield a more definitive conclusion.

The average deflections predicted by the proposed equation yielded excellent correlation with measured values. The scatter around the data fit line was less than 1.5 percent deflection. The approximation of the ideal correlation line was very good. The predictions by the AWWA method did not correlate well with measured values.

Maximum deflection prediction by the proposed method was very good. With the limited data available for this study, the anticipated results of a maximum deflection correlation would be an overprediction of deflection. The proposed equation reflected this anticipated behavior. The AWWA method yielded a reasonable correlation with maximum measured deflection; however, several data points were distributed on the unconservative side of the ideal correlation line.

PRACTICAL APPLICATIONS

Short-Term Deflection

Measurement of vertical pipe deflection is a frequently-used, quality control technique. These measurements are taken after backfilling has been completed to grade elevation. It is common practice to limit short-term deflection to levels of two or three percent less than the allowable long-term value.

Equation 14 can be used to calculate short-term deflection to obtain an estimate of expected maximum initial deflection during pipeline installation. The purpose of estimating short-term deflection is to determine what level of control may be required during installation to achieve a quality installation.

For short-term deflections, the time effects represented by creep retention factors for both the pipe and soil properties are set equal to one in Equations 8, 12 and 14. The short-term vertical load is estimated as a fraction of the full prism load by calculating the arching reduction factor, given by Equation 3A. Equation 14 can be used for predicting average short-term deflections.

Long-Term Deflection

Long-term deflection requires application of creep retention factors for both the pipe and soil properties. The long-term soils load is assumed to be equal to the full prism weight acting on the pipe after all soil arching has dissipated. These parameters are given by Equations 3A and 8. Typical average long-term deflections can be realistically predicted by Equation 14, and upper limit deflections can be estimated by Equation 16.

Limitations

The method presented has been developed empirically using principles of pipe-soil interaction. The correlation studies presented include fiberglass pipe data. The behavior is likely to be similar for a variety of flexible pipe materials. This method can be applied to flexible pipe products other than fiberglass. Correlation studies are continuing to expand the confirmed range of application of the method. Based on current studies, Equations 14 and 16 have been found to be valid for pipe stiffnesses greater than 600 N/M². Good engineering judgment must be used when selecting representative soil properties. When burial depths are in excess of 10 meters, soil properties may be approximated by extrapolation of Table 3. Depths exceeding 15 meters should be very carefully evaluated. These limitations are provided to restrict the use of the method to a range of pipe and soil stiffness and installation condition for which the method is considered to be applicable.

SUMMARY AND CONCLUSION

Prediction of vertical deflection of buried flexible pipelines has been reconsidered in view of recent research findings. A semi-empirical approach has been taken to introduce modifications to the original Spangler method. These modifications include the following parameters which govern the response of buried flexible pipe.

1. Pipe Stiffness: Dependent on time via stiffness retention data. Sensitivity to pipe stiffness is introduced into each element of deflection.

2. Soil Stiffness: Defined as a function of trench width, embedment and native soil modulus and time. Soil modulus is determined by the confined compression test and related to soil type, density and burial depth. Pipe stiffness dependency is introduced via a pipe-soil interaction coefficient.

3. Soil Load: The weight of the soil prism above the pipe is represented as long-term vertical soil load. Short-term load is obtained via the Marston Theory of Arching as a reduction of the prism load.

4. Trench Configuration: In addition to trench width, a composite soil modulus is calculated which introduces the effect of split embedment installations into the soil stiffness term.

5. Haunch Support: Nonuniform haunch support is demonstrated by a bedding factor since variable haunch support is found to occur with a high degree of regularity.

6. Non-Elliptical Deformation: The ratio of vertical to horizontal deflection is not assumed to equal one. Values based on measured data are provided as input to the proposed method which are dependent on soil type, density and pipe stiffness.

7. Deflection Caused by Construction: Initial ovalization that occurs during installation is included as a function of pipe stiffness, soil type, and soil density.

8. Deflection Caused by Installation Variability: Variability resulting from field construction practice is introduced as a difference between achieved versus specified soil compaction. A variability term is included based on parametric studies of density sensitivity of the model developed.

An equation for predicting vertical pipe deflection has been developed, considering these governing parameters, to better characterize short-term and long-term response. The following expression of maximum vertical deflection is a result of these considerations.

$$\delta_{VP} = \frac{k \left(\Delta V / \Delta H \right) \left(C_L \gamma^{H+W} \right)}{8 C_{TP} EI/D^3 + 0.061(0.6) \zeta C_I E_S} \times 100 - \delta_{VO} + \delta_{VA} \tag{16}$$

Average long-term deflection can be predicted by eliminating the installation variability term. When compared to actual field measurement data, this method demonstrates improved correlation and reliability to predict long-term vertical pipe deflection. Comparison of the newly-developed approach versus that in Appendix A of ANSI/AWWA C 950-88 for the same field measurement data base shows that correlation of Equation 16 is significantly better.

RECOMMENDATIONS

To extend the confidence in the proposed deflection prediction method, further correlation studies should be completed. Issues of initial deflection prediction and installations using cohesive soils should be resolved through additional studies.

ACKNOWLEDGEMENTS

Dr. A. P. Moser, Dr. R. K. Watkins, Dr. L. R. Anderson of Utah State University for directing the soil box testing on full-scale pipe and FEA studies.

Dr. S. Torp in Oslo, Norway and R. R. Bishop of Carlon for initiating and collecting the data.

Owens-Corning Fiberglas Corporation for sponsoring the programs that allowed the additional technology to be developed.

REFERENCES

[1]   Marston, A., "The Theory of External Loads on Closed Conduits in the Light of the Latest Experiments", Engineering Experiment Station, Bulletin 96, 1930.
[2]   Spangler, M. G., "The Structural Design of Flexible Pipe Culverts", Iowa Engineering Experiment Station, Bulletin 153, 1941.
[3]   Watkins, R. K. and Spangler, M. G., "Some Characteristics of the Modulus of Passive Resistance of Soil: A Study of Similtude", Highway Research Board Proceedings, Vol. 27, 1958.
[4]   Hoeg, K., "Stress Against Underground Structural Cylinders", Journal of Soil Mechanics and Foundations Division, ASCE, Vol. 94, No. SM4, July 1968.
[5]   Galili, N. and Shmulevich, I., "A Refined Elastic Model for Soil-Pipe Interaction", ASCE International Conference on Underground Plastic Pipe, New Orleans, Louisiana, March, 1981.
[6]   Duncan, J. M., Byrne, P., Wong, K. S., "Strength, Stress-Strain and Bulk Modulus Parameters for Finite Element Analysis of Stresses and Movements in Soil Masses", University of California, Department of Civic/Engineering, Report No. UCB/GT/80-91, August 1980.
[7]   Zarghamee, M. S., Tigue, D. B., "Soil-Structure Interaction of Flexible Pipe Under Pressure", Presented at the Transportation Research Board Annual Meeting, January 1986.
[8]   Leonhardt, G., "Die Erdlasten bei Uberschutteten Durchlassen", Die Bautechunik, 56(11), 1979.
[9]   Leonhardt, G., "Einflub der Bettungssteifigkeit auf die Tragfahigkeit und die Verformungen von Flexiblen Rohren", Strasse Bruecke Tunnel, Volume 3, 1972.
[10] ATV - Regelwerk, A127, "Rightlinie Fur die Statische Berechnung von Entwasserungskanalen und-Leitungen", December 1984.
[11] Howard, A. K., "Modulus of Soil Reaction (E'), Values for Buried Flexible Pipe", Report REC-ERC-77-2, Bureau of Reclamation Engineering and Research Center, Denver, Colorado.
[12] Krizek, R. J., et.al., "Structural Analysis and Design of Pipe Culverts", National Cooperative Highway Research Program, Rèport 116, 1971.
[13] Bishop, R. R. and Lang, D. C., "Design and Performance of Buried Fiberglass Pipes - A New Perspective", ASCE National Convention, Pipeline Division, San Francisco, California, October 1984.
[14] Duane, J., Robinson, R., and Moore, C. A., "Culvert-Soil Interaction Finite Element Analysis", ASCE Journal of Transportation Engineering, Vol. 112, No. 3, May 1986.
[15] Lang, D. C. and Howard, A. K., "Buried Fiberglas Pipe Response to Field Installation Methods", ASCE International Conference on Advances in Underground Pipeline Engineering, Madison, Wisconsin, August 1985.

[16] Moser, A. P., Bishop, R. R., Shupe, O. K., Bair, D. R., "Deflection and Strains in Buried FRP Pipes Subjected to Various Installation Conditions", Presented at the Transportation Research Board 64th Annual Meeting, January 1985.

[17] Moser, A. P., Clark, J. C., Bair, D. R., "Strains Induced by Combined Loading in Buried Pressurized Fiberglass Pipe", ASCE International Conference on Advances in Underground Pipeline Engineering, Madison, Wisconsin, August 1985.

[18] Sharp, K., Anderson, L. R., Moser, A. P., Bishop, R. R., "Finite Element Analysis Applied to the Response of Buried FRP Pipe Due to Installation Conditions", Presented at the Transportation Research Board 64th Annual Meeting, January 1985.

[19] Spangler, M. G. and Handy, R. L., Soil Engineering 3rd Edition, Harper and Row, Publishers, Inc., 1973.

[20] Hartley, J. D. and Duncan, M., "E' and Its Variation with Depth", ASCE Journal of Transportation Engineering, Vol. 113, No. 5, September, 1987.

[21] Brown, F. A. and Lytton, R. L., "Design Criteria for Buried Flexible Pipe", ASCE National Convention, Pipeline Division, San Francisco, California, October 1984.

[22] Chuo, K. M. and Lytton, R. L., "Viscoelastic Approach to Modeling Performance of Buried Pipes, "ASCE Journal of Transportation Engineering",Vol. 115, No. 3, May, 1989.

[23] Carlstrom, B., "Calculation of Circumferential Deflections and Flexural Strains in Underground GRP Pipes Used for Non-Pressure Applications", Europe '82 Conference, Basel, Switzerland, Paper 3.

[24] Howard, A. K., "Diametrical Elongation of Buried Flexible Pipe", ASCE International Conference on Underground Plastic Pipe, New Orleans, Louisiana, March 1981.

[25] Torp, S., Paul, P. N., Molin, J., "Deflection and Strain in Buried GRP Pipelines: Experiences from the Middle East", Europipe '83 Conference, Basel, Switzerland, June 1983.

Installation

Kleovoulos G. Leondaris

FIELD EXPERIENCE WITH FIBERGLASS PIPE IN THE MIDDLE EAST

REFERENCE: Leondaris, K.G., "Field Experience with Fiberglass Pipe in the Middle East", Buried Plastic Pipe Technology, ASTM STP 1093, George S. Buczala and Michael J. Cassady, Eds., American Society for Testing and Materials, Philadelphia, 1990.

ABSTRACT: Glassfiber Reinforced Plastic (GRP) pipes were first introduced into the Middle East in the mid 1970's. High temperatures prevail, high and saline groundwater tables exist in coastal areas, and corrosive soils abound making GRP the ideal pipe material for water and sewer lines, cooling systems for power and desalination plants, storm-water networks, and other applications.

KEYWORDS: deflection, stiffness, field experience, installation, Middle East, Glass Fiber Reinforced Plastic (GRP) Pipe

Today, GRP pipes are produced in Saudi Arabia, Dubai (U.A.E.), Kuwait and Egypt, for the Middle East markets. Thousands of kilo-meters of GRP pipe have been installed so far in varying soil conditions and with different design and loading criteria. The overwhelming majority of these pipes had a Specific Tangential Initial Stiffness (STIS = EI/D3) of 1250 N/m2 and 2500 N/m2.

The quality of pipe installations has been monitored by taking initial and final deflection readings. The paper is a detailed study of GRP pipe performance in relation to different types of soils, cover depths, levels of groundwater table, and types of backfill materials used.

Deflection readings are presented positive (increase in vertical diameter) and negative (decrease in vertical diameter).

The paper also evaluates special installation practices necessary for particularly poor native soils.

Kleovoulos G. Leondaris is Manager, Marketing & Field Engineering, International Pipe Operations, Owens-Corning Fiberglas Corporation, Athens Tower, Building A, Athens R-11527, Greece.

The experience gained on over a thousand kilometers of GRP (or FRP) pipe demonstrates that low stiffness GRP pipes perform very satisfactorily when installed with reasonable care.

GRP PIPE INSTALLATIONS IN SAUDI ARABIA

The discovery of vast oil and gas reserves enabled the Kingdom of Saudi Arabia to transform itself into a dynamic 20th century society in a relatively short time. Traditional Saudi Arabian towns witnessed unprecedented building booms with rapidly growing industries, residential housing and the introduction of international technology.

While much of that technology has been directed to capturing the Kingdom's huge gas and oil supplies, it has also been used to improve the quality of water supply. New methods of treatment, including recycling of wastewater, have made groundwater, generally high in saline, usable for agriculture and industry as well as personal consumption. A number of power and desalination plants were also built along the coasts of the Red Sea and the Arabian Gulf making Saudi Arabia one of the largest users of desalination water in the world. Of course, the Kingdom's rapid and successful growth also demanded a large network of water and sanitary transmission pipes.

The answer, in many cases, has been Glassfiber Reinforced Plastic (GRP) pipe.

The vast majority of the GRP pipes in the Kingdom of Saudi Arabia have been supplied by Amiantit Fiberglass Industries Ltd. (AFIL), a joint venture established in 1977 between the Saudi Arabian Amiantit Company of Dammam and Owens-Corning Fiberglas Corporation of Toledo, Ohio.

GRP PIPE INSTALLATIONS IN YANBU

Yanbu on the Red Sea and Jubail on the Arabian Gulf were chosen for the two most ambitious and advanced industrial programs ever undertaken in the Kingdom of Saudi Arabia.

By Royal Decree No. M/75 in July, 1975, a Royal Commission was created to implement the basic infrastructure plans necessary in both areas. Major projects were planned and executed such as: power and desalination plants, various industrial facilities, housing complexes, roads, potable water, irrigation, drainage and sewerage networks, etc..

All these projects utilize significant quantities of pipes of different diameters and pressure capabilities. GRP pipes were specified in some projects as the only acceptable material. In others, together with alternatives.

The vast majority of these pipes were locally produced by AFIL. Table 1 gives a breakdown of GRP pipes supplied to Yanbu by Amiantit Fiberglass.

TABLE 1 -- FRP Pipes Supplied By Amiantit
Fiberglass To Various Projects In Yanbu, Saudi Arabia

Project - Client/Engineer Contractor	Application	Lineal Meters	Year
Yanbu 1077, Abu Al-Enain & Jastaniah	150,200,250 mm, H-120 Blue color	15,450	1988
Yanbu PPH-1014, RC/Parsons, H. Zosen	400 mm, H-060	3,000	1980
Yanbu 1005, RC/Parsons, Ret-ser	350,400,500,600,700,800, 900,1100,1300,1400,&1600 mm 1800,2000 mm, H-010 800,900,1000,1000 mm, H-060	47,900 13,225	1980 1980
Yanbu PIC-G-1631, DITCO	600 mm, H-010	204	1984
Yanbu Repairs Work, Korean Express Co.	400,500,600,800,900,1400, 1500,2000 mm, H-010 400,500 800 mm, H-010 with fittings	300 84	1984
Yanbu PIC-A-1042, Abu El-Enain & Jastaniah	500,600,700,800,1000, 1100,1200 mm, H-030	4,284	1986
Yanbu PIC-A-1025, and 1680, Intrafor-Co	300,350,400,600,700,800, 900 mm, H-150	24,024	1986
Yanbu PIC-A-1042, Shairco	1200 mm, H-010	108	1986
Yanbu 1077, Abu Al-Enain	500,800 mm, H-060 350 mm, H-120	171	1988
Yanbu PIC-A-1200-08,RC/Gibbs &Hill/Dong Ah	600,700,900,1100,1400,170 2700,2900,3000 mm	18,763	1981
Yanbu Steam Generating Power Plant,RC/Parsons SICOM SIMCO	350 mm, H-120	156	1982

Project - Client/Engineer Contractor	Application	Lineal Meters	Year
Yanbu PIC-B- 1153 & PIC-B- 1168, Hazama Gumi	800,1400,2100,2400,2500, 2700,2900,3000,3400,3700 mm	8,650	1985
Royal Commis- sion Warehouse Yanbu, Riza Investment Co.	600,750,800,900,1100,1200 1400,1700,2000,2300,2400, 2600,2700,2900,3000,3200 3400,3700 mm, H-120	324	1985
Yanbu 1085, ENDECO	400,600,1000,1400,2900 3000 mm, H-060 600 mm, H-010 aboveground	324 3,528	1989 1989
Water & Sewer Treatment Plant Yanbu, Royal Commission for Jubail and Yanbu	600,800 mm, H-120	48	1989
Yanbu 1013, RC/Parsons/ Keang Nam	350,400,500,600,800 1,000 mm, H-120	21,696	1980
Yanbu PIC-A- 1021,RC/Parsons/ T.G.M.	800,1000 mm, H-120	380	1981
Yanbu PIC-B- 1241,RC/Parsons/ Keang Nam Ent.	350 mm, H-150 and fittings	3,840	1982
Yanbu PIC-A- 1006,RC/Parsons/ Intes	400,450,500,600,1000 mm	2,000	1983
Yanbu PIC-A- 1042, Abu Al- Enain & Jastaniah	500 mm, H-090	3,493	1986

It is interesting to note that in all these projects, specifications called for pipes with minimum stiffness 1250 N/m2 STIS.
Pipes were installed in accordance to the instructions of the pipe
manufacturer and the quality of installation was monitored by
checking initial and/or final deflection of the pipes.

An analysis of GRP pipe installations in Yanbu is of particular interest since Yanbu is quite representative of soil and
ground conditions encountered in most coastal areas of Saudi
Arabia -- sandy soils and high water table.

We will now review some projects in Yanbu area, emphasizing
installation conditions and corresponding deflection results.

Delta Housing Project, PIC-A-1676

The project involved 600 mm gravity sewer pipes. Native
soil was loose sand and groundwater table was well above the
pipe crown. To be compatible with native soil, crushed gravel/
sand mix was used as backfill material. Proper compaction
(70% Relative Density) was achieved by water jetting. Lightweight
plate compactors were also used. Select backfill was brought up
to 70% of pipe diameter. The trench was then filled with compacted sand to ground level. Cover depth was 4.0 to 4.5 m above pipe
crown. Fig. 1 shows a typical trench cross-section and gives
minimum dimensions for different diameters.

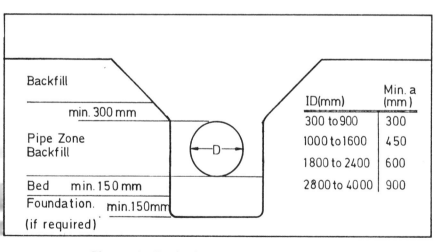

ID(mm)	Min. a (mm)
300 to 900	300
1000 to 1600	450
1800 to 2400	600
2800 to 4000	900

Figure 1. Typical Trench Dimensions

Initial vertical deflection readings varied between 0.80%
and 1.55% comparing well with maximum allowable value of 3%.

Site Development Project, PIC-A-1025

The following pipes were used in this project: Pressure Sewer (6 bar), 350, 400, 600, 700, 800 and 900 mm in diameter.

Potable water (15 bar), 300, 400 and 500 mm in diameter.

Irrigation water (15 bar), 400 and 600 mm in diameter.

Native material was sand and in places rock. Water table was mostly below pipe invert and in some cases, above pipe crown. Pipes were always installed in dry trench, as pumps were used to lower the water level. Well-graded sand was used as bedding and backfill material. It was compacted by water jetting in 200 mm layers and extended to 300 mm over the pipe crown. A minimum 90% SPD was achieved in the pipe zone area. Cover depth varied between 1 and 3.2 m above pipe crown.

Deflection readings were taken five to seven months after installation and varied between 1.0% and 3.90% comparing well with maximum allowable long-term deflection of 5.0%.

Yanbu, PIC-A-1005

This project required gravity and pressure (6 bar) sewer pipes. Diameters were 350, 400, 500, 700, 1100, 1300, 1600, 1800 and 2000 mm.

This was a huge project involving a total of 48 Kms of pipe. Management of construction was carried out by Parsons. Ret-ser, a large Chinese company, was the contractor.

Native soil was stable coarse sand with 6" to 10" rounded boulders embedded in it. Minimum cover over pipe crown was 1.2 m while maximum was 3.5 m.

Backfill material was well-graded coarse sand compacted in layers (90% SPD) by roller compactors. Initial deflection readings varied between 0.90% and 2.25% which compares well with maximum allowed values of 3%.

Yanbu, PIC-B-1153/1168

The project called for large diameter GRP pipes, 600, 2100, 2700, 2900, 3000, 3400 and 3700 mm in diameter. These were the largest diameter pipes ever produced in the Middle East. They were joined with double bell couplings and were used in a cooling water system for the industrial area of Yanbu. The pipes were designed for an operating pressure of 5.6 bars. Minimum stiffness was 1250 N/m2.

The native soil was sand and groundwater was above pipe invert. Backfill material in the pipe zone area (up to 300 mm above pipe crown) was 20 mm uniformly graded, crushed gravel and maximum cover depth above the pipe crown was 3.0 m.

Following installation, all the pipes were inspected internally and deflection readings were taken. Initial deflection readings were between 0.10% and 1.52% well below 3% maximum allowable value.

The pipelines have been operating since 1987. In mid 1989, the line was shut down, drained and inspected internally. Even though deflection readings were not taken, the pipes visually were in excellent condition. Please note that this project was, in fact, the second phase of an ambitious plan to provide cooling water to the various industries in Yanbu. Phase 1, project PIC-A-1200, was designed by Gibbs & Hill and constructed by the Korean company, Dong Ah. It involved 19 Kms of 6 bar pressure pipes of various sizes -- 600, 700, 900, 1150, 1400, 1700, 2700, 2900, 3000, 3200 3400 and 3700 mm in diameters. This project represents the world's largest (money wise) installation of GRP pipes. Installation conditions were similar to those of project PIC-B-1153. The pipe-lines have been in operation for seven years now and no problems have been reported.

PIPE INSTALLATIONS IN AL KHARJ

Al Kharj is an area in the Central Province of Saudi Arabia about 175 Km north of Riyadh. The development of this region led to the execution of many projects in the last years. Table 2 details GRP pipes supplied to Al Kharj projects by Amiantit Fiberglass in detail.

TABLE 2 -- FRP Pipes Supplied by
AFIL To Various Projects in Al Kharj

Project - Client/Engineer Contractor	Application	Lineal Meters	Year
Sewer Projects In Al Kharj:			
Al Kharj Sewage Disposal System,	500,600,700,800,900,1000 and 1200 mm, H-120	14,856	1984
Al-Marafik Const.	500, 800, 1200, 1450 mm, H-150 with fittings	24,300	1984
Al Kharj Project,	600, 1200mm, H-010	264	1985
Sambu	300, 800 mm, H-10.5	1,912	1985
Al Kharj Airbase Pkg 102, J&P	400, 500, 1200 mm, H-010	2,655	1988
Al Kharj 202 Al-Henaki Trading	400, 500 mm, H-030	4,994	1989

Project - Client/Engineer Contractor	Application	Lineal Meters	Year
Water Projects in Al Kharj:			
Al Kharj Airbase Pkg 202, J&P	350, 400, 500, 600, 900 1000 mm, H-120	32,725	
Al Kharj 202 Al-Henaki Trading	350, 400, 500, 600, 800, 900 mm,H-120	2,393	1989
Al Kharj 202 Al-Henaki Trading	80, 100, 150, 200, 250, 300 mm	13,951	1989

One of these projects, the sewage disposal system, used 15 Kms of gravity and 15 bar pressure pipes of the following sizes: 500, 600, 700, 800, 900, 1000, 1200 and 1400 mm.

Pipe stiffness was 1250 N/m2.

The native material was hard soil, basically coarse sand with rock formation. Ground-water table was below pipe invert in contrast to the coastal areas referred to earlier.

Pipe zone backfill material was well-graded crushed stone with maximum particle size equal to 10 mm. Maximum cover depth over the pipe crown was 7 m.

Initial deflection readings varied between 0.15% and 0.62% indicating a high quality installation was achieved. Maximum allowable initial deflection was 3%.

GRP PIPE INSTALLATIONS IN DUBAI, U.A.E.

Another Middle East Pipe producer, Gulf Eternit Industries Ltd. (GEI), operates in Dubai, U.A.E. Gulf Eternit has been very active in producing high quality GRP pipes and other products. This company has been very successful in promoting the use of GRP in the Gulf area.

One major project where GRP pipes were used is the New Dubai Sewage Treatment Plant. This multi-million scheme will handle 130,000 cubic meters of raw sewer per day at its first phase. The future second phase will increase daily capacity to 200,000 cubic meters.

This project was necessary as the old treatment plant was designed for a peak population of 60,000 whereas the current population of Dubai is around 400,000.

Associated with the construction of the treatment plant was the installation of over 100 Kms of large diameter GRP pipes. The pipes were used to carry raw sewage to the new treatment plant, as well as treated effluent from the plant to the city (used for irrigation).

The sketch in Figure 2 shows the New Dubai Sewage Treatment Plant (STP) located 15 Km from Dubai City, pumping stations S,E, C,G and X, as well as the pipelines laid between.

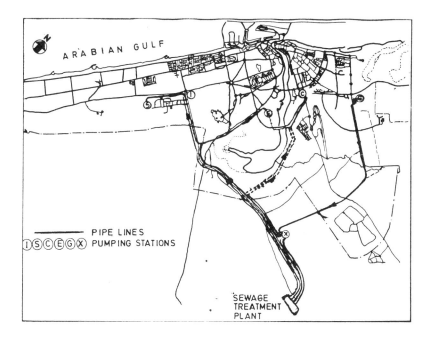

Figure 2 - New Dubai Sewage Treatment Plant

The project was designed by the German joint venture GWE-OMS and the first phase of construction was carried out by Keang Nam Enterprises Ltd., part of Daewoo Corporation.

GRP pipes, couplings, and fittings were manufactured in Dubai by Gulf Eternit Industries S.A. under the close supervision of VEROC Technology A/S (a subsidiary of Owens-Corning Fiberglas). The pipe used in this project ranged from 400 mm to 2000 mm with over 50 Kms being 1000 mm in diameter.

More than 80% of the 100 Kms of pipes were designed to with stand a 6 bar working pressure and on-site hydrotesting of 9 bars. About 2 Kms were designed for a 2.5 bar working pressure whereas the balance were gravity pipes. Stiffness class was minimum 2500 N/m2 for all pipes

As far as pipe installation is concerned, the consultant's specification was in line with ASTM D3839, Standard Practice for Underground Installation of Flexible Reinforced Thermosetting Resin Pipe and Reinforced Plastic Mortar Pipe. At the same time it depended heavily on the pipe supplier's recommendation. In fact, in order to ensure that installation was in strict accordance with the manufacturer's recommendations, the contractor had to make available to the engineer for the duration of the contract, two experienced pipeline inspectors who were full time employees of the GRP pipe manufacturer.

In addition to the two inspectors, a qualified engineer from the GRP pipe manufacturer was made available to provide specialist technical advice on pipe laying problems. Obviously, the objective of all these precautions was for a successful execution of this high prestige project in Dubai.

The pipelines were installed over a very large area. As a result, the native soil conditions varied substantially, thus, effecting the installation procedures followed. Another variable was the level of groundwater depending on the proximity of the line to the creek.

In general, native materials were coarse sand medium to very dense and cohesive soils of medium to hard consistency. These soils, combined with proper dewatering, were adequate support for standard installation procedures.

The type and stability of native soils was established by conducting a detailed soil survey including bore-holes and exploratory excavation.

Where native soils were not structurally adequate to properly support the pipe (min. SPT*Blow Counts 15), special installation instructions were followed, thus, allowing the use of one stiffness class (2500 N/m2) of pipe for the entire project.

Select backfill material, used for bedding and backfilling the pipe zone area, was well graded gravel with maximum particle size equal to 19 mm. An important factor in selecting this backfill material was compatibility with the native soil in order to prevent the pipe zone backfill material from being washed away or migrating into the native soil. Where incompatible materials had to be used, they were separated by filter cloth which completely surrounded the bedding and pipe zone materials as illustrated in Figure 3.

Ground-water table in Dubai is generally high, particularly in areas close to the Arabian Gulf and the creek. It also shows considerable tidal fluctuations. In all cases, GRP pipe installation was done in dry trench conditions. Dewatering was achieved using well points and the groundwater was pumped into the creek.

* Standard Penetration Test in Accordance with ASTM D1586.

Figure 3 - Select Backfill Material Separated by Filter Fabric

Minimum cover depth over the pipe crown was 1.5m, whereas maximum was set at 6.0 m. When cover depth exceeded 6 m (in certain areas it reached 11 m), pipes were encased in concrete. At major road crossings, pipes were protected from heavy traffic loads by concrete slabs placed at least 300 mm over the pipe crown.

As stated earlier, in most cases the native soils were quite stable and standard installation procedures were adequate. The trench was excavated so that the minimum distance between pipe and trench wall at the pipe springline was:

300 mm for pipe I.D. 300 - 900 mm
450 mm for pipe I.D. 1000 - 1600 mm
600 mm for pipe I.D. larger than 1600 mm

In certain sections of C,S and E lines, pipes were installed in soft cohesive soils with an unconfined compressive strength between 26 and 100 kN/m2 or in loose granular soils with a blow count in accordance with ASTM D1586 Standard Method for Penetration Test and Split Barrel Sampling of Soils between 3 and 10. In such cases, the weakness of native soil was overcome by increasing the trench width to between 2.25 and 3.0 times the pipe diameter.

In many cases, the trench bottom was soft or loose and had to be improved before laying the pipes. This problem was solved by over-excavating the soft sections and providing a proper foundation using compacted select backfill material surrounded by filter cloth. The thickness of this foundation varied between 150 and 500 mm, depending on the situation.

In parts of the E-influent line, the native soil in the pipe zone area was very soft clay, totally inadequate to provide support to the 1000 mm pipe. The solution was either to use a stiffer pipe or follow special installation procedures. The engineer decided to use the same pipes (2500 N/m2 STIS) with special installation instructions, even though, a stiffer pipe would have been as cost effective. Special installation procedures were developed for this case as follows:

- Minimum trench width was five times pipe diameter.

- Minimum bedding thickness was 150 mm.

- Under the bedding, a foundation with minimum thickness equal to 500 mm was provided.

- Select backfill material was used for both bedding and foundation.

- Filter cloth was used to totally surround the foundation, bedding and pipe zone area. The filter cloth was folded over the top of the pipe zone area.

- Maximum pipe length was 6.0 m.

- Trench boxes were used to support the trench walls. The boxes were pulled in steps to allow for compaction of the select backfill material against the trench native walls.

The trench boxes were 3.0 m long sections with adjustable width varying between 1.0 and 5.0 m. Different widths were achieved by inserting 1.0 m long beams at each end. The whole unit was lifted easily with either a crane or an excavator using a chain hooked at the four corners of the section. Figure 4 details the use of trench boxes.

Figure 4. Trench boxes used in unstable soils.

The quality of installation was confirmed in several way.

Pipelines larger than 600 mm in diameter were inspected internally to make sure that no flat areas or bulges were formed and that proper gaps were left at every joint.

Each pipe joint larger than 600 mm in diameter was hydro-tested at 1.5 times the operating pressure prior to backfilling the line. A portable hydraulic joint test equipment supplied by VEROC Technology was used for testing these joints. Leaks were observed on three occasions only due to twisted rubber gaskets. This was an excellent result considering the large number of joints tested (over 8000). Initial hydrotesting was carried out in sections from chamber to chamber (200 to 450 m long). This was followed by final hydrotesting in larger sections (3.0 to 8 Kms). By testing longer sections this time, valves, fittings and in general, chamber connections, were also tested. Testing of such connections was not part of initial field testing. Finally, deflection readings were taken at four different stages in order to ensure compliance to the requirements of the consult-ants specification (maximum allowed initial deflection of 3%, maximum allowed long term deflection 5%).

Firstly, deflection was checked after select backfill was compacted to 300 mm above the pipe crown. The purpose of this exercise was to give a quick indication of how well the pipe was installed, and if necessary, take corrective action while it was still easy to do so. The contractor was urged to place and compact the select backfill material in such a way as to slightly ovalize the pipe in the vertical direction. This initial ovalization was verified by the first deflection check and varied between 0.5% and 1%, well within acceptable limits. It was observed that the same compactive effort caused slightly larger ovalization in large diameter pipe. This was due to the fact that smaller diameter pipes were stiffer than large ones even though minimum design stiffness was 2500 N/m2 in all cases.

Secondly, deflection was checked after trench backfilling
was completed to ground level and the dewaterting system was still
operating.

A third deflection reading was taken after dewatering was
stopped and ground water reached equilibrium. This was the true
initial deflection reading which the specification directed was not
to exceed 3%. An analysis of these four sets of deflection readings is quite interesting. There is a definite gradual increase of
deflection as we move from the first set of readings to the last.
However, this increase is fairly small due to the very good
installation procedures followed.

On the average, deflection readings taken after pipe zone
backfill was completed and after the trench was totally backfilled
but groundwater was kept below the pipe invert, show positive
deflections. That is, pipe is still slightly ovalized upwards.
When groundwater was allowed to reach its normal level, pipes were
either still ovalized or slightly deflected (0.1% to 0.6%). Final
deflection readings were taken at least six months after installation. In fact, most of the readings were taken one year after
installation.

Final deflection readings indicated that pipes were mostly
round, some still ovalized in the vertical direction, and others
deflected only about -0.2% of pipe diameter. Maximum deflection
recorded was -1.14% in C-line (1650 mm diameter).

Dubai Contract DS 52
Pipe Deflection Records
Chainage 2+778.71 to 3+001 'E' Line

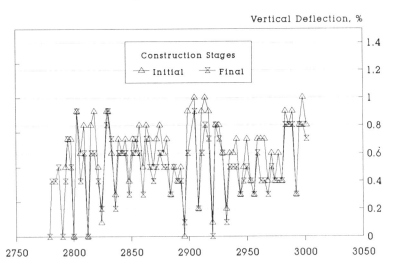

Figure 5. "E" Line with 2.12 m Average Depth of Cover

Figure 5 presents the results of deflection measurements taken on a portion of the 1000 mm pipe in "E" line. Although the data only portrays 225 m of pipe, the results are very representative of the entire line. Deflection measurements were taken at different stages of construction. Here initial values, taken just after completion of backfilling to grade and the return of the natural groundwater are compared to final values measured eight months later. Because of the favorable initial ovalization of the pipe created during compaction of the pipe zone backfill material, the initial deflections with 2.12 m of cover averaged 0.6%. That is, the vertical diameter is still elongated upwards. After eight months, the deflections have increased slightly. The average vertical deflection is 0.4%. This demonstrates the effectiveness of using a wide trench to minimize the negative effect on pipe performance in weak native soils.

The deflection values for chainage 7+099 to 7+1447 in line "C-9" have been graphed in Figure 6 for all four measurements. Consequently, one can visually see the effects of construction on the pipe's deflection performance. The first set of measurements, labelled "To Crown", shows the initial ovalization induced in the 1650 mm pipe. This only averaged 0.3% and was quite uniform in a fairly large diameter pipe, even after mechanically compacting the backfill. As expected, deflections increased (i.e. vertical diameter decreased) with the addition of 1.845 m of backfill over the pipe. Now, average deflection was -0.3%. When the groundwater was allowed to return to its natural elevation, the pipe responded by rebounding. The buoyancy effect of the water effectively reduced the weight of the soil cover over the pipe, without significantly effecting the lateral soil support (i.e. soil modulus). At this stage, deflections only averaged 0.15%. However, after 10 months when the final deflection measurements were taken, the effects of the water on the submerged soil side fill's structural properties becomes apparent. Deflection measurements were now averaging -0.85%.

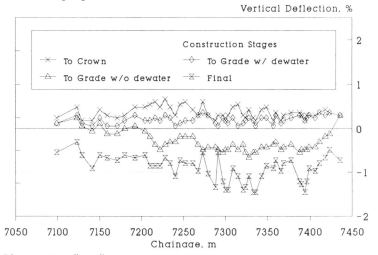

Figure 6 - "C-9" Line with 1.845 m Average Depth of Cover

CONCLUSION

The above mentioned GRP pipe installations are representative of the whole Middle East region, not only as far as soil and groundwater conditions are concerned but the stiffness of GRP pipe used.

Pipe deflection readings and other installation checks clearly indicate that when properly installed, GRP pipes perform very well. This is widely recognized and GRP pipes are steadily gaining acceptance in the Middle East as well as elsewhere in the world.

F.J.M. Alferink

SOME EXPERIENCE WITH 30 YEARS OLD BURIED (uPVC) PIPES FROM THE
VIEWPOINT OF STRESS AND STRAIN

REFERENCE: Alferink, F.J.M., "Some experience with 30
years old buried (uPVC) pipes from the viewpoint of
stress and strain", Buried Plastic Pipe Technology, ASTM
STP 1093, George S. Buczala and Michael J. Cassady, Eds.,
American Society for Testing and Materials, Philadelphia,
1990.

ABSTRACT: This contribution reports on the experience
with (uPVC) pipes used for gas distribution in the Netherlands. An extensive research program has been carried out
in order to determine the behaviour of buried PVC pipes,
some of which have been in service for over 30 years.
This research program covers both theoretical and experimental studies, the latter being the subject of this
contribution. Over the last three years a lot of pipe
deformation measurements were carried out on pipes buried
under different circumstances concerning soil type,
burial depth and loading conditions.
The results clearly show that buried PVC pipes behave
quite well from the viewpoint of stress and strain. The
results of measurements on old pipes when compared with
results from measurements on newly installed pipes and
from different stress calculation methods, also indicate
that the buried pipes are not in a situation of constant
stress, but in a constant strain situation. This supports
the idea of using the short-term modulus of elasticity
for calculation of the pipe deflection instead of the
long-term value.

KEYWORDS: plastics pipes, deflections, soil types,
strain, long-term behaviour.

F.J.M. Alferink is a research engineer at
VEG-GASINSTITUUT n.v., P.O. Box 137,
7300 AC Apeldoorn, the Netherlands

INTRODUCTION

An extensive gas distribution network having a total length of
about 90,000 km is in operation in the Netherlands. Twenty four
percent of the Dutch gas grid consists of rigid PVC. These pipes
were installed during the period from 1956 to 1974. Initially, they
carried town gas obtained from gasification of coal.
Since 1964 these pipes have been carrying natural gas. After the
early seventies rigid PVC was no longer installed. Impact modified
PVC was used instead, having a slightly lower stiffness, but a significantly better resistance against impact and stress cracking.
The total amount of PVC pipes in the distribution grid is about
50%.
Low pressure distribution was chosen in the Netherlands for reasons
of safety and installation convenience. The maximum pressure in PVC
pipes is 200 mbar (2.10^4 Pa). So from the mechanical point of view,
these pipes can be considered as pressureless. In order to get a
better understanding about the behaviour of buried flexible pipes
and to get real values for the deformation of pipes which have been
in service for a long time, it was decided to carry out an extensive
measuring program into the deformation of these pipes.
For this purpose a special device has been developed enabling the
whole shape of buried pipes with diameters varying between 110 and
200 millimeter, SDR 41, to be measured. The pipes were buried under
verges, sidewalks and roads, at depths varying between 30 centimeter
and 120 centimeter, in different soil types.
From the measurements the deflection and the tangential strain were
calculated. It was also possible to calculate the tangential stress
in the buried pipe, by remeasuring the pipe after digging up, and by
using a viscoelastic constitutive law.

MEASUREMENTS

The measurements are divided into two series, the first series
consisting of measurements with the objective to describe the field
conditions, and the second series having the objective to determine
the pipe deformation.
Pipe deformations were measured both for the buried pipes and for
the pipes that had been dug up.
To determine the field conditions, the following measurements were
performed:
- height of soil cover
- soil density
- minimum, maximum and in-situ porosity of the soil
- graine size distribution
- water content of the soil.
- plasticity index of the soil (in the case of clay only)
- pipe location (verge, road, sidewalk).

Part of these field conditions have been used for classification purposes.

The deformations of the pipes were measured by means of the DEFLEC, developed by VEG-GASINSTITUUT n.v. (1). This device enables measuring of the whole shape of buried pipes with diameters varying between 110 millimeter and 200 millimeter (Fig. 1).

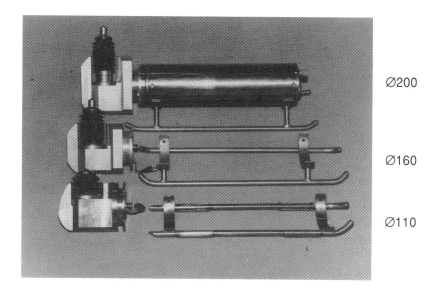

⌀200

⌀160

⌀110

Figure 1: DEFLEC⌀200 with interchangeable head's and skids for⌀160 and⌀110 mm.

The deformation measurements were carried out at different locations in the Netherlands.
At every location a buried pipe length of about 12 meters was made free of gas, after two manholes had been dug and after the length of pipe had been isolated.
The pipe deformation was then measured by pushing the DEFLEC into the pipe to its first position. Here the equipment measured the position of the sensor and the distance between pipewall and the axis of rotation.
Thus 40 positions (angle of rotation) and distances for every cross-section were measured and stored on a computer disc. At every location about ten cross-sections were measured.
After the measurements of the buried pipe were completed, the pipe was dug up, causing the pipe to recover from the loading which was initiated during installation, by soil load and by overburden loads, like traffic loading, during its time of operation.

CALCULATION OF STRAIN

In order to calculate the strain from the displacement measurements a method was used as shown in Figure 2. The curvature is calculated through three points on the pipe's circumference. In this case the curvature was calculated from the originally measured values. The angle between two points is about 9 degrees.

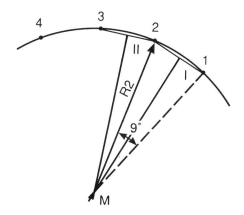

Figure 2: Calculation of tangential strain.

When a finer distribution is required more points should be measured or an interpolation polynomial should be used which allows for a good fit of the measured values. The disadvantage of the latter method is that the curvature might be influenced slightly by the choice of the polynomial.

That's why in this study the curvature has been calculated by using the originally measured values, although it was realised that this approach might result in a slight underestimation of the real strains.

After the curvature was estimated, the strain was calculated by the following formula:

$$\varepsilon_i = \frac{S}{2} \left(\frac{1}{R_o} - \frac{1}{R_i} \right) \cdot 100 \qquad (1)$$

where:

ε_i = tangential strain at point i   (%)
s = wall thickness   (mm)
R_i = curvature at point i   (mm)
R_o = radius of undeformed pipe   (mm)

Then these discrete values were fitted by using a cubic spline, as shown in Figure 3, after which the maximum strain value was determined. These maximum values were stored on disc to be used for a statistical evaluation.

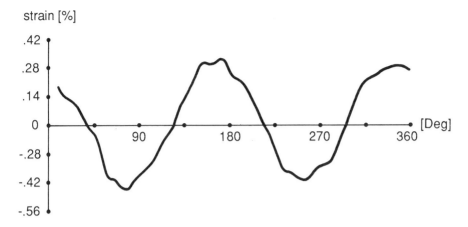

strain [%]

Figure 3: Tangential wall strain as function of rotation angle.

CALCULATION OF DEFLECTION

The deflection of the cross-section of pipes is defined as the relative diameter increase or relative diameter decrease. In the case of buried pipes, the former is generally associated with horizontal pipe deflection, while the latter is associated with vertical deflection. The deflections have been calculated as follows; first the measured values were fitted by using a cubic spline. Then by scanning, the maximum and minimum diameter were determined, from which the deflections were calculated by using the following formula:

$$f_{min} = \frac{D_{nom} - D_{min}}{D_{nom}} \cdot 100 \qquad (2)$$

$$f_{max} = \frac{D_{max} - D_{nom}}{D_{nom}} \cdot 100 \qquad (3)$$

where:

f_{min} = vertical pipe deflection (%)
f_{max} = horizontal pipe deflection (%)
D_{max} = maximum pipe diameter (mm)
D_{min} = minimum pipe diameter (mm)
D_{nom} = nominal pipe diameter (mm)

The nominal diameter was also calculated from the measurements, by determining the arch length along the circumference.

CALCULATION OF TANGENTIAL STRESS

As mentioned before, the deformation of the pipe was measured before and after digging up. The difference between these two deformations, called the recovery, is a measure for the stress in the pipe just before digging up.
The recovery of the pipe will be immediate, as long as the material can be considered to behave in a linear-elastic way. If so, the time passing between the two measurements is of no importance. PVC, however, does not behave like an elastic material; it behaves visco-elastically. This means, that the recovery will be time-dependent, which makes the calculation of the stress more complex. First of all, the loading condition before and in any case, after installation should be determined.
According to our measurements on a few newly installed pipes, and from the results of others (2), plastics pipes reach their final state of deformation within a few years after installation. From this point, the pipe is in a stage of stress relaxation, so in a constant strain condition (see Fig. 4).
At the time t = 0 in Figure 4, so the time at which the pipe is dug up, the stress in the pipe has decreased to a value of σ_o, due to stress relaxation, under a constant strain of ε_o.
Unloading of the pipe can be modelled by superposition of the opposite value of σ_o at time t = 0. So from that moment on, basically two processes should be considered, the first process being further stress relaxation from σ_o to a lower value in time, and the second process being creep from ε_o to ε_t under a constant stress of σ_o. However, the effect of the first process can be negleted, because the rate of a stress relaxation process that is going on for more than 10 years, approaches zero.
So in summary, the pipe unloaded by digging up can be considered to be in a constant stress condition, under a stress of σ_o, which is equal to the stress in the buried pipe just before digging up. Now the stress in the pipe can be calculated by first measuring ε_o (loaded pipe), and by measuring ε_1 (unloaded pipe) at time t_1. By using the effective creep modulus of elasticity for time t_1 the stress is calculated by using the following formula, considering a plane strain condition:

$$\sigma o = \frac{E(t1)}{1 - v^2} \left(\varepsilon(t1) - \varepsilon(o) \right)$$ (4)

where:

σo = tangential stress   (MPa)
$E(t1)$ = Young's modulus at $t = t_1$ (MPa)
V = Poisson's ratio   (-)
$\varepsilon(t1)$ = strain at t=t1   (-)
$\varepsilon(o)$ = strain at t=0   (-)

In view of the fact that the creep rate rapidly slows down within the first few minutes after unloading, and for reasons of convenience, t1 was chosen at 30 minutes after dig-up.

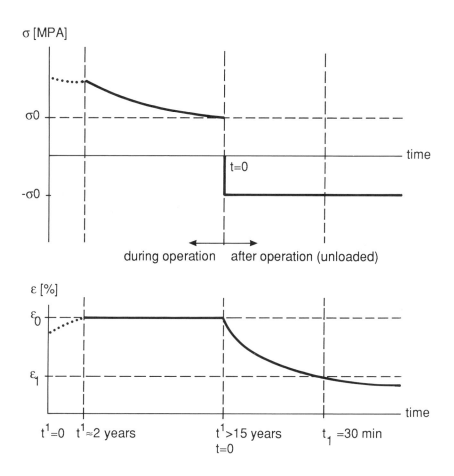

*Figure 4: Stress and strain in a buried pipe
as a function of time.*

STATISTICAL EVALUATION AND CLASSIFICATION

One of the objectives of this study is to get values for the
maximum strain, stress and deflections to be expected in buried
PVC pipes, having been in operation for a long time. Another
objective is to determine the influence of some parameters involved in soil-pipe interaction such as the effect of soil type,
pipe diameter and laying depth on pipe deformation.
The major handicap in achieving this last objective is that there
is no information available about the quality of installation of
the individual pipes that have been in service for such a long
time. From other experimental studies on newly installed pipes (2),
and from a theoretical study (3) the effect of installation was

shown to be the most important one.
Nevertheless, when a lot of data is getting available, then it may
be assumed that careless and careful installation qualities will
be distributed more or less equally among the data.
Classification of the measurements was carried out after the deflections and the strain had been calculated for each cross section.
Plotting the maximum strain values for each cross-section in a normal probability plot clearly showed that the data did not fit such
a distribution.
By applying a Weibull probability distribution the data fit was
much improved (Figure 5). So in order to calculate 95% confidence
levels used for estimating the maximum values to be expected, such
a distribution was used.

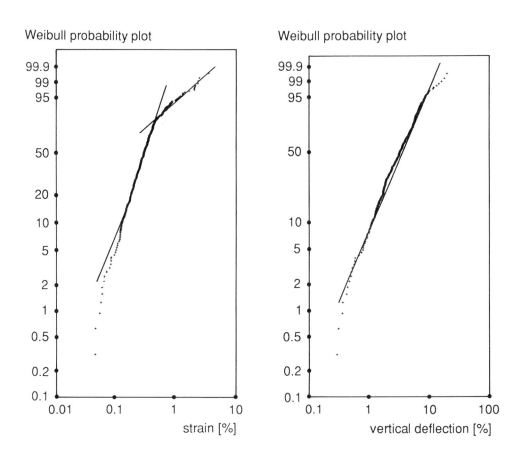

*Figure 5: Results of strain and deflection calculations for
all the measurements.*

In Figure 5 it can be observed that sometimes the measured strain values do not fit to one distribution class, but can be described by two distribution classes. The lower strain distribution class represents the gross soil-pipe interaction process. The higher strain values, however, are caused by local point-loading effects. The stress in the pipe was calculated on the basis of the lower strain values, representing the gross soil-pipe interaction.

EFFECT OF SOIL TYPE AND PIPE DIAMETER

After the measurements were classified according to soil type the maximum values to be expected, based on a 95% confidence level were estimated.
The classification was carried out by using the grain size distribution.
A detailed classification of the type of soil was obtained by using the classification chart of the triangular type as developed by the Bureau of Public Roads in the United States of America.
This resulted in three main soil-type groups. The first group, sand, contains the results of 20 locations. The second group, clay, comprises 12 locations, and the third, miscellaneous group comprises 9 locations.

The results of the measurements and deflection calculations performed on the basis of the foregoing classification are summarized in Table 1.
The results given in this Table are the 95% probability values obtained from a Weibull distribution and by regression analysis.

TABLE 1 -- Results of measurements and calculations for the maximum deflections to be expected

Soil type	Pipe diam. (mm)	Deflection vertical (%)	Deflection horizontal (%)	Number of measurements
Sand	110	10.2	9.9	78
	160	6.9	7.0	31
Clay	110	5.6	7.2	66
	160	6.8	6.9	26
Miscellaneous	110	12.2	10.5	58

The results in this Table indicate that the deflections for pipes having a diameter of 160 millimeter buried in sand are slightly lower than the deflections for the 110 mm pipes. Furthermore, the pipes embedded in clay show lower values, than those buried in sand.

The results of strain and stress calculations are summarized in Table 2.

For the stress calculation an effective Young's modulus (30 minutes) of 3030 MPa obtained from a creep test and a Poisson's ratio
of 0.35 were used.

TABLE 2 -- Results of stress and strain calculations
(maximum values to be expected)

Soil type	Pipe diam. SDR 41 (mm)	$\varepsilon t(o)$ (%)	$\varepsilon t(30)$ (%)	σ (MPa)	m (-)
Sand	110	0.61	0.37	8.4	4.5
	160	0.40	0.26	4.8	4.6
Clay	110	0.50	0.30	6.8	5.2
	160	0.43	0.23	6.9	4.3
Miscellaneous	110	0.52	0.39	4.5	8.8

Table 2 also summarises the so-called shape factor m. This
value is obtained using the following formula:

$$\varepsilon(o)_{max} = m. f_{max} \cdot \frac{S}{D} \qquad (5)$$

where:

$\varepsilon(o)$ = maximum tangential strain (%)
m = shape factor (-)
f_{max} = maximum deflection (%)
S = wallthickness (mm)
D = nominal outside diameter (mm)

The m-value was obtained by applying formula 5 for each cross-
section measured. After sorting of the data into the defined
groups, the m-value was estimated for each group by linear regression between ε and $f_{max} \cdot \frac{S}{D}$.
In case of an elliptical deformation, the value of m is 3. In
design codes a value of 6 is often proposed (4).
As can be seen from the Table, this value is conservative for
most cases.
The results in Table 2 clearly show that for pipes buried in sand
strain and stress have the highest values for pipes having a diameter of 110 mm.
This may be caused by the fact that horizontal support of the soil
besides the pipe depends on the absolute deflection of the pipe and
not on the relative deflection. A larger pipe size will have the
same absolute horizontal pipe deflection at much lower relative deflection values than smaller-sized pipes.
The pipe diameter effect on strain is considerably lower for pipes
buried in clay than for pipes embedded in sand.
For the pipes buried in clay, however, no effect of pipe diameter
on stress has been observed. This means that in clay there is less

horizontal support than in sand (clay behaves more viscously than sand).
On the other hand, clay will also create, by its viscous behaviour, a more uniform radial pressure along the outside pipe circumference. This results in rather low pipe deflections, and stress and strain levels.
The m-value is rather high, which is caused by the effect of point-loading.
The stress values given in Table 2 have been calculated using the recovery behaviour of the pipe after digging-up.
In this calculation a short-term Young's modulus has been used. If it is assumed, however, that the dug-up pipe is already in a constant strain situation immediately after installation the stress in the pipe can also be calculated using the strain of the buried pipe and a long-term (30 years) relaxation modulus and Poisson's ratio, which are 1300 MPa and 0.15, respectively.
The results of these calculations compared to the results of Table 2 are summarized in Table 3.

TABLE 3 -- Results of stress calculations according to the "recovery" method (30 min.) and the "stress relaxation" method (30 years)

Soil type	Pipe diameter mm	Max. values to be expected	
		σo-30 min MPa	σo-30 year MPa
Sand	110	8.4	8.1
	160	4.8	5.3
Clay	110	6.8	6.6
	160	6.9	5.7
Miscellaneous	110	4.5	6.9

The calculations using a stress relaxation process over a period of 30 years show that the stresses are quite close to the values based on calculations using a short-term modulus (recovery method).
This supports the idea that buried pipes can be considered to be in a constant strain situation.
Otherwise the latter calculation method (30 years' modulus) would have resulted in clearly lower stress values.

CONCLUSIONS

The results of measurements and calculations carried out on 30-years-old buried uPVC pipes ($\emptyset$ 110 mm, 160 mm, SDR 41) show that soil-pipe interaction cannot be described by one process only. Real (macroscopic) soil-pipe interaction results in relatively low de-

flections, strains and stresses. But a second process may also occur, namely pipe-structure interaction, like point loading effects.
It was shown that the strain can be estimated from the maximum deflection by using a certain value for the shape factor m. A value of 6 for m proved to be sufficient for uPVC pipes (SDR 41) as shown in this study and by others (4).

Furthermore the results show that for pipes buried in sand the outstanding behaviour of plastic pipes is mainly caused by the continuous horizontal support of the soil besides the pipe.
For pipes buried in clay, however, the viscous behaviour of the soil, especially at high moisture levels, is of prime importance. This results in a more or less hydrostatic loading of the pipe.

Stress calculations indicate that the buried pipes can be considered to be in a state of constant strain shortly after installation. This means that shortly after installation the pipe will have the highest stresses; these stresses will decrease in time by stress relaxation.
These stresses calculated are below the critical stress values at which failure or stress cracking may occur.
Although the pipes have been buried under rather severe circumstances e.g. having low soil covers and subjected to traffic loading, they do not deform excessively nor do they develop high stresses.

The differences found between the various soil groups and pipe diameters should be interpreted carefully, because installation quality, which is usually unknown, may be the most important parameter. Therefore, it was decided to perform a number of well-controlled tests on some test sites. The important parameters, like soil type, burial depth, installation quality, etc. will be controlled as much as possible. It is expected that results of these measurements will be available within a few years.

REFERENCES

(1) Alferink, F.J.M. and Wolters, M.
    Deformations in buried flexible pipes
    International Conference Plastics Pipes VII,
    University of Bath, September 1988, UK, pp. 27/1 - 27/10.

(2) De Putter, W.J. and Elzink, W.J.
    Deflection studies and design aspects on PVC sewer pipes
    Int. Conf. on Underground Plastics Pipes, ASCE, New Orleans,
    USA, 1981.

(3) Alferink, F.J.M., Munting, T.G., Oranje, L. and Wolters, M.
    Stresses and strains in buried flexible pipes
    Eleventh Plastic Fuel Gas Pipe Symposium,
    San Francisco, October 1989, USA, pp. 406-420.

(4) Janson, L.E.
    Plastic pipes for water supply and sewage disposal
    Stockholm, 1989, pp. 93.

Harold Kennedy Jr., Dennis D. Shumard, and Cary M. Meeks

THE DESIGN OF UNDERGROUND THRUST RESTRAINED SYSTEMS
FOR POLYVINYL CHLORIDE (PVC) PIPE

REFERENCE: Kennedy, H. Jr., Shumard, D. D., and
Meeks, C. M., "The Design of Underground Thrust
Restrained Systems for Polyvinyl Chloride (PVC)
Pipe", Buried Plastic Pipe Technology, ASTM STP
1093, George S. Buczala and Michael J. Cassady,
Eds., American Society for Testing Materials,
Philadelphia, 1990

ABSTRACT: Direct shear tests were performed on soil
to polyvinyl chloride (PVC) pipe surfaces to study
pipe-to-soil friction. Soil type and moisture
content were varied. The resulting data were used
to formulate design parameters for PVC pipe thrust
restrained systems installed in a wide range of
soil types. Soil types are identified by ASTM
D2487, "Classification of Soils for Engineering
Purposes". Design methods, equations, and
recommendations are presented for horizontal bends,
tees, vertical offsets and other combinations of
pipe and fittings.

KEYWORDS: thrust restraint, polyvinyl chloride
(PVC) pipe restraint, pipe-to-soil friction,
polyvinyl chloride (PVC)-to-soil shear strength

INTRODUCTION

During the fall and winter of 1988-1989 approximately
300 direct shear tests were conducted on soils and
pipe-to-soil surfaces to study the friction, i.e., the shear
resistance, at the interface between the pipe and soil. The
major goal of these tests was to study the interface shear
strength of polyvinyl chloride (PVC) pipe and various soils.
The results of the tests were used to formulate reasonable

Mr. Kennedy is vice president - research and
development and Messrs. Shumard and Meeks are product
engineers at EBAA Iron, Inc., Eastland TX 76448.

245

pipe-to-soil friction parameters. These parameters were then used in existing equations to formulate design recommendations for restrained pipelines. The study has led to the publication of a thrust restraint design handbook for PVC pipe. The design equations and recommendations used in the handbook are included in this paper.

BACKGROUND

PVC pipelines utilize joints with rubber gaskets for connections between individual pipe and also pipe-to-fittings. An unbalanced force exists at any change in direction in the pipeline. It is necessary to restrain the pipeline at changes in direction to prevent joint separation. This restraint is often provided by devices designed specifically to prevent individual joints from separating. Thrust restraint design involves calculation of the length of pipeline to be provided with joint restraint on each side of a change in direction.

Many papers have been published on the design of thrust restraint for pipelines. Thrust blocks as well as restrained joint systems have been studied analytically resulting, for the most part, in conservative design approaches. The design approach presented here is the result of model studies on a restrained joint system in 1969-1970 [1], full scale tests on 12" horizontal bend systems in 1981-1982 [2], and subsequent analytical work. The 12" tests on 90 and 45 degree bends were used to evaluate numerous design equations. Those which were obviously inadequate were eliminated from consideration. The resulting design approach is a modification of the equation proposed by Rodger Carlsen [3].

The soil parameters for the design procedure were selected from design values recommended by recognized soil engineering authorities [4]. The pipe-to-soil friction values were taken from a publication of direct shear tests done on steel surfaces by Potyondy [5]. These values have proven very conservative for the design of hundreds of installations since 1982 [6].

During 1987-1988, the need for a thrust restraint design procedure for PVC pipe became increasingly apparent. Since no test data were available on PVC surfaces in contact with soils, it was decided to perform a series of direct shear tests on actual pipe-to-soil interfaces, to evaluate the friction, and to determine realistic values for design.

THEORY

The friction force acting to oppose movement of a pipeline is the shearing strength of the pipe-to-soil interface and is related to the internal shear strength of the soil. The internal shear strength of the soil can be expressed by the Coulomb equation below.

$$S = C + N \, Tan(\phi) \qquad (1)$$

where S = shear strength of the soil,
       C = cohesion intercept of the soil,
       N = normal force,
       ϕ = angle of internal friction of the soil.

The pipe-to-soil interface shear strength can also be expressed in the Coulomb form, similar to Potyondy [5].

$$Sp = (Fc)C + N \, Tan(F\phi\phi) \qquad (2)$$

where Sp = shear strength of the pipe-to-soil interface,
       Fc = proportionality constant relating the cohesion intercept of a direct shear test on the pipe-to-soil interface and the cohesion intercept of the soil itself,
       $F\phi$ = proportionality constant relating the friction angle of a direct shear test on the pipe-to-soil interface and the friction angle of the soil itself,
       C = cohesion intercept of the soil (zero intercept of a plot of shear strength vs. normal force),
       ϕ = Angle of internal friction of the soil,
       N = Normal force (force acting perpendicular to the plane of shearing).

The determination of the proportionality constants Fc and Fϕ for design purposes was the object of this study. The proportionality constant Fc was found by dividing the pipe-to-soil interface cohesion intercept by that of the soil at the same moisture content. Likewise, the Fϕ constant is found by dividing the friction angle of the pipe-to-soil interface by the friction angle of the soil at the same moisture content.

TEST PROCEDURE

Potyondy and others have performed steel-soil interface shear tests in small direct shear boxes using flat samples. However, in order to flatten a PVC pipe sample, the surface would be damaged. It was therefore necessary to test actual pipe surfaces. A large box, 12.7 cm x 12.7 cm (5 in.x 5 in.), was used for the direct shear tests on the soil and an upper section was constructed to fit the curvature of a

30.48 cm (12 in.) pipe. This allowed actual surfaces of
30.48 cm (12 in.) C900 PVC pipe to be mounted in the same
loading apparatus as the direct shear test on the soil. The
same normal and axial loads were used in the direct shear
tests on the soil and pipe-to-soil interface.

Preliminary tests were conducted on several different
mixtures of soil components and pipe-to-soil interfaces to
determine general parameters. During this preliminary
testing it was determined that the shear strength of the
pipe-to-soil interface varied considerably with moisture
content and soil type. Although shear strength also varied
with initial density of the soil, it was decided to
eliminate loose soil from the investigation because the
reduction in passive resistance of the soil is drastic when
loose soils are encountered. Dumped, unconsolidated, native
soils are not recommended as bedding for restrained joint
pipe systems. It was therefore decided to study only soils
consolidated above their critical densities (below critical
void ratio) with identical compactive effort applied.

After the preliminary study it was decided to use a
stress controlled procedure on both the soil and
pipe-to-soil interface. All samples were consolidated to an
estimated density above 80% Standard Proctor. While the same
procedure was used to compact each sample, no attempt was
made to measure the actual density of the soil being
tested.

SOIL

Soils were classified by Atterberg limits and sieve
analyses using the Unified Soil Classification System, ASTM
D2487, "Classification of Soils for Engineering Purposes".
Soil type was varied by mixing a uniform fine grained sand
with a manufactured soil readily available. This
manufactured soil was composed of sodium montmorillonite
and calcium montmorillonite combined with a small percentage
of wood flour and sea coal. Both the sand and the clay were
readily available because they were used in the foundry
process. Mixing the two components yielded a wide range of
soil properties.

Five mixed soils and two natural soils were tested.
Plasticity Index ranged from zero (assumed for sand) to 138
for the 100% montmorillonite, wood flour, sea coal mix.
Classification by Atterberg limits of all mixtures were
located slightly above but in close proximity to the "A"
line in ASTM D2487. The manufactured soils tested were
classified SP, SC, and CH. The natural soils were classified
as CL.

TEST SPECIMENS AND DATA AQUISITION

Test specimens were prepared by mixing the two dry components in various percentages by weight and adding the prescribed water content. The mixture was used in a series of direct shear tests on the soil to determine the angle of shearing resistance and the cohesion intercept. The same soil was then used in tests on the pipe-to-soil interface, using the same normal loads with the curved shear box described above. After completing the multiple shear tests at one moisture content, a sample was taken for the determination of the actual moisture content of the specimen. The moisture content was then varied in the next soil sample. A minimum of three and as many as five different moisture contents were tested for each soil mixture.

Peak values of shear strength were recorded for each direct shear test. Individual shear tests were conducted for a minimum of three normal loads for each soil and soil-to-pipe interface. The normal loads ranged from 6.9 $\times 10^{-3}$ to 2.8×10^{-2} MPa (1 to 4 psi). This range of vertical loading corresponds to the vertical prism load on a pipe at ordinary depths of cover. The peak values of shear stress were used to plot the Mohr-Coulomb failure line, allowing the determination of the friction angle and cohesion intercept for the soil and pipe-to-soil interfaces under near identical conditions.

Values of the friction angle of the soil as well as the friction angle for each pipe-to-soil interface were determined for each moisture content. A similar determination was made for the cohesion intercept for the soil and pipe-to-soil interfaces. Plots were made of both friction angle and cohesion intercept versus moisture content for the soil and pipe-to-soil interfaces. Best fit curves were made for the data.

RESULTS

For sand, the variation of friction angle with moisture content was small. For sand-clay mixtures and 100% clay the variation with moisture content was large and therefore very important. The friction angles increased to a maximum and then decreased. Decrease in the friction angles continued until saturation was reached, i.e., the friction angle of the soil reached zero. An example can be seen in Figure 2.

The cohesion intercept for the soil and for the pipe-to-soil interface also varied with moisture content. However, a cohesion intercept remained, even when the soil reached saturation.

Figures 1 through 4 are plots of friction angle vs.

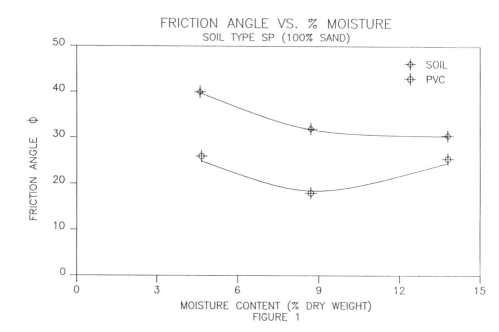

FRICTION ANGLE VS. % MOISTURE
SOIL TYPE SP (100% SAND)

FIGURE 1

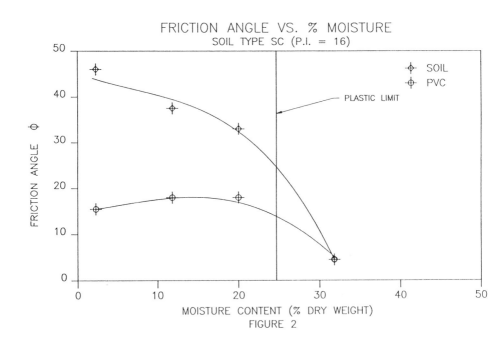

FRICTION ANGLE VS. % MOISTURE
SOIL TYPE SC (P.I. = 16)

FIGURE 2

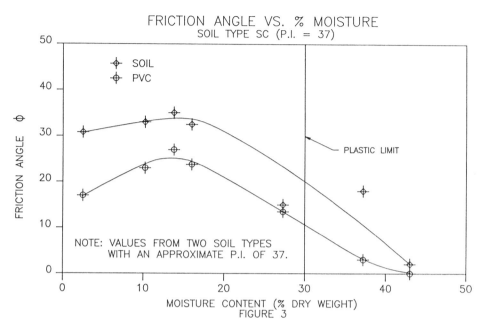

FRICTION ANGLE VS. % MOISTURE
SOIL TYPE SC (P.I. = 37)

PLASTIC LIMIT

NOTE: VALUES FROM TWO SOIL TYPES
WITH AN APPROXIMATE P.I. OF 37.

MOISTURE CONTENT (% DRY WEIGHT)
FIGURE 3

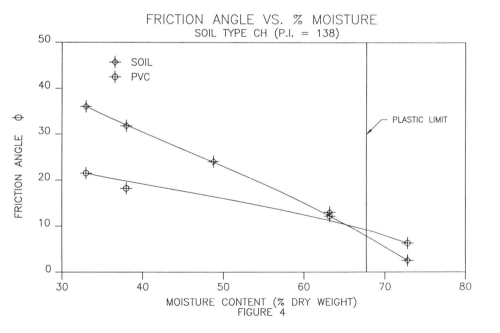

FRICTION ANGLE VS. % MOISTURE
SOIL TYPE CH (P.I. = 138)

PLASTIC LIMIT

MOISTURE CONTENT (% DRY WEIGHT)
FIGURE 4

moisture content for both the soil and the PVC pipe-to-soil interfaces. Note that the friction angle of the pipe-to-soil interface generally follows the pattern of the friction angle of the soil.

Although the friction angle was very important, it was not the only component of the interface shear strength. The cohesion intercept plays an important role, especially in soils with a significant clay content. The cohesion intercept of the soil, and soil-to-PVC pipe vs. moisture content were plotted and are shown in Figures 5 through 8. Note that the cohesion intercept of the interface did not follow that of the soil as closely as the friction angle.

The moisture content in the tests on the fine sand induced a small amount of apparent cohesion. Although very small compared to the sand-clay mixtures, the value was recorded and plotted along with the other data.

Friction angles and cohesion intercepts were determined for each soil type at its plastic limit. Friction angle at the plastic limit was then plotted versus plasticity index on semi-log paper along with effective friction angles generated by Kenney [7], for natural clays. As can be seen in Figure 9, the data was near the lower limit of values obtained for these naturally occurring clays and encompassed almost the complete range of plasticity index for these clays. Another comparison of interest was made. While Potyondy's mixed soils encompassed a much smaller range of plasticity index, the range of friction angles in this study was almost identical.

It was concluded that the measured values of friction angle for the soil and pipe-to-soil interface, yield conservative values for design when taken at moisture contents equivalent to the plastic limit.

It should be noted that many clays (below the water table) exist in their natural state with moisture contents considerably above the plastic limit. Under these conditions, the soil shear strength becomes independent of normal load. For pipelines in these areas special considerations have to be made.

In these cases the " $\phi=0$ principle" must be used and the undrained shear strength of the soil should be substituted for cohesion intercept. Restrained pipelines in these soils should be bedded in granular material.

DETERMINATION OF DESIGN VALUES

Soils were grouped using the Unified Soil Classification System into six catagories ranging from coarse grained to fine grained soils, covering inorganic soils only. Heavy clays and heavy silts as well as organic

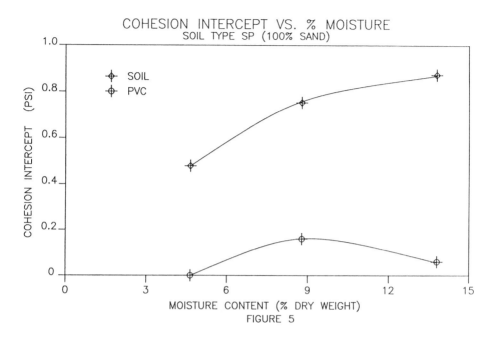

COHESION INTERCEPT VS. % MOISTURE
SOIL TYPE SP (100% SAND)

FIGURE 5

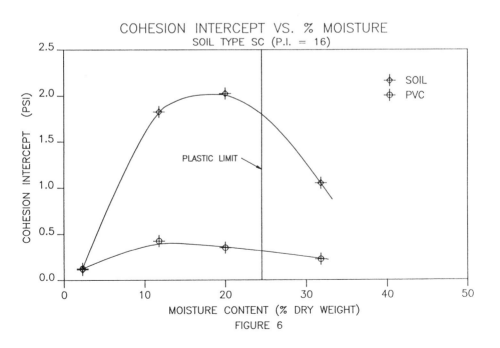

COHESION INTERCEPT VS. % MOISTURE
SOIL TYPE SC (P.I. = 16)

FIGURE 6

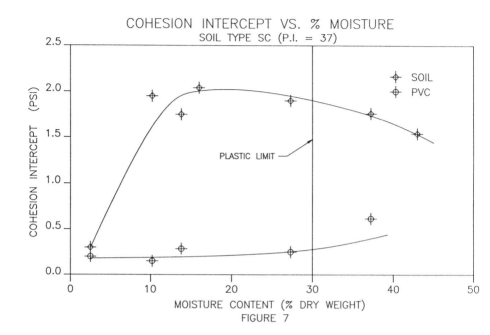

FIGURE 7

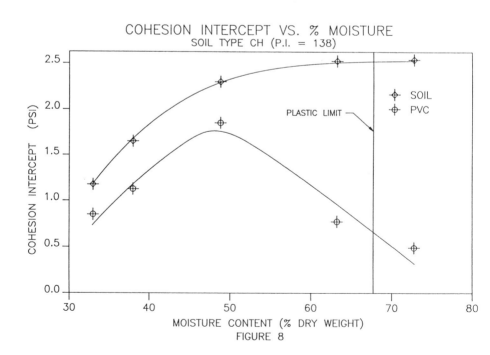

FIGURE 8

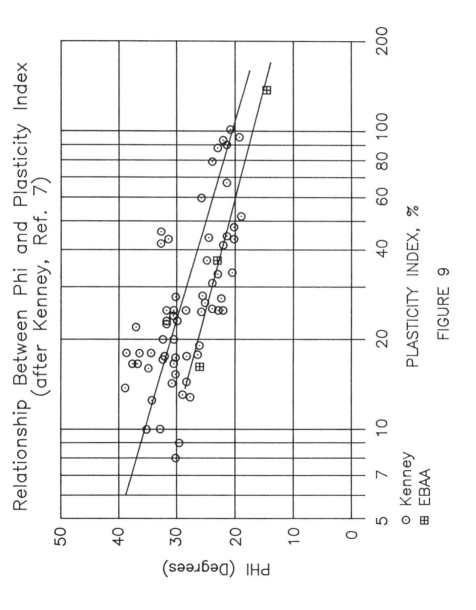

Relationship Between Phi and Plasticity Index (after Kenney, Ref. 7)

PHI (Degrees)

PLASTICITY INDEX, %

⊙ Kenney
⊞ EBAA

FIGURE 9

clays, organic silts and peat were classed as not recommended for bedding restrained pipe.

Values of ϕ for granular, noncohesive soil in the lower range of recommended design values of medium dense soils given by B.K. Hough, 1957 [4] were selected as design values for the suggested procedure.

Values of Fϕ and Fc were determined from the results previously described. Fϕ and Fc values for design were based on the general soil classifications of the soils tested. The values for silts from Potyondy's work for steel were used as a guide for PVC. This seemed to be a general relationship for the other soils as well.

Friction angles of the PVC pipe-to-soil interface were less than the internal friction angle of the soil. The values of Fϕ were taken at the plastic limit for design recommendations in unsaturated soils, i.e., above the water table. Design values of Fϕ were 0.7 for all granular soils; 0.6 for GC, SC, and ML soils, and 0.5 for CL soils.

Values of Fc were determined for moisture contents equivalent to the plastic limit of the soils. Design values for PVC pipe were 0.2 for GC and SC soils and 0.3 for CL soils.

As stated above, the friction angle of the PVC pipe-to-soil interface went to zero in saturated fine grained and cohesive soils. Therefore the design procedure recommends that restrained pipelines in these soils should be bedded in granular material. The value of interface shear strength should be calculated using the bedding material properties. The passive resistance should be calculated using the in situ soil properties. The "ϕ= 0 principle" applies and the undrained shear strength of in situ soil is substituted for cohesion.

THE USE OF RESTRAINED JOINTS.

In the case of a horizontal bend, as shown in Figure 10, the resultant thrust force is given by the equation:

$$T = 2PA \sin(\theta)/2 \qquad (3)$$

Where: T = unbalanced force at the bend,
       P = internal pressure,
       A = cross sectional area of the pipe based on the outside diameter,
       θ = angle of bend.

By restraining certain joints at bends and along the pipeline, the resultant thrust force is transferred to the surrounding soil by the pipeline itself. In a properly

designed pipeline using restrained joints the bearing
strength of the soil and the frictional resistance between
the pipe and the soil balance the thrust force.

RESISTANCE TO THE UNBALANCED THRUST FORCE, T.

Resistance to the unbalanced thrust force, T, is
generated by the passive resistance of the soil as the bend
tries to move, developing resistance in the same manner as a
concrete thrust block. In addition to the passive
resistance, friction between the pipe and soil also
generates a considerable resistance to joint separation.

Figure 10, is a free body diagram of a restrained
horizontal pipeline-bend system, designed to resist the
unbalanced thrust created by the change in direction.
Notice that the thrust, T, is resisted by the passive
resistance, Rs, as well as the frictional resistance, Fs,
along a length, L, on each side of the bend. L is the
required length of pipe to be restrained. Note that every
joint, within L, whether pipe joint or fitting joint, must
be restrained on both sides of the bend. On small diameter
pipelines L is often less than a full length of pipe and
therefore, with planning, only the fitting has to be
restrained.

As shown in Figure 10, the following equation can be
used to calculate L.

$$L = Sf \frac{(PA) \tan(\theta/2)}{Fs + 1/2Rs} \qquad (4)$$

Where: Fs = pipe-to-soil friction,
          Rs = bearing resistance of the soil along the
               pipeline,
          θ = angle of bend,
          Sf = Safety factor.

Pipe-to-soil Friction, Fs

The friction force, acting to oppose movement of the
pipeline is the shear strength of the pipe-to-soil
interface. For design purposes, the pipe-to-soil friction,
Fs, can be found using equation 5.

$$Fs = Ap(FcC) + W \tan(F\phi\phi) \qquad (5)$$

Where:
          Ap = area of pipe surface bearing against the soil,
               (In the case of horizontal bends, 1/2 the pipe
               circumference),
          W = normal force per unit length
          W = 2We + Wp + W
          We = vertical load on the top and bottom surfaces of

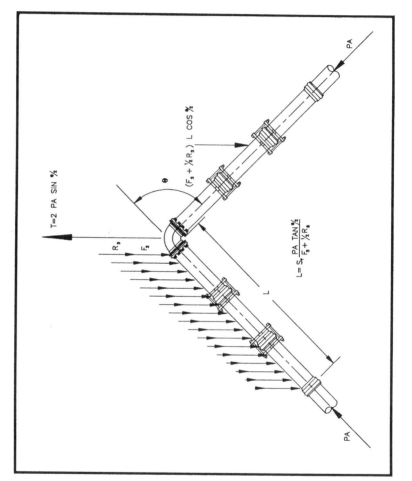

FIGURE 10   FREE BODY DIAGRAM OF A RESTRAINED HORIZONTAL BEND SYSTEM.

the pipe, taken as the prism load,
Wp + Ww = weight of pipe plus weight of water
     Fc = proportionality constant relating the cohesion
          intercept of the pipe-to-soil shear strength
          and the cohesion intercept of the soil shear
          strength
     Fϕ = proportionality constant relating the angle of
          pipe-to-soil friction to the friction angle of
          the soil.

Friction values should be based on the soil used for
bedding the pipe and not the native soil unless that soil is
also used as bedding material.

Bearing Resistance, Rs

In addition to friction along the pipe, the resultant
thrust is also resisted by the passive pressure of the soil
as the pipe tends to move into the surrounding soil. The
passive pressure of the soil is generated by the movement.
The maximum resistance to this movement can be calculated
with the Rankine passive pressure formula. The amount of
movement required to generate the resistance depends upon
the compressibility of the soil. In general, soils having a
Standard Proctor Density of 80% or greater require very
little movement to generate the maximum passive resistance
of the soil. However, because the compressibility of the
soil can vary greatly between the prescribed trench types,
the design value of passive pressure should be modified by
an empirical constant, Kn. Considerable ecconomies can be
obtained, in the number of restrained joints required, by
specifying Trench Type 4, or 5, (Fig. 11) for restrained
joints.

Rankine's passive pressure formula:

$$Pp = Hc\ Np + 2Cs(Np)^{1/2} \qquad (6)$$

Where: Pp = passive soil pressure,
            = soil density (backfill density for loose soil
              bedding and native soil density for compacted
              bedding),
         Hc = Mean depth from the surface to the plane of
              resistance (centerline of pipe),
         Cs = cohesion of soil,
         Np = $Tan^2(45 +\phi/2)$,
         ϕ = internal friction angle of the soil.

Therefore: Rs = KnPpD   (7)

Where: Rs = bearing resistance of soil,
           Kn = trench constant,
           D = Pipe outside diameter.

DESIGN VALUES FOR SOIL PARAMETERS

TABLE 1
PROPERTIES OF SOILS USED FOR
BEDDING TO CALCULATE Fs and Rs

Soil Group[a]	Fc	C[b]	Fϕ	ϕ	γ[c]	Kn Trench Type (Fig.11)		
	(psf)				(pcf)	3	4	5
GW & SW	0	0	0.7	36	110	.60	.85	1.00
GP & SP	0	0	0.7	31	110	.60	.85	1.00
GM & SM	0	0	0.6	30	110	.60	.85	1.00
GC & SC	.20	225	0.6	25	100	.60	.85	1.00
CL	.30	250	0.5	20	100	.60	.85	1.00
ML	0	0	0.6	29	100	.60	.85	1.00

[a] Note: Soil Classification symbols are those specified in ASTM D2487. [b] 47.88 psf = 1 N/M^2. [c] 16.0184 psf = 1 K/M^3.

Note: Soils in the CL and ML groups must be monitored closely since moisture content is difficult to control during compaction. Free draining soils are much better pipe bedding. Soils in the MH. CH, OL, OH, and PT groups are not generally recommended for pipe bedding.

Pipelines laid in highly plastic soils subject to high moisture contents are usually bedded in granular material. If the bedding material has a higher bearing value than the native soil the value of Fs should be calculated using the bedding and the value of Rs should be based on the native soil. The values below are for saturated in situ soil, types CL, ML, CH and MH with the pipe bedded in sand or gravel with a minimum Standard Proctor density of 80% or greater. Undrained shear strength values should be used for cohesion in the ϕ = O principle. Actual values of the vane shear test (AASHTO T223 -76, "Field Vane Shear Test in Cohesive Soil"), unconfined compression test (ASTM D2166, "Test Methods for Unconfined Compressive Strength of Cohesive Soil"), or the standard penetration test (ASTM D1586, "Method for Penetration Test and Split-Barrel Sampling of Soils") should be used when available. A competent soils engineer should be contacted for pipelines in wetlands, river bottoms, etc.

TABLE 2
IN SITU VALUES OF SOIL PROPERTIES FOR Rs

SOIL GROUP	C=Su[b]	γ[c]	Kn Trench Type (Fig.11)		
	(psf)	(pcf)	3	4	5
CL	450	100	.60	.85	1.00
CH	400	100	.40	.60	.85
ML	300	100	.60	.85	1.00
MH	250	100	.40	.60	.85

(Note: ϕ = 0 principle for undrained shear, C = Su = undrained shear strength of soil.   is for undisturbed soil.) [b] 47.88 psf = 1 N/M^2. [c] 16.0184 psf = 1 K/M^3.

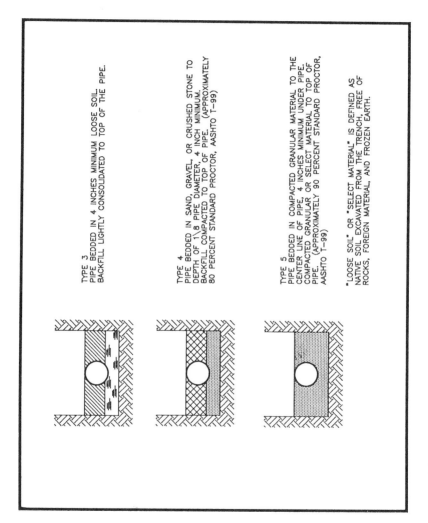

TYPE 3
PIPE BEDDED IN 4 INCHES MINIMUM LOOSE SOIL.
BACKFILL LIGHTLY CONSOLIDATED TO TOP OF THE PIPE.

TYPE 4
PIPE BEDDED IN SAND, GRAVEL, OR CRUSHED STONE TO
DEPTH OF 1\8 PIPE DIAMETER, 4 INCH MINIMUM.
BACKFILL COMPACTED TO TOP OF PIPE. (APPROXIMATELY
80 PERCENT STANDARD PROCTOR, AASHTO T-99)

TYPE 5
PIPE BEDDED IN COMPACTED GRANULAR MATERIAL TO THE
CENTER LINE OF PIPE, 4 INCHES MINIMUM UNDER PIPE.
COMPACTED GRANULAR OR SELECT MATERIAL TO TOP OF
PIPE. (APPROXIMATELY 90 PERCENT STANDARD PROCTOR,
AASHTO T-99)

"LOOSE SOIL" OR "SELECT MATERIAL" IS DEFINED AS
NATIVE SOIL EXCAVATED FROM THE TRENCH, FREE OF
ROCKS, FOREIGN MATERIAL, AND FROZEN EARTH.

FIGURE 11 BEDDING FOR RESTRAINED PIPE SYSTEMS.

TABLE 3,
BEDDING SOIL PROPERTIES FOR Fs

	Fc	C	Fϕ	ϕ
GP & SP	0	0	0.70	31

(Bedding assumed to be granular soil of at least the GP and
SP groups consolidated to a minimum of 80% Standard Proctor
Density.)

OFFSETS, TEES, REDUCERS, DEAD-ENDS, AND BEND COMBINATIONS

Offsets

Offsets made with restrained bends and pipe are
becoming more and more commonplace. Some precautionary
notes are warranted here. As stated for horizontal bends,
the resultant thrust at a bend varies greatly with the angle
of the bend. This thrust varies from 1.414(PA) for 90
degree bends to .398 (PA) for a 22-1/2 degree bend, a
reduction of 72%. Vertical down bends and offsets should be
made with small angle bends if possible. All piping systems
move slightly at bends as pressure is brought to bear.
Therefore the joint restraint systems should be resilient,
responding to that movement in a manner to prevent the
concentration of load in a small area of the bend. If a
system is used such that deflection of the fitting causes
load concentrations, the end result will be a highly
stressed fitting bell. This can and has resulted in failure
of the fitting or pipe. Thrust restraint systems are
available for this situation, responding to deflection by
continually redistributing the load around the circumference
to protect the pipe and fitting.

Therefore use a resilient restraint system with as
small angle bends as possible. Restrain all of the joints
in the offset and use equation 8 to determine the length,
L, on the high side of the vertical down bend and equation 9
for the length, L, on the low side of the vertical offset.

High side of vertical offset:

$$L = Sf (PA \tan \theta/2)/Fs \qquad (8)$$

Low side of vertical offset:

$$L = Sf (PA \tan \theta/2) / (Fs + 1/2Rs) \qquad (9)$$

Tees

Tees can be restrained utilizing the pipe-to-soil
friction along the branch and the passive resistance of the

soil on the run, on each side of the tee. Since the equation involves two "L's", the simplest method is to specify that at least a distance, Lr, be installed on each side of the run of the tee using a single length of pipe without joints. With this amount specified, the amount of pipe to be restrained on the branch can be easily calculated. In some branch reducing tees, wherein the run size is much larger than the branch, it is unnecessary to restrain the run joints on the tee.

$$Lb = Sf\ (PAb - RsLr)/Fsb \qquad (10)$$

Where:

Lb = Length of restrained pipe on the branch,
Ab = Cross sectional area of the branch pipe,
Lr = single pipe length on each side of the run of the tee,
Fsb = Frictional resistance along the branch using the entire circumference of the pipe for Ap,
Sf = Safety factor.

Reducers

An unbalanced force exists at reducers caused by the differential area between the large and small pipes. The thrust can be resisted by the frictional resistance of either the large side of the reducer or the small side of the reducer. If a specified length, Ls, on the small side is free of bends, valves, tees or other fittings the small pipe will take the load in compression and no restraint is necessary. However, if, Ls is not free of unbalanced thrust then restraint must be used in the large pipeline side for a length, Ll.

$$Ls = Sf\ P(Al-As)/Fsb \qquad (11)$$

Where:

Ls = Length of straight, unobstructed pipe on the small side of the reducer.
Al = Area of large pipe,
As = Area of small pipe,
Fsb = Frictional resistance based upon the entire circumference of the small pipe,
Sf = Safety factor.

$$Ll = Sf\ P(Al-As)/Fsb \qquad (12)$$

Where:

Ll = Length of restrained pipe on the large side of the reducer,
Fsb= Frictional resistance based upon the entire circumference of the large pipe.

Dead Ends

Dead ends usually require the ability to extend the
pipeline at a later date. The restrained length required can
be found from the following equation. However in some larger
pipelines it may be more economical to pour an axial thrust
collar across the trench. Please note that the extreme end
thrust caused by a dead end cannot ordinarily be restrained
by a concrete thrust collar without a thrust collar
attachement to the pipe.

The restrained length is:

$$L = Sf \ (PA)/Fsb \qquad (13)$$

Where:
L = restrained length
Fsb = Frictional resistance based on full
circumference of the pipe.

Bend Combinations

Often, two or more bends are located in close
proximity to each other. For example, a 90 degree
bend located close to a 45 degree bend results in a
change in direction of flow of 135 degrees. The
resultant thrust is much greater for the combination
than for either of the fittings considered separately
and consequently requires a much greater length to be
restrained. Therefore, when possible, no bend should
be located within the calculated length, L, of any
other bend or combination.

CONCLUSIONS

Pipe-to-soil interface shear strength is a major
component of the design of thrust restraint systems. Its
value depends upon soil type, moisture content, soil
compaction, and surface roughness. The pipe-to-soil
interface shear strength of actual pipe samples was found to
be considerably different from the original assumptions made
for design.

REFERENCES

[1]   Kennedy, H. Jr. and Wickstrom, W.; "Model
       Studies on Number of Pipe to be Restrained at 90
       Degree Bend and 45 Degree Bend", Unpublished
       Report, United States Pipe and Foundry Company.,
       February 9, 1970.

[2]   Kennedy, H. Jr. and Conner, R. C.; "Thrust Restraint of
      Buried Ductile-Iron Pipe at Horizontal Bends", American
      Cast Iron Pipe Co., Presented at AWWA Distribution
      Systems Symposium, 1982.

[3]   Carlsen, Rodger J.; Thrust Restraint for Underground
      Piping Systems", Ductile Iron Pipe News, CIPRA, Spring
      1975.

[4]   Lambe, T. William and Whitman, Robert V.; Soil
      Mechanics, Series in Soil Engineering, Massachusetts
      Institute of Technology, John Wiley and Sons, New
      York.

[5]   Potyondy, J.G.; " Skin Friction Between Various Soils
      and Construction Materials", Geotechnique, London,
      England, Volume II, No. 4, December, 1961, pp 339-353.

[6]   DIPRA; "Thrust Restraint Design for Ductile-Iron Pipe",
      Second Edition, 1986.

[7]   Kenney, T.C., 1959: Discussion, Proc. ASCE, Vol. 85,
      No, SM3, pp. 67-79.

BIBLIOGRAPHY

Kennedy, H. Jr., Shumard, D. D., and Meeks C. M.; "PVC Pipe
Thrust Restraint Design Handbook"; EBAA Iron Inc., P.O. Box
877, Eastland TX 76448.

Kennedy, Harold Jr., Shumard, Dennis D., and Meeks, Cary M.;
"Investigation of Pipe-to-soil Friction and its affect on
Thrust restraint Design for PVC and Ductile-Iron Pipe";
Proceedings of 1989 Distribution Systems Symposium; American
Water Works Association, 6666 West Quincy Avenue, Denver CO
80235.

UNI-BELL Plastic Pipe Association; Handbook of PVC Pipe,
Design and Construction, Dallas, Texas.

Larry J. Petroff[1]

REVIEW OF THE RELATIONSHIP BETWEEN INTERNAL SHEAR RESISTANCE
AND ARCHING IN PLASTIC PIPE INSTALLATIONS

REFERENCE: Petroff, L. J., "Review of the Relationship Between
Internal Shear Resistance and Arching in Plastic Pipe
Installations," Buried Plastic Pipe Technology, ASTM STP 1093,
George S. Buczala and Michael J. Cassady, Eds., American
Society for Testing and Materials, Philadelphia, 1990.

ABSTRACT: A soil mass shifts load from its weaker components
to its stronger components such that the load is distributed
in proportion to the stiffness of the components. This
characteristic often results in the shifting of load around
weak zones or cavities in the soil. Such redistribution of
stresses in a soil mass is called arching. Arching is usually
thought of in regard to pipe and tunnels. However, arching is
a consequence of a more general property of soil known as
internal shearing resistance. Slope stability, the bearing
capacity of shallow footings, and the pressure distribution on
retaining walls depend on the internal shear resistance. This
paper discusses internal shearing resistance, the movement
required to mobilize it, its measurement, typical values for
common trench soils, and how it relates to the load applied to
a buried pipe.

KEYWORDS: arching, buried pipe, shear resistance, earth
pressure

The past quarter century has seen the introduction and widespread
use of plastic pipes. Early on, plastic pipes were manufactured in
small diameters, the largest sizes being less than 12 inches. The
design of this pipe for underground installation consisted mainly of
the calculation of pipe deflection using Spangler's Iowa Formula. As
manufacturers and design engineers learned more efficient methods for
making pipes with larger diameters and lighter cross-sections, two
other buried design considerations became significant: (1) compressive
wall crushing from ring thrust loads and (2) local wall buckling. Both
of these considerations depend on the earth pressures acting on the

[1]Engineering Supervisor, Spirolite Corp.,
3295 River Exchange Dr., Suite 210, Norcross, GA, 30092.

pipe. To obtain the most efficient design the earth pressures must be accurately known.

In designing small bore plastic pipes it is commonplace to assume that the overburden load applied to the crown of the pipe is equal to the weight of the prismatic element of soil projecting above the pipe, often referred to as the prism load. The prism load is a handy convention for calculating deflection, but the actual load applied to a plastic pipe may be considerably less than the weight of the overburden soil. This reduction in the applied load is due to arching. To account for arching, pipe designers can use the Marston method for calculating loads. But, many designers of plastic pipes use the prism load routinely, even when considerable arching occurs. This paper will show the soil mechanics concepts that link arching, earth pressure parameters and loads on buried pipe, so that designers can better understand the cases where arching will exist and where it can be safely considered in design.

PRISM LOAD

The simplest case for determining the vertical earth load on a horizontal surface in a mass of soil occurs when the soil has uniform stiffness and weight through-out with no large voids or buried structures present. Then, the vertical earth pressure acting on a horizontal plane at depth H is equal to the geostatic stress:

$$P = WH \qquad (1)$$

where P = vertical soil pressure, W = unit weight of soil, and H = height of soil mass above the horizontal surface. The prism load per unit length of pipe is defined as the geostatic stress times the outside diameter of the pipe.

If a structure is present in the soil mass, there will be a redistribution of load toward the stiffer zones. Consequently, it is unlikely that the vertical load on a horizontal plane at the top of the structure will equal the prism load. The load may be greater or less, depending on the structure's stiffness relative to the soil. The variance from the prism load can be considerable as demonstrated in pipe whose crowns are so corroded as to expose the overburden soil. In this case the prism load does not act on the crown of the pipe; in fact, there is no vertical load at this point.

Theoretically, the prism load occurs on a buried pipe only when the pipe has the equivalent stiffness of the surrounding soil. More commonly, the pipe and soil are not of the same stiffness and therefore the pipe either sees more or less than the prism load, depending on the relative stiffness between pipe and soil. When the pipe is less stiff than the soil, as in the case of most flexible pipe, the soil above the pipe redistributes load away from the pipe and into the soil beside the pipe. The load "flows" away from the pipe. This is illustrated in Fig. 1, which shows the distribution of soil pressure around a buried pipe. The arrows indicate the general path along which the load flows.

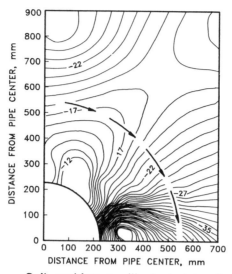

FIG. 1 —— Soil arching as illustrated by the horizontal
stress distribution for 18 in. HDPE pipe in soil box (Yapa [1]).

The path suggests that the soil forms an arch. Terzaghi points out
that the notion of an arch forming is credited to Bierbaumer in 1913,
although the mechanics of load reduction were worked out by Engesser in
1882 and Kotter in 1899 [2]. More recently, Handy suggests that the
horizontal flow of load from the backfill into the side fill resembles
a catenary, in other words the soil hangs from the trench walls [3].
Arching can be defined as the difference between the applied load and
the prism load. The term "arching" is usually taken in the sense to
imply a reduction in vertical load. When the pipe takes on more
vertical load than the prism load, "reverse arching" is said to occur.

ORIGIN OF STRENGTH IN SOIL

   As demonstrated in the previous section, the actual load applied to
a buried pipe varies from the theoretical prism load because of the
redistribution of stresses. Redistribution is possible because the
soil mass possesses shear strength, which enables it to resist
distortion much like a solid body. Shear strength, or shear
resistance as it is often called, arises from the structure of the
soil's fabric. Soil is an assemblage of (1) mineral particles such as
silica or aluminum silicates, (2) water, and (3) air. Mineral
particles can range in size from the large, such as boulders, to the
microscopic, such as the colloidal particles making up clay. The size
of the individual soil particles or grains is significant in
determining the soil's behavior. Very small (colloidal) size soil
particles are capable of adsorbing large quantities of water, as much
as 10 times their own weight. These particles attract each other to
produce a mass which sticks together. This property is called cohesion
or plasticity. Soils containing such particles are referred to as
"cohesive" and include clayey soils. Cohesion gives clayey soils
resistance to shear. The strength of clayey soils is dependent on the

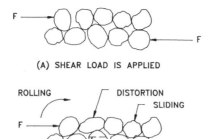

(A) SHEAR LOAD IS APPLIED

(B) SHEAR DEFORMATION

FIG. 2 —— Grain movement during shear.

amount of water within the soil.   As the content of water increases, the shear resistance decreases.   On the other hand assemblages of larger particles such as silts, sands, or gravels do not exhibit plasticity.   Their strength is almost unaffected by water.   These soils are called "cohesionless" or "granular."   Normally, cohesionless soils have high shear resistances.   When a mass of cohesionless soil is sheared individual grains either roll, slide, fracture, or distort along the surface of sliding as shown in Fig. 2.   Likewise, many cohesive soils contain grains of sand or silt, so they can exhibit significant shear resistance.

The grain size, shape, and distribution will affect the shear resistance.   In general, soils with large grains such as gravel have the highest strengths.   Rounded grains tend to roll easier than angular, or sharp, grains which may interlock.   So, the angular grains resist shear better.   Well graded mixes of grains, that is soils which have a good representation of grains over a wide range of sizes, tend to offer more resistance than uniform graded soils, which contain similar size grains.   Aside from the grain characteristics the density has the greatest affect on shear resistance.   For instance, in a dense soil there is considerable interlocking of grains and a high degree of grain to grain contact.   When shear occurs in a dense mass, the volume of the soil along the surface of sliding expands (dilatancy) as the grains are displaced.   This requires a high degree of energy. Therefore, the soil mass has a high resistance.   In a loose soil, shearing causes the grains to roll or to slide, which requires far less energy than dilatancy.   So, loose soil has a lower resistance to shear.

ANGLE OF INTERNAL FRICTION

Shear resistance can have both frictional and non-frictional components.   The frictional component being that resistance which is affected by increasing the pressure normal to the surface of sliding and can be likened to the static friction between two rigid bodies. The resistance to shear of cohesionless soils such as sand and gravel is determined by the frictional component.   The non-frictional component is related to cohesion or the bond between individual particles. Thus, it is not affected by increasing the normal pressure. Most clayey soils possess both a frictional and a non-frictional component. The difference between frictional and non-frictional forces is illustrated by the different ways vertical cuts in sand and vertical cuts in clay behave. If we attempt a vertical cut in sand, the grains slide downward to the trench bottom. The grains collect in a pile, with the side of the pile having a definite slope. The angle of the slope is called the angle of repose. Whereas, a cut in clay will stand vertical due to cohesion until it collapses. After collapsing, it does not exhibit an angle of repose.

Frictional resistance in soils is developed the same way as frictional resistance between rigid bodies. The shear force, T, required to overcome the static resistance between two rigid bodies along their surface of contact is given by the following equation:

$$T = N\tan(u) \tag{2}$$

where N = the force normal to the surface of sliding, tan(u) = the coefficient of friction, and u = the angle of friction. Coulomb developed an analogous equation for shear resistance in a soil mass. For a cohesionless soil, Coulomb's equation is:

$$s = p\tan(\emptyset) \tag{3}$$

where s = the shear stress, p = the normal pressure along the surface of sliding, and $\emptyset$ = the angle of internal friction, which is sometimes simply called the "friction angle". Coulomb's equation shows that the resistance to sliding within a mass of soil depends on the pressure acting normal to the surface of sliding and on the soil's friction angle. The friction angle represents the soil's "internal" (within the soil mass) frictional resistance, which is a property of the soil's fabric and density as discussed above. For instance, dumped or loose sand has a friction angle around 30 degrees. (The friction angle for dumped sand is usually equal to the angle of repose.) When the density of sand is increased by compaction the friction angle increases and therefore so does its shear resistance.

Non-frictional resistance, such as the cohesion in clay, is accounted for by an additional term, c, which does not depend on the normal pressure. Coulomb's equation for soils with cohesion is:

$$s = c + p\tan(\emptyset) \tag{4}$$

The shear resistance of soil can be determined by several laboratory methods. Perhaps, the simplest is the direct shear test. Soil is placed in a shear box which consists of two rigid frames as shown in Fig. 3. A load head is placed on top of the soil to confine the soil and to apply normal pressure. A horizontal force applied to the top frame shears the top half of the soil over the bottom half. Usually, the test is run at several normal pressures. The resulting horizontal force or peak shear stress at failure is plotted in terms of the normal pressure as shown in Fig. 4. Generally, the graph of shear stress versus normal pressure is a straight line. The slope of this line is defined as the friction angle. If the soil's density is changed and

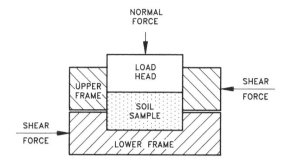

FIG. 3 —— Direct shear box.

the test is repeated, a line with a new slope is obtained as shown in Fig. 4. For cohesionless soils the straight line can be extended back to the origin. For cohesive soils the line will be straight, but it will not pass through the origin. The value of shear resistance when no normal pressure is applied is the cohesion term, c, in Coulomb's equation.

Table 1 gives typical values for the friction angle taken from Bowles [4]. No friction angles are shown for clay because they depend on the load history and the drainage condition of the clay. Depending on these conditions the friction angle can range from 0 to 30 or more degrees.

Wetting or submerging a soil will reduce its friction angle. In cohesionless soils the reduction may only be one or two degrees, but in clay it can be much larger.

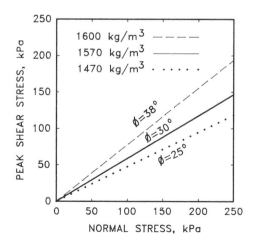

FIG. 4 —— Shear strength of a dry sand (Direct Shear).

TABLE 1 -- TYPICAL VALUES FOR ANGLE OF INTERNAL
FRICTION (Bowles [4])

Soil	Density	Friction Angle, deg.
Gravel	Medium	40-55
	Dense	35-50
Sand	Loose	28-34
	Dense	35-46
Silt & Silty Sand	Loose	20-22
	Dense	25-30

MOBILIZATION OF SHEAR RESISTANCE

Shear resistance in a soil mass develops or is "mobilized" with the initiation of movement along a surface of sliding. Even slight movements may mobilize high resistances. As movement continues more resistance is mobilized. This continues up to a point when the maximum limit to resistance is reached. At this point movement may continue, but there is little change in resistance. The soil is said to have reached a state of limiting equilibrium. As a practical matter, most soils in a state of limiting equilibrium can undergo large shear deformations without significant loss of shearing resistance.

EARTH PRESSURE COEFFICIENTS

The earth pressure exerted on a buried structure depends on the state of the soil placed against the structure. There are three earth pressure states to consider: (1) at-rest, (2) passive, and (3) active. The at-rest state occurs naturally in the ground, whereas the passive and active states are brought about by movement of the soil which ultimately mobilizes the limiting resistance of the soil.

The pressure existing at a point in a fluid is the same in all directions. This is not the case at a point within a soil mass. Seldom does the horizontal (or lateral) pressure equal the vertical pressure. The horizontal pressure can be found by multiplying the vertical pressure times an earth pressure coefficient, K.

$$K = p_h/p_v \qquad (5)$$

where p_h = horizontal pressure and p_v = vertical pressure. The value of the earth pressure coefficient depends on the state of the soil mass, ie. whether it is at-rest, passive, or active.

Consider the condition existing in a mass of cohesionless, backfill soil located beside a vertical wall. If the wall is perfectly rigid, the horizontal pressure exerted by the wall on the soil mass is no different than the horizontal pressure existing in a large, level soil mass, such as the ground. That horizontal pressure is:

$$p_o = K_o WH \qquad (6)$$

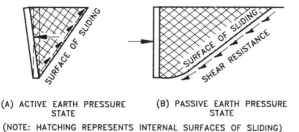

(A) ACTIVE EARTH PRESSURE   (B) PASSIVE EARTH PRESSURE
          STATE   STATE
(NOTE: HATCHING REPRESENTS INTERNAL SURFACES OF SLIDING)

FIG. 5 —— Active and passive earth pressure states.

where p_o = horizontal earth pressure at-rest, W = unit weight of the soil, H = depth , and K_o = coefficient of at-rest earth pressure. The value of K_o depends on the soil type and its load history, but typically for granular backfill, that is not heavily compacted, K_o ranges from 0.4 to 0.6. The earth pressure coefficient is said to be "at-rest" because the wall is perfectly rigid and will admit no movement. Undisturbed ground is considered to be in the at-rest state.

If the wall were to move, the earth pressure exerted on it would change. Consider such a case. If a force is applied to the wall in such a manner as to push the wall into the soil, then the soil will offer resistance to the movement by virtue of its internal shear resistance. Because the force is acting on the soil, this resistance is referred to as "passive" resistance. As the force increases, movement along shear planes as shown in Fig. 5 eventually mobilizes the soil's maximum shear resistance and brings the soil to a state of limiting equilibrium. The pressure required to bring the soil to this state is the soil's "passive" earth pressure, p_p, as given by:

$$p_p = K_p WH \qquad (7)$$

where K_p = coefficient of passive earth pressure, which depends on the friction angle of the soil. It is given by the following:

$$K_p = \tan^2(45 + (\emptyset/2)) \qquad (8)$$

Typical values for K_p range from 2 to 14 for cohesionless soils [4]. Therefore, the passive earth pressure is much higher than the at-rest pressure. For instance, the force required to pull a buried anchor from the ground is found using the passive pressure. Likewise, the minimum depth of cover required to prevent flotation of an empty, shallow buried pipe can be determined knowing the passive pressure.

In the passive case, the soil mass was acted on by a force; the reverse may also occur. The soil mass may actively exert a force on a surface. Again, consider the wall. The active state occurs when the wall deflects away from the soil. When this occurs, the soil undergoes slippage along many internal surfaces as shown in Fig. 5. The slippage mobilizes shear resistance, which carries part of the soil weight and reduces the pressure against the wall from the at-rest pressure. The pressure on the wall, or the active earth pressure, p_a, is given by:

$$p_a = K_a WH \qquad (9)$$

where K_a = the coefficient of active earth pressure:

$$K_a = \tan^2(45 - (\emptyset/2)) \qquad (10)$$

Typical values for K_a range from 0.25 to 0.35. The soil in the active state exerts less pressure against a wall than soil in the at-rest state. It is common practice with cohesionless backfills to use the active earth pressure for the design of retaining walls that can tolerate slight movements. Seldom is clay designed for the active state because of soil creep.

The amount of wall movement required to produce the active state is small. Typically, for granular soils the movement at the top of the wall only need be 0.1% to 0.2% of the wall's height. Thus, for a 20 ft. high wall a movement of one-quarter to one-half inch can reduce the pressure applied to the wall by as much as fifty (50) percent.

ARCHING

Terzaghi experimented with arching by placing a trapdoor underneath a bin of sand as shown in Fig. 6 [2]. As the trapdoor was lowered the sand followed and the downward movement mobilized shear resistance along the surface of sliding. This action transferred some of the weight of the sand above the trapdoor to the sand above the bin floor. There followed a decrease in pressure against the door and a corresponding increase in pressure against the floor of the bin. The total pressure against the bottom of the bin (floor and trapdoor) remained constant. Therefore, the load redistributed over the trapdoor and formed an arch. Terzaghi found that the pressure against the trapdoor was equal to the weight of the prism of soil above it less the shear resistance mobilized along the surface of sliding. He gives the following for the pressure, p, on the trapdoor in a cohesionless soil:

$$p = \frac{BW}{2K\tan(\emptyset)} \left(1 - \exp(-K\tan(\emptyset)\frac{2H}{B})\right) \qquad (11)$$

where W = unit weight of soil, $\emptyset$ = angle of internal friction, H = depth of cover, B = width of trapdoor and K = an experimentally

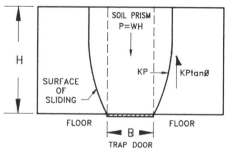

FIG. 6 -- Terzaghi's trapdoor in soil bin experiment.

determined earth pressure coefficient equal to the ratio of horizontal to vertical pressure in the backfill above the trapdoor. The Ktan(Ø) terms in Eq 11 represent the shear resistance. The shear resistance in a cohesionless soil is a function of the normal pressure multiplied by tan(Ø) as shown in Eq 3. In the soil bin the normal pressure on the surface of sliding is the horizontal pressure. Therefore, as the ratio K increases more load will be carried in shear. Terzaghi established a value of one for K. Other values have been proposed. For instance, Handy suggests using a value of $K = K_a$ for conservative design [3].

The amount of arching or load reduction on the trapdoor is equal to the amount of shear resistance developed along the surface of sliding. The following example illustrates this. Consider a bin of loose sand that is 3 meters high with a 1 meter trapdoor. Before lowering the door, the load on the door is 5700 kg (12,566 lbs.). After lowering, the load on the trapdoor can be calculated using Eq 11. This gives a load of 1600 kg (3527 lbs.). The difference equals the arching, which is 4100 kg (9039 lbs.). This value is also equal to the shear resistance along the surface of sliding. This can be demonstrated by calculating the shear resistance directly from Eq 3. For this example, only a gross approximation can be obtained for the shear resistance, since the length of the surface of sliding must be estimated as well as the horizontal pressure on it. The following is assumed; the shear surface has a length of 2 door widths (which equals 2 m), K equals one, and the average vertical pressure in the volume of soil that is sliding equals half the weight of that volume ($2 \ m^3$) divided by 1 m^2, which gives (2 m) * ($1900 \ kg/m^3$)/2 = 1900 kg/m^2. In Eq 3, let the normal pressure be equal to the horizontal pressure on the surface of sliding, which equals K times the average vertical pressure, and let Ø = 30 deg. Equation 3 gives a shear stress of ($1900 \ kg/m^2$) * tan(30) = 1097 kg/m^2. The shear resistance mobilized is equal to 4 m^2(1097 kg/m^2) or 4388 kg (9674 lbs.). In light of the simplifying assumptions it can be seen that the shear resistance (4388 kg) and the arching (4100 kg) are equal.

Terzaghi's experiment revealed an important consequence of arching. The surface of sliding in a sand bin extends for only two to three door widths above the trapdoor. This means that as the height of sand in the bin is increased the pressure on the trapdoor reaches a limit. Once the limit is reached, adding sand has virtually no effect on the pressure on the trapdoor.

ARCHING IN PIPE TRENCH

Mobilization of an active state in the backfill above a buried pipe can be brought about by downward movement of the backfill. This can occur as the result of deflection of a flexible pipe, settlement or compression of the deeper layers of the backfill, settlement beneath a pipe, or placement of soft, compressible material in the trench above the pipe. As stated in a previous section, wall movements that are a fraction of a wall's height are sufficient to mobilize an active state in the soil behind the wall. Likewise, small vertical deflections in buried pipe can produce significant load reductions.

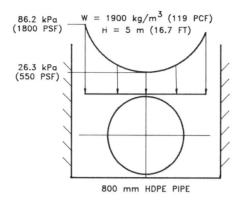

FIG. 7 —— Vertical earth pressure at pipe crown per Gaube.

As flexible pipe undergoes vertical ring deflection, it acts like Terzaghi's trap door. The soil tries to follow the pipe downward, but the soil's movement is resisted by shear resistances (frictional forces and cohesion) along the trench walls. As this occurs, slippage surfaces form throughout the backfill mass. Movement along these slippage surfaces is resisted by the soil's shear resistance, which brings about an active state in the soil above the pipe. Through this action, part of the weight of the backfill soil is carried over into the trench walls. Therefore, the amount of force exerted on the pipe by the backfill is less than the weight of the backfill soil mass, or the prism load. Field measurements illustrate this load reduction. For instance, Lefebvre et al. instrumented a 15.5 m (51 ft.) metal arch culvert in a 13.4 m (44 ft.) embankment on the Vieux Comptoir River [5]. They reported vertical soil pressures at the crown equal to about one-fourth of the prism load. Total crown movement, which mobilized this reduction, was 9 cm (3.5 in.) or 0.7% of the embankment height.

When arching occurs, the soil directly adjacent to the sides of the pipe experiences an increase in vertical pressure due to the load transferred from the backfill. This increase in pressure is much like the pressure increase seen by the floor of Terzaghi's bin. In the case of a flexible pipe it further stiffens the sidefill soil's resistance to horizontal deflection of the pipe. Gaube has measured the vertical pressure distribution across the crown of an HDPE pipe [6]. See Fig. 7.

The arching described above resulted in a decrease in load. If the pipe is stiffer than the surrounding soil, it may attract more load since it deflects less than the soil beside it compresses. This is more likely to be the case with a rigid pipe than a flexible pipe.

Pipe designers often question the permanence of arching. Many use the prism load "to be safe". Terzaghi states that "since arching is maintained solely by shearing stresses in the soil, it is no less permanent than any other state of stress in the soil which depends on the existence of shearing stresses, such as the state of stress beneath the footing of a column [2]." He goes on to say that if shearing stresses were not permanent footings would settle indefinitely. As shown previously retaining walls are designed using an active or arched load. The stability of slopes is determined by the shear resistance of the soil. Shear resistance may be reduced by strong vibrations or soil creep. In shallow cover applications, vibrations from traffic loading may reduce arching but typically at these depths the earth load is so small that other parameters control the pipe design. Pipe located near large vibrating machines should be designed for the prism load. Creep will reduce arching. Generally, in the design of retaining walls with cohesionless backfill, creep is not considered. However, it is considered for clayey backfills. The usual assumption for clay is that the full at-rest pressure is reached on the wall at some point in time. For buried pipe, soil creep is accounted for in the design method by selecting conservative design parameters, as will be discussed later.

METHODS OF LOAD CALCULATION

In 1930 Marston published a design method for determining loads on buried pipe that accounts for arching. His work was based on experiments and field measurements. His method is widely accepted and can be found in ASCE Manual No. 60 [7]. Marston assumed that the vertical pressure on a pipe in a trench was analogous to the vertical pressure on Terzaghi's trap door. So, Terzaghi's equation, Eq 11, can be used to obtain the loads on a buried pipe. Bulson shows that this can be done by replacing B, in Eq 11 with the trench width, Bd [8]:

$$p = \frac{B_d W}{2K \tan(\emptyset)} \quad (1 - \exp(-K \tan(\emptyset) \frac{2H}{B_d})) \qquad (12)$$

where the terms in Eq 12 are the same as in Eq 11, except H = height of cover above the pipe. There is no cohesion term in Marston's equation. He assumed that significant cohesion would not develop in the pipe trench. Nor does Eq 12 account for the pipe flexibility. However, the customary method is to assume that the load on a rigid pipe is equal to the pressure, p, in Eq 12 times the trench width, Bd. Whereas, for a flexible pipe the load is equal to the pressure, p, times the pipe diameter. This gives a somewhat smaller loading on flexible pipe.

Since Marston's equation and Terzaghi's equation have the same form, many analogies can be drawn between arching in a pipe trench and arching in a soil bin. Two important ones will be discussed here: (1) the effect of the horizontal pressure in the backfill on the vertical pressure applied to the pipe or trapdoor and (2) the limit to the load reaching the pipe or trapdoor.

The horizontal pressure in the backfill is an important parameter in determining the amount of arching. The higher the value chosen for K in Eq 12, the greater the arching. Terzaghi determined that K equaled one, but Marston assumed that the horizontal pressure would equal the active pressure and concluded that $K = K_a$. It turns out that Marston's assumption may be conservative. Bulson cites laboratory studies which indicate Marston underestimated K for dense cohesionless soils to such an extent that the actual pressure may only be half that predicted. Likewise, field measurements support this notion. Lefebvre reported values for K of approximately 0.6 near the crown and 1.2 higher in the

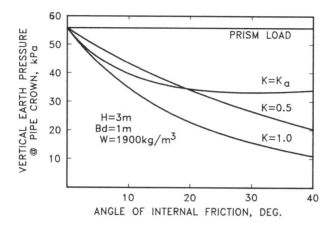

FIG. 8 —— Vertical earth pressure versus angle
of internal friction per Marston's equation, Eq 12.

fill. Wetzorke has proposed the following for K: 1.0 for dense sand,
0.5 for loose sand and clay, and 0.11 for saturated clay [8]. Figure
8 shows the sensitivity of the vertical pressure to the value of K.

Terzaghi showed that as the depth of the bin increased the load on
the trapdoor approached a limit. This suggests that the same phenomena
may occur in a pipe trench, at least in cohesionless soils. Figure 9
is a plot of the vertical crown pressure calculated with Eq 12 versus
depth of cover for a flexible pipe in a trench of 1 m width. The load
approaches a limit. The depth at which the load reaches the limit
depends on the value of K. Whether a limit is actually reached or not
in practice has not been established. Adams et. al. report that the

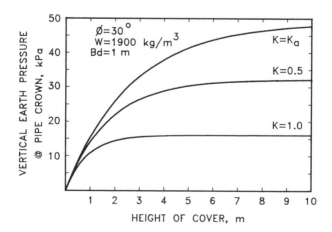

FIG. 9 —— Vertical earth pressure versus
depth of cover per Marston's equation, Eq 12.

vertical pressure on a 600 mm (24 in.) HDPE pipe in a 30.5 m (100 ft.) embankment continued to increase up to the completion of the fill, although the resulting pressure was only a fifth of the prism load [9].

All soils creep. However, most pipe designers ignore creep, when the backfill is cohesionless. This is a conservative design approach for plastic pipe, which tends to creep at a faster rate than cohesionless soils [10]. Clayey soils, especially when saturated, exhibit significantly more creep than cohesionless soils. When clayey soils are subjected to loads of 50% or more of their peak shear strength, considerable creep occurs. When a clay backfill is initially placed over a pipe, mobilization of shear resistance occurs and arching may be high. In the backfill where stress concentrations exist such as along surfaces of sliding, the stress level in the clay may approach a significant portion of its peak value. At these points of stress concentration creep occurs, which allows movement of the backfill soil toward the pipe with a corresponding load increase. With the passage of time more creep occurs. Because most clays have some frictional resistance the prism load is usually not reached. But a conservative approach should be taken for design. Therefore, a low friction angle is usually assumed for clays when using Marston's equation. Typically the values assumed for an ordinary clay is 11 deg. and for a saturated clay 8 deg.

Other methods for load calculation include the ATV method [11] and the TAMPIPE method [12]. The ATV method finds the load on a pipe by multiplying the Marston load by an additional factor which accounts for the redistribution of stresses around the pipe due to the relative stiffness between the pipe and the soil. When calculating loads on flexible pipe the ATV method usually gives lower loads than the Marston method.

The Marston method considers only the vertical earth pressure. Typically, a designer would apply the Marston load to the pipe as a normal (radially directed) load. In fact, the distribution of normal pressure around the pipe is not uniform. Where arching occurs, the horizontal pressure on the pipe may exceed the vertical pressure. Both Adams and Lefebvre measured horizontal pressures equal to about half the prism load at the pipes' springlines, which is a pressure greater than the vertical pressure. Additional pressure occurs due to the tangential shear between the soil and the pipe's surface. This more complicated stress distribution can be analyzed using a finite element code. In lieu of this, the resulting stress in the pipe can be found using TAMPIPE. For instance, TAMPIPE shows that pipe placed in highly compacted embedment and entrenched in stiff insitu soil may see total loads as little as a fifth of the prism load [13].

When the Marston load is used in conjunction with the modulus of soil reaction, E', the designer must be careful in selecting the appropriate value for E'. For instance, Howard has made an extensive determination of E' values [14]. He measured pipe at 113 installations. Howard backcalculated E's using Spangler's equation. Since he did not know the exact soil load on the pipe, he assumed the prism load. Based on the discussions in this paper, it is a reasonable assumption that arching did occur on those installation, thus his E' value in part accounts for arching. Whenever Howard's E' values are used in design, it is appropriate to use the prism load. Whenever E' values suggested by Spangler are used, it is appropriate to use the Marston load.

SUMMARY

Designers commonly assume that the maximum earth load a plastic pipe will see is the prism load. This assumption is unnecessarily conservative as arching occurs. Arching is as permanent as any other form of shear resistance. Arching may be considerable. In dense cohesionless soils, only a fraction of the prism load is applied to the pipe. The Marston load with a value of K such as K_a is a practical choice for a conservative design load.

REFERENCES

[1]   Yapa, K. A. S. and Lytton, R. L., "An Analysis of Soil-Box Tests on HDPE Pipes," Texas A&M University, Research Report, 1989.
[2]   Terzaghi, K., Theoretical Soil Mechanics, John Wiley & Sons, New York, 1943.
[3]   Handy, R. L., "The Arch in Soil Arching," Journal of Geotechnical Engineering, ASCE, 111 (3), 1985, pp. 302-318.
[4]   Bowles, J. E., Foundation Analysis and Design, McGraw-Hill Book Co., New York, 1982.
[5]   Lefebvre, G. et. al., "Measurement of Soil Arching above a Large Diameter Flexible Culvert," Canadian Geotechnical Journal, 13, 1976, pp. 58-71.
[6]   Gaube, E. et. al., "High Density Polyethylene Sewage Pipes - Results of Tests to Determine the Time Dependence of Soil Pressure and Deformation," Kunststoffe, Vol. 61, 1971, pp. 765-769.
[7]   Gravity Sanitary Sewer Design and Construction, ASCE, No. 60, 1982.
[8]   Bulson, P. S., Buried Structures, Chapman and Hall, London, 1985.
[9]   Adams, D. N., Muindi, T. M., and Selig, E. T., "Performance of High Density Polyethylene Pipe Under High Fill," University of Massachusetts, Geotechnical Report No. ADS88-351F, 1988.
[10] Petroff, L. J., "Stress Relaxation Characteristics of the HDPE Pipe-Soil System," Proceedings of the International Conference on Pipeline Design and Installation, Las Vegas, NV, 1990.
[11] ATV Regelwerk, Arbeitsblatt 127, Abwassertechnische Vereinigung e. V. (ATV), Markt (Stadthaus) 5202 St. Augustin 1, West Germany.
[12] Chua, K. M. and Lytton, R. L., "A New Method of Time-Dependent Analysis for Interaction of Soil and Large-Diameter Flexible Pipe," 66th Annual Meeting, Transportation Research Board, Washington, DC, 1987.
[13] Chua, K. M. and Petroff, L. J., "Soil/Pipe Interaction of Large Diameter Profile Wall-HDPE Pipe,"Managing Corrosion with Plastics, Vol. 7, NACE, Houston, TX, 1985.
[14] Howard A. K., "Modulus of Soil Reaction Values for Buried Flexible Pipe," Journal of Geotechnical Engineering, ASCE, 103, GT 1, 1972, pp. 33-46.

Timothy J. McGrath, Richard E. Chambers, Phillip A. Sharff

RECENT TRENDS IN INSTALLATION STANDARDS FOR PLASTIC PIPE

REFERENCE: McGrath, T.J., Chambers, R.E., and Sharff, P.A. "Recent Trends in Installation Standards for Plastic Pipe," Buried Plastic Pipe Technology, ASTM STP 1093, George S. Buczala and Michael J. Cassady, Eds., American Society for Testing and Materials, Philadelphia, 1990.

ABSTRACT: ASTM Committees are nearing the end of long projects to revise and upgrade Standard Practices D 2321 and D 3839 for the installation of buried thermoplastic and fiberglass pipe, respectively. These revisions include improved and expanded guidance for selection of soils, control of construction procedures and compaction of backfill. The Appendices to D 2321 provide a commentary on installation issues that are critical to the long term performance of flexible pipe and specific guidance to engineers on important topics to be considered in preparing project specifications. The task group working on D 3839 is considering modifications to the deflection prediction equation that will treat flexible pipe deflection in a broader, more rational way than previous versions. This paper highlights the important revisions being addressed in these standard practices and provides background on their development.

KEYWORDS: Compaction, Culverts, Deflection, Installation, Plastic Pipe, Soil Density, ASTM D 2321, ASTM D 3839

INTRODUCTION

ASTM Standard D 2321 has just been revised, culminating several years of work by ASTM Committee F17.62. This standard, with the new title "Standard Practice for Underground Installation of Thermoplastic Pipe for Sewers and Other Gravity-Flow Applications", was first published in 1964 and may be the most cited installation standard for plastic pipe. The revisions include broader guidance on the use of various soil classes as pipe embedment, on various aspects of ground water control and on other important issues. Most of the information is not new from a research point of view. The achievement of the revisions is that they bring known information about the interaction of pipe and soil into an understandable, application-oriented standard.

Another Task Group in Subcommittee D20.23 is actively pursuing revisions to ASTM Standard D 3839 "Standard Practice for Underground Installation of Fiberglass Pipe". This standard provides much the same information for fiberglass pipe as D 2321 does for thermoplastic pipe, and many of the anticipated revisions

Messrs McGrath, Chambers and Sharff are Associate, Principal and Staff Engineer respectively at Simpson Gumpertz & Heger Inc., 297 Broadway, Arlington, MA 02174.

will include information similar to D 2321; however, D 3839 maintains an Appendix that offers guidance in predicting the deflection in buried flexible pipe subjected to earth and live loads. This equation is used to select appropriate soils and densities for given installation conditions. The revisions being considered will broaden the range of conditions that can be addressed with the deflection prediction equation.

This paper first considers the important general principles of behavior of flexible pipe, and then presents the new revisions that have been or are being considered for incorporation into D 2321 and D 3839 and finally, how these changes can contribute to a better pipe installation. Although the title of this paper, and the standards that serve as its primary focus, include the term "plastic pipe", the principles discussed are applicable to all flexible pipe.

REQUIREMENTS FOR PIPE PERFORMANCE

The main subject of this paper is standard installation practices for buried thermoplastic and fiberglass pipe. These practices do not directly specify pipe materials or performance guidelines, yet, as discussed below, the soil placed and compacted around a pipe plays an integral role in its performance. Key structural aspects of pipe performance that should be considered when developing installation specifications include:

Constructability - The installer must have the means available to him to achieve a quality installation. Groundwater must be controlled, the pipe must withstand handling and installation forces and the specified soils must have the necessary stiffness to support the pipes.

Stability - The pipe must maintain its shape and strength to resist imposed loads over the installation design life.

PIPE-SOIL INTERACTION

Investigation of the behavior of buried pipe is often called the study of "soil-structure interaction". We do not use that phrase here because it implies that the pipe constitutes the sole structural element of the pipe and soil system. Most buried pipe (flexible and rigid) are not capable of performing as an independent structure under earth loads without the benefit of uniform bedding and lateral soil support; thus, it is the interaction of the pipe with the surrounding soil that forms a structure. We will refer here to "pipe-soil interaction" to emphasize the importance of both the pipe and the soil in providing a viable structure.

The emphasis on pipe-soil interaction suggests that design of pipe for earth loads is an interactive process of designing both the pipe and the soil envelope around it. If engineers allow different types of pipe in a specification, they must separately consider, for each type of pipe, if the backfill specification is appropriate. The pipe and the soil together constitute an engineered system.

Flexible Pipe

What constitutes a flexible pipe installation is often discussed among experts and various definitions have been put forth. This paper will work with the imprecise definition of a flexible pipe as a pipe with relatively low flexural stiffness, such that when vertical earth load is applied the pipe will deflect downward vertically and outward horizontally. This outward horizontal deflection mobilizes passive lateral soil support for the pipe, in turn preventing further downward vertical deflection. Thus, pipe deflection is controlled more by soil stiffness than pipe flextural stiffness. This definition emphasizes pipe-soil interaction as discussed above.

One of the benefits of this flexible behavior is that the quality of the installation can be readily checked via a deflection test after installation is complete.

Soil Density and Soil Stiffness

One of the chief purposes of a standard practice for installation of flexible pipe is to provide guidance to the pipeline installer in providing suitable, uniform soil support such that the pipe will perform successfully as a conduit, culvert, sewer or any other purpose for which it was designed. The primary soil property that provides this support is stiffness. In engineering terms soil stiffness is measured as a modulus. The most common modulus used to define soil stiffness around a buried pipe is the modulus of soil reaction or "E' ". This is the semi-empirical term used in Spangler's equation to calculate pipe deflections (1,2) and is not a true material property.

The importance of soil stiffness is not always evident in practice because installation specifications rely on measurement of soil density for field quality control and no direct correlation between percent of Proctor density and soil stiffness exists. There is even considerable confusion about density specifications, i.e. the differences between standard Proctor, modified Proctor and relative density. This is also important but is outside the scope of this paper.

The varying relationship between soil stiffness and soil density can be demonstrated with the recommended values of E' developed by Howard (3). Howard proposed a table that related values of E' to soil density for various general types of soil. This table has since been reproduced in many publications and adopted by some standards. Although normally listed in tabular form, Figure 1 is a graphic presentation of the relationships proposed by Howard. Examination of Figure 1 shows that a fine grained soil with less than 25 percent coarse particles must be compacted to 95 percent of maximum standard Proctor density to achieve the same soil stiffness (i.e. the same value of E') as crushed stone that is simply dumped around a pipe.

COMPACTIVE ENERGY

One of the major economic decisions in the design of buried pipe installations is the cost of achieving the required soil stiffness (E'). A processed granular backfill material will cost more to purchase, but, as shown in Figure 1 will provide high soil stiffness (as measured by E') at a low percent of maximum Proctor density, i.e. with relatively little compactive effort. A fine-grained backfill may be very inexpensive as a material (it may be the same material that was excavated from the trench) but may require compaction to a significantly higher percent of maximum Proctor density to achieve the same level of soil stiffness. The increased amount of compactive energy required relates directly to an increase in installation costs.

The relative cost of compaction, as indicated by required compactive energy is demonstrated in Figure 2, by Selig (4). This figure shows the percent of maximum density achieved in various types of soil as a function of the percentage of the total energy specified in AASHTO T-99 (ASTM D698, also known as the standard Proctor test). Figure 2 shows that a coarse grained soil can be installed

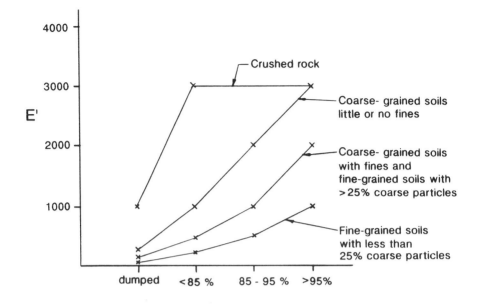

% of Maximum Standard Proctor Density

Figure 1 - Recommended Values of E' [3].

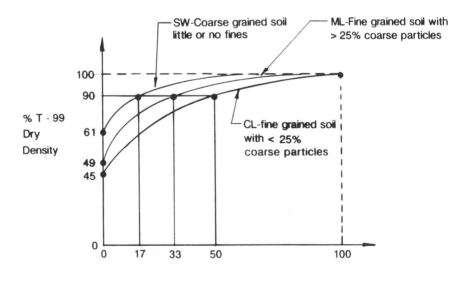

% T - 99 Effort (Standard Proctor)

Figure 2 - Effect of Soil Type on Relationship Between
Compaction Level and Compaction Effort [4].

to 90 percent of maximum standard Proctor density with about one third the compactive energy required for a fine grained soil with less than 25% fines. Furthermore, Figure 1 then shows that at about 90 percent of maximum density the coarse grained soil will have an E' of 2000 psi (13.8 MPa) while the fine grained soil will have an E' of 400 psi (2.75 MPa). Thus with a coarse grained soil five times the soil stiffness is achieved with one third of the compactive effort.

Table 1 is a combination of Figures 1 and 2 that demonstrates the relative amount of energy required to achieve a specific value of E'. This approximates the relative cost of achieving soil stiffness. As discussed later, D 2321 indirectly considers a value of 1000 psi (6.89 MPa) as the recommended minimum value of E' for typical installations. Table 1 shows that about seven times the energy is required to achieve this soil stiffness in a silty clay compared to a gravelly sand. This is a significant difference in installation requirements and cost; yet, even this may be optimistic, since the comparison is made at optimum moisture contents. If the soil moisture contents are at other than optimum then the relative cost of compacting fine grained soils is even higher.

The above discussion shows that Table 1 may be used in a general fashion to determine the relative installation cost of a soil for comparison with the cost of the soil itself to reach an economic decision about overall costs. It may often be the case that using "expensive" granular backfill is economical when compaction costs are considered.

Another indirect consideration in evaluating costs is the cost of quality control. The increased level of effort required to achieve high soil stiffness in fine grained soils will also generally mean increased levels of monitoring and inspection for quality assurance purposes.

OTHER GENERAL ASPECTS OF FLEXIBLE PIPE BEHAVIOR

The above discussion relates to the general behavior of flexible pipe. There are a great number of situations that occur in the field that can influence that behavior and these must be addressed in any comprehensive installation standard.

Native in-situ soil stiffness versus pipe embedment stiffness: Passive soil support at the sides of a pipe is derived from the stiffness of the pipe embedment soil and, when buried in a trench, the stiffness of the native in-situ soil in the trench wall. Thus, the trench width and the relative stiffness of the pipe embedment and native in-situ soils must be considered.

Water: The presence of excessive water is typically one of the biggest problems faced during installation. Sometimes the absence of water will make it difficult to achieve specified compaction levels in fine grained soils.

Bedding and haunching: The bedding, the soil on which the pipe rests (Figure 3) should provide a firm base for the pipe that will not settle beyond acceptable limits. The bedding should also isolate the pipe if the native soil is too hard. Because of difficulty of access, the soil in the haunch zone (Figure 3) generally needs to be compacted by hand, making it the most difficult area in which to achieve uniform soil properties.

The manner in which these items are addressed in D 2321 and D 3839 is discussed below.

Table 1

Energy [1] Required to Achieve Soil Stiffness

Soil Type[2]	Modulus of Soil Reaction - E', (psi)[3]			
	400	1000	2000	3000
Gravelly sand (SW)	≤5	10	17	30
Sandy Silt (ML)	25	33	40	≥100
Silty Clay (CL)	50	70	≥100	≥100

Notes:

1. Energy expressed as a percent of the energy specified in AASHTO T-99 (standard Proctor test)

2. Soil types correspond to Howard soil descriptions as follows:

    Gravelly Sand (SW) = coarse grained soil ≤12%, fines
    Sandy Silt (ML) =   fine grained soil, liquid limit ≤50 and more than 25% coarse grained particles
    Silty Clay (CL) = fine grained soil, liquid limit ≤50 and less than 25% coarse grained particles.

3. 1 psi = 6.89 kPa

IMPROVEMENTS TO ASTM D 2321

Soil Types

The revised version of D 2321 expands the description of soil types by not only describing soils in physical terms, such as gradation, but also in practical terms such as:

●   compactibility - This is a subjective term used to describe the relative amount of compactive energy that must be supplied to produce acceptable levels of soil stiffness (as was demonstrated in Table 1). Guidance is also provided on appropriate types of compaction equipment required for different classes of soil.

●   susceptibility to migration - When an open-graded soil is placed next to a soil with substantial fines, the movement of water can carry the fines into the voids in the open coarse graded material. This mixing can result in a net loss of soil volume that will in turn result in a loss of soil support to the pipe and an increase in pipe deflection. This is the major concern in using open-graded backfill materials that are otherwise very compactible. Migration

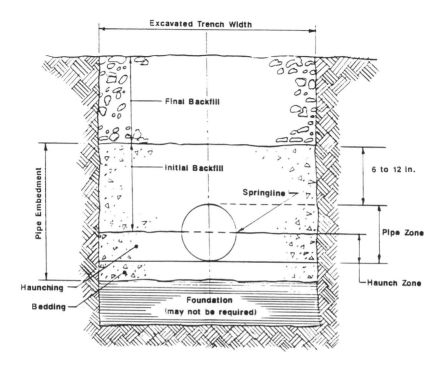

Figure 3 - Trench Cross Section Showing Terminology
(From ASTM D 2321-89)

can be controlled by geotextile filter fabrics or by specifying appropriate gradations for adjacent materials. Appendix X.1 to D 2321 provides criteria for evaluating the gradation of adjacent materials.

Trench Construction

In the area of trench construction, the revised version of D 2321 provides more background and more definitive information in several areas related to trench construction.

Dewatering - As noted above the control of groundwater is one of the most difficult tasks encountered in installing buried pipe and an area with many pitfalls. D 2321 provides guidance in areas where the installer should pay particular attention.

• The water table should be kept below the bottom of the pipe trench excavation to such an extent as to provide a stable and sound foundation and bedding and to allow proper assembly of the pipe and proper compaction of most soil types.

Table 2

D 2321 Guidelines for Use of Class III Soil Around Buried Pipe

General Recommendations and Restrictions	Do not use where water conditions in trench may cause instability.
Foundation	Suitable as foundation and for replacing over-excavated trench bottom as restricted above. Do not use in thicknesses greater than 12 in. total. Install and compact in 6 in. maximum layers.
Bedding	Suitable only in dry trench conditions. Install and compact in 6 in. maximum layers. Level final grade by hand. Minimum depth 4 in. (6 in. in rock cuts).
Haunching	Suitable as restricted above. Install and compact in 6 in. maximum layers. Work in around pipe by hand to provide uniform support.
Initial Backfill	Suitable as restricted above. Install and compact to a minimum of 6 in. above pipe crown.
Embedment Compaction	Minimum density 90% Std. Proctor. Use hand tampers or vibratory compactors. Maintain moisture content near optimum to minimize compactive effort.
Final Backfill	Compact as required by the engineer.

• Dewatering should be completed in such a fashion that fines are not removed from soils that will stay in place (e.g. water coming out of pumps should not be muddy), since this may cause subsequent loss of support to the pipe. It also should be completed to assure that soils behind sheet piling do not wash into the trench, leaving voids that will have a deleterious effect on the pipe after the sheeting is removed.

Backfilling - D 2321 and D 3839 both provide sketches to classify the most important areas of the backfill (Figure 3). D 2321 now includes a table that notes special considerations for each of the soil classifications when used in each of the major backfill areas. The part of this table that pertains to what D 2321 calls Class III soils (coarse grained soils with more than 12 percent fines) is reproduced in Table 2. The table notes such important matters as the compactibility in wet and dry conditions, minimum and maximum lift thicknesses and recommended types of compaction equipment. The table also recommends a minimum percent of standard Proctor density at which a given soil type should generally be used. This minimum is based on providing a soil stiffness (E') of at least 1000 psi (6.89 MPa), which is considered adequate for typical pipe installations. This should always be given careful consideration when designing for specific conditions.

Commentary

Appendix X.1 to D 2321 is a commentary on the major variables that affect pipe-soil interaction and pipe behavior. The commentary emphasizes the importance of verifying the quality of installation during and after construction whenever possible. This includes monitoring soil density as embedment is placed and compacted around the pipe, and/or monitoring pipe deflection levels with a deflectometer or a properly sized go no-go mandrel after the backfill is placed in its entirety. The commentary notes the desirability of waiting thirty or more days after construction before measuring deflections. This allows for a period of soil consolidation and stabilization after backfilling. On large projects, designers should consider deflection checks after installing the first sections of pipe. This practice can detect problems with backfilling procedures before they are multiplied over an entire project.

Guidelines for Specifications

D 2321 now includes an Appendix X.2 titled "Recommendation for Incorporation in Contract Documents". This Appendix provides a list of sections of the standard where information provided is general in nature and where more specific information should be provided whenever available.

This recognizes that the number of soil types, water conditions, burial depths, construction methods and pipe materials are such that a standard practice such as D 2321 can not address all possible situations.

IMPROVEMENTS TO D 3839

The proposed revisions to D 3839, the standard practice for installing fiberglass pipe, are still being developed. The general guidance that will be provided in the body of the standard will be similar to that in D 2321 because both standards are concerned with the behavior of flexible pipe. This section will focus on perceived problems with the methods that are currently used to estimate field deflections in the Appendix to D 3839 and modifications being considered to address those problems.

Backfill Soil and Native Soil

When a pipe is installed in a trench both the native material in the trench and backfill material will contribute to the soil support provided to the pipe. This is a function of the stiffness of the two individual materials, the width of the trench and the diameter of the pipe. The Spangler equation for calculating deflection offers no method to treat this situation. Leonhardt (5) developed a simple interaction equation that considers the important parameters and computes a factor(ζ) that, when multiplied by the E' value for the trench backfill, produces a composite value of E'. The Leonhardt expression is:

$$\zeta = \frac{1.662 + .639\left(\frac{B}{D} - 1\right)}{\left(\frac{B}{D} - 1\right) + \left[1.662 - .361\left(\frac{B}{D} - 1\right)\right]\frac{E_s}{E_3}} \qquad \text{Eq. (1)}$$

Where E_s is the value of E' for the trench backfill, E_3 is the value of E' for the native in-situ material and B/D is the ratio of the outside pipe diameter to the trench width. The derivation of this factor assumes that when a trench is five pipe diameters or wider that the native material has no effect on pipe behavior. This Leonhardt factor can be used to help engineers decide how wide a trench must be when the native materials are very soft.

Behavior of Soils

When making analytical models of soils there is always a problem in determining properties because soils are both variable and nonlinear. As was noted, the modulus of soil reaction, E', is semi-empirical in nature and there is no practical test method available to determine a value for this parameter for a specific type of soil. The use of E' implies a linear relationship between load and deflection in a buried pipe, even though actual soil behavior and therefore, buried pipe behavior, is nonlinear. Since the standard values of E' (Figure 1) were determined over a wide range of soil types and over typical ranges of soil depths they are suitable for typical installation conditions; however, if a designer wants values for E' for a specific soil type, or wishes to consider the effects of depth of backfill there is currently no guidance offered in D 3839. The task group revising D 3839 is considering methods to handle these situations as discussed here.

One dimensional modulus: The one dimensional modulus (M_s), also called the constrained modulus, is a measure of the stiffness of a material when subjected to a uniaxial stress and no lateral strain is allowed. It is related to Young's modulus by the equation:

$$M_s = \frac{E(1 - \nu)}{(1 + \nu)(1 - 2\nu)}$$

Eq. (2)

Where E is Young's modulus and ν is Poisson's ratio.

Values for the one dimensional modulus can be determined from a test where the soil is compacted to the desired density in an appropriately sized consolidation ring and then subjected to an increasing amount of vertical stress while monitoring vertical strain. The results can be used to plot a stress strain curve, as shown in Figure 4. The one dimensional modulus for any given vertical stress is the instantaneous slope of the curve at that stress as shown at stress p_s in Figure 4. An average value for M_s over a given range of loads can be computed as the slope of the secant from the starting point of the range to the end point of the range, as shown for stress 0 to p_v in Figure 4.

Research by two of the authors (6) has shown that the one dimensional modulus may be substituted directly for values of E' in the Spangler equation. This allows a designer to run a test on a specific soil to determine soil stiffness. By applying load incrementally and using appropriate values of M_s, as shown in Figure 5, the designer can also approximate nonlinear soil behavior. This method uses lower values of M_s for shallower burials and higher values for deeper burials, even though the soil may be the same type and compacted to the same density. This reflects the effect of greater soil confinement at greater depths of fill.

This increase in soil stiffness with increasing burial depth is an important aspect of buried pipe behavior that designers and installers should bear in mind when specifying pipe installations. Pipe deflection will occur at a higher rate during the early stages of backfilling when the soil is confined by a lower stress.

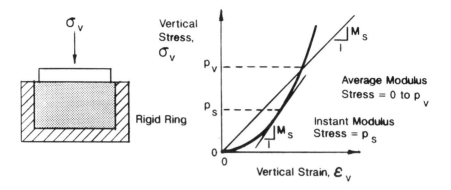

Figure 4 - One - Dimensional Compression Test

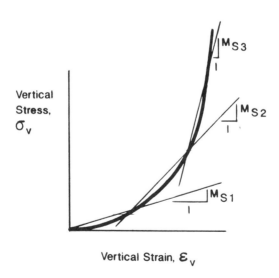

Figure 5 - Approximation of Non - Linear Soil Behavior

The use of the one dimensional modulus gives designers the opportunity to evaluate the actual soils on a project and to better understand how deflections will develop with increasing fill height.

DEFLECTION VARIABILITY

Deflection calculations result in a single value of deflection for a single set of input parameters. In the field, however, deflections will vary along the length of a pipeline as a result of inherent variability in soil, moisture, construction practices and other factors. We call variability "Installation Deflection" because it is a function of the overall installation practice and cannot be predicted by the Spangler equation.

Variability cannot be measured in a laboratory test because it is by nature a result of field conditions. Attempts to measure it in the field are limited by the number and types of installations that are available for measurement. Until more specific guidance is available designers should consider that deflection variability is generally reduced when:

- pipe with higher stiffness is used.

- pipes are installed in soils that do not require a lot of energy to compact to suitable levels of soil stiffness, (i.e. coarse grained soils)

- specified compaction levels are lower.

The last two items suggest that variability is increased with increased passes of compaction equipment, but no studies have been made to verify this.

The most important aspect of variability is that it occurs while the pipe is being installed and while backfill is being placed to the top of the pipe. Spangler's deflection equation (1) only predicts deflection that occurs as fill is placed above the crown. The label "Installation Deflection" helps to emphasize this and increase concern for control of construction practices.

SUMMARY

The reorganization and revision of D 2321 and D 3839 were undertaken to provide engineers with a wider range of information on important installation issues, and to provide it in a format that is easily referenced and readily incorporated into specifications. The revisions reflect a better understanding of the interaction of pipe and soil, and of practical construction problems that can affect that interaction. The standards should help specifiers develop a better understanding of the situations that may occur in the field and prepare for those situations with more detailed specifications. Careful advance study of conditions on any project will generally pay dividends in a more precise specification and fewer field problems.

REFERENCES

[1]   Spangler, M.G., "The Structural Design of Flexible Pipe Culverts", Iowa Engineering Experiment Station, Bulletin 153, 1941.

[2]   Watkins, R.K. and Spangler, M.G., "Some Characteristics of the Modulus of Passive Resistance of Soil: A Study of Similitude", Highway Research Board Proceedings, Vol. 27, 1958.

[3]   Howard, A.K., "Modulus of Soil Reaction (E') Values for Buried Flexible Pipe", Report REC-ERC-77-2, Bureau of Reclamation Engineering and Research Center, Denver, Colorado.

[4]   Selig, E.T., "Soil Parameters for Design of Buried Pipelines", Proceedings of the Conference on Pipeline Infrastructure, American Society of Civil Engineers, New York, NY, 1988, pp 99-116.

[5]   Leonhardt, G., "Die Erdlasten be Uberschutteten Durchlassen", Die Bautechunik, 56(11), 1979, pp 361-368.

[6]   Chambers, R.E., McGrath, T.J., Heger, F.J. Plastic Pipe for Subsurface Drainage of Transportation Facilities, NCHRP Report 225, Transportation Research Board, Washington, D.C. 1980.

Rehabilitation

John E. Collins, Jr.

LARGE DIAMETER SEWER REHABILITATION BY SLIPLINING IN FLORIDA

REFERENCE: Collins, J. E., Jr., "Large Diameter Sewer
Rehabilitation by Sliplining in Florida," Buried Plastic Pipe
Technology, ASTM STP 1093, George S. Buczala and Michael J.
Cassady, Eds., American Society for Testing and Materials,
Philadelphia, 1990.

ABSTRACT: Case history of two large diameter sanitary sewer
rehabilitation projects in Florida. The projects, located
in Jacksonville and Tampa, involved the rehabilitation of
deteriorated reinforced concrete pipe (RCP) by sliplining
new centrifugally cast fiberglass pipe into the interior of
the existing RCP. The projects are described from the
planning stage; including an evaluation of replacement
versus rehabilitation; the design stage, which specified a
pipe product that would be a structural- and corrosion-
resistant alternative to replacement and the construction
stage, ensuring not only conformance to installation
requirements but product compliance through laboratory
testing.

KEY WORDS: rehabilitation by sliplining, structural and
corrosion protection, centrifugally cast fiberglass pipe

Many of the nation's large diameter sanitary sewers installed 20
or more years ago were constructed either with unlined/ unprotected
or coal-tar, epoxy-coated concrete sewers. Today many of the
municipalities who installed these concrete sewers are faced with
the inevitable task of total replacement or rehabilitation of these
concrete sewers due to the corrosive effects of hydrogen sulfide on
the concrete. In the State of Florida the production or generation
of hydrogen sulfide in the sewers is more significant than in most
other areas of the country due to the ideal, warm climate and the
fact that the relatively flat terrain dictates the use of gravity
piping with minimum slopes and requires more pumping and repumping
of wastes. Each of these factors, temperature, flat slopes, and
depth of flow, pressure mains and long detention times of wastes

John E. Collins, Jr., is a Senior Project Manager at Reynolds,
Smith and Hills, 6737 Southpoint Drive South, Jacksonville,
Florida 32216.

contribute to an oxygen-depleting and sulfide-producing environment which ultimately releases hydrogen sulfide to the pipes. The hydrogen sulfide generated in the wastewater stream causes severe deterioration of the concrete and steel reinforcing bars. The effects of the hydrogen sulfide eventually will cause catastrophic failure of these pipes. To combat this problem, many municipalities are now in the process of restoring these concrete pipes so that such major sewer system failures do not occur.

In the cities of Jacksonville and Tampa, it was recognized that a failure of the major interceptor or trunk sewers could seriously affect the health and safety of the citizens of the area as well as the city's economy. Based upon the foregoing, each of the two cities initiated programs in the early 1980s to investigate the conditions of these important infrastructure links and to establish appropriate rehabilitation or replacement requirements.

This paper will discuss two case histories on large diameter sewer system rehabilitation projects in the State of Florida. The two projects, 16th Street Trunk Sewer Rehabilitation - Phase III in Jacksonville, and Main Outlet Interceptor Rehabilitation Project in Tampa, involved the rehabilitation of 1,280 m (4,200 LF) of 1,829-mm (72-inch) diameter reinforced concrete pipe and 3,292 m (10,800 LF) of 1,219-mm and 1,372-mm (48- and 54-inch) diameter reinforced concrete pipe, respectively, by sliplining centrifugally cast fiberglass liner as manufactured by HOBAS. The planning, design, and construction-related services were provided by Reynolds, Smith and Hills, of Jacksonville and Tampa, Florida. The Jacksonville project was completed by Hall Contracting Corporation and the Tampa project was constructed by Kimmins Contracting Corporation. The projects are described from the planning stage, including an evaluation of replacement versus rehabilitation; the design stage, specifying a pipe product that would be both structurally sound and corrosion resistant; and the construction stage, ensuring not only conformance to installation requirements but product compliance through laboratory testing.

PLANNING STAGE

The planning stage in Jacksonville involved the development of a cost-effective program for rehabilitation or replacement of the trunk sewer. This program involved four basic steps:

o   Evaluation of trunk sewer capacity requirements;
o   Evaluation of present sewer conditions;
o   Assessment of rehabilitation/replacement methods; and
o   Formulation of the program plan.

The program involved the study of approximately 5,578 m (18,300 LF) of 1,676-, 1,067- and 2,134-mm (66-, 72-, and 84-inch) diameter reinforced concrete sewer pipe (RCP). Due to the enormous capital expenditures required for replacement or reconstruction of the entire length of this project, the project was divided into six phases. Since the most recent project, Phase III (1,280 m of 1,829-mm diameter) was completed in the last quarter of 1989, this paper will address the success of the Phase III project.

The hydraulic capacity analysis of the trunk line was completed to ensure that the rehabilitated or replaced line would be adequate to carry the required present and future wastewater flow rates. In addition, since many rehabilitation alternatives required removing the line from service, a detailed understanding of by-pass pumping requirements was included in the analysis of the rehabilitation alternatives. The hydraulic capacity analysis was completed by:

1.   Flow monitoring existing trunk sewer to determine
     average and peak flow conditions;
2.   Performing sewer hydraulic capacity comparison
     between existing concrete pipe and new
     rehabilitated pipe (reduced diameter); and
3.   Determining future flow requirements.

Based on this data, it was determined that by-passing facilities installed to divert flow during construction should be designed to handle an average daily flow of 20 MGD and a peak flow of 50 MGD. This meant installing temporary or portable pumps capable of pumping approximately 35,000 GPM plus a major network of aboveground force mains.

Table No. 1 presents the results of the sewer hydraulic capacity comparison between the existing reinforced concrete pipe and a proposed plastic/fiberglass liner pipe or other rehabilitation methodology which reduces the internal diameter. From this evaluation, it was determined that the existing hydraulic capacity of the sewer could be maintained even though the internal diameter (I.D.) was reduced a total of 153 mm to a final I.D. of 1,676 mm. The use of an inserted pipe material such as a fiberglass or plastic liner would, of course, reduce the cross-sectional area of flow but, due to the extremely smooth flow characteristics of fiberglass, the friction factor or roughness coefficient is lower which results in the flow carrying capacity of the new 1,676-mm diameter pipe to be greater than the original 1,829-mm diameter RCP.

A review of the future flow projections for this line found that the existing 1,829-mm RCP or a new 1,676-mm fiberglass/plastic pipe would be adequate to transport flows through this reach.

Table 1 also presents the comparison of the flow capacities of the existing and proposed (planning stage) pipe with the actual pipe installed. The liner pipe, centrifugally cast fiberglass pipe manufactured by HOBAS, had a nominal diameter of 1,676 mm; but due

TABLE 1 -- Sewer Hydraulic Capacity Analysis

City of Jacksonville - 16th Street Trunk Sewer Rehabilitation - Phase III

Existing Internal Diameter	Slope %	Capacity (N=0.013)	Proposed (c) Internal Diameter	Capacity (N=0.010)	Actual (d) Internal Diameter	Capacity (N=0.10)
1,829 mm	0.092 (a)	83.2 MGD	1,676 mm	85.8 MGD	1,708 mm	90.4 MGD
1,829 mm	0.050 (b)	61.3 MGD	1,676 mm	63.2 MGD	1,708 mm	66.4 MGD

(a) - Maximum slope on project.
(b) - Minimum slope on project.
(c) - Based on fiberglass/plastic pipe with n = 0.010.
(d) - Based on actual pipe provided on project;
      Hobas centrifugally cast fiberglass pipe.

to an oversized internal diameter, this pipe was actually 1,708 mm in diameter which provided a flow capacity increase of almost nine percent over the original 1,829-mm diameter RCP and five percent over the capacity of the proposed 1,676-mm diameter.

The next step in the planning program was the evaluation of the structural conditions of the existing sewer. An actual walk-through inspection was performed to measure the depth of concrete or rebar deterioration and core samples of the existing pipe were obtained to perform a detailed structural analysis. The core sampling was used to verify present compressive strength of the concrete, the thickness of the existing concrete pipe, and the depth of corrosion on the inside face of the pipe. The evaluation of this field data determined that the line was in an advanced stage of deterioration and had become severely weakened by the corrosion in the pipe. In the Phase III project, which had an original concrete thickness of 178 mm (seven inches), the concrete loss ranged from 64 mm to 114 mm at the spring line and 38 mm to 89 mm at the crown. The inner layer of reinforcing steel was completely deteriorated. It was estimated that the flexural capacity of the outer cage of reinforcement was just above the capacity required to carry the backfill and live loads on all reaches. The life expectancy of much of this line section was estimated to be, at best, only five years.

In Tampa, the initiation of a planning program in the early 1980s was prompted by various collapses of the city's large diameter concrete interceptors. This program involved the determination of the structural condition of the existing interceptor sewers, an evaluation of the existing sewer line capacity and, ultimately, the development of a program for replacement or rehabilitation. The study to determine the structural stability of the city's interceptor sewers began with an evaluation of techniques for determining the actual condition of the large diameter concrete sewers. Methods such as ground-penetrating radar, ultrasound, and electromagnetic pulse were analyzed. Ultimately, a methodology was developed, referred to as "Sonic-Caliper measurement," to determine the internal dimensions of the sewers. This technique used the travel-time measurement of a sonic signal to determine the distance from a sonic transmitter to the target.

The actual investigation work involved the movement of an instrument raft through the interceptor sewer system. The end result of this effort was longitudinal plots showing original pipe diameter, water level during inspection, measured crown location, and the amount of debris measured in the invert. The study verified the concrete loss in the pipe at an accuracy of 13 mm. Lines found to be in the most critical condition were prioritized for rehabilitation or replacement.

One of the interceptor sewers that was the subject of this study was the Main Outlet Interceptor which was comprised of 2,347 m of 1,372-mm diameter reinforced concrete pipe and 945 m of 1,219-mm diameter reinforced concrete pipe. This line was found to have serious structural problems and rehabilitation/replacement was imminent. Concrete loss in the original 127-mm wall of the 1,372-mm diameter RCP was up to 76 mm in the crown of the pipe.

Prior to selecting a rehabilitation or replacement option, a hydraulic capacity analysis was conducted to ensure that the renovated/replaced sewer interceptor could adequately transport the required wastewater volumes. Table 2 shows that, if a smooth fiberglass/plastic pipe is used, the existing interceptor sewer could be reduced in diameter by 152 mm; 1,372 mm reduced to a 1,219 mm, and 1,219 mm reduced to a 1,067 mm; and still maintain 95% and 91%, respectively, of the two pipes' original flow-carrying capacity. It was determined from this analysis that the flow capacities of the new 1,219- and 1,067-mm diameter pipe would be acceptable for a rehabilitation option such as sliplining. Table 2 also presents the flow comparison of the actual liner pipe installed on the project versus the original and proposed (planning stage) pipes. From this table, it is evident that the use of the HOBAS centrifugally cast fiberglass pipe with an oversized internal diameter (for example, the nominal 1,219-mm diameter was actually 1,245 mm) produced much better results than originally proposed. Instead of a five percent flow loss compared to the original capacity of the 1,372-mm diameter RCP, the HOBAS pipe provided no loss of hydraulic capacity. In comparison to the original 1,067-mm diameter RCP, the centrifugally cast fiberglass pipe improved to only a four-percent loss of hydraulic capacity compared to the previous estimate of nine percent. If factors such as Manning's "n" of 0.009 is taken into consideration, as recommended by the manufacturer, then the nominal 1,067 mm would actually be considered to have no loss of hydraulic capacity.

In both Jacksonville and Tampa, the process for determining the procedure for repairing or replacing the deteriorated sewers involved the evaluation of various rehabilitation techniques and total replacement. Of the many rehabilitation alternatives available, the sliplining of new liner pipe into the existing reinforced concrete pipe was found to be the most favorable option. Basically, the sliplining alternative involved the placement of a new fiberglass/plastic pipe, which is structurally sound and corrosion resistant, into the interior of the deteriorated sewer pipe. The new liner pipe extends the life of the old sewer and eliminates excessive excavation work associated with a new installation.

The sliplining option, a proven method for rehabilitating deteriorated sewers, has been actively utilized in the smaller diameter sewer rehabilitation programs over the last two decades. Due to the recent development of larger diameter plastic/fiberglass products in the U.S., sliplining of large diameter concrete or brick sewers with these new liner pipes has become a viable method for rehabilitation. The greatest advantage of utilizing the slipline process in these two cities was that the relining operation could take place without diverting or by-passing the wastewater flow. Due

TABLE 2 -- Sewer Hydraulic Capacity Analysis

City of Tampa - Main Outlet Interceptor

Existing			Proposed (c)		Actual (d)	
Internal Diameter	Slope %	Capacity (N=0.013)	Internal Diameter	Capacity (N=0.010)	Internal Diameter	Capacity (N=0.10)
1,372 mm	0.065 (a)	32.6 MGD	1,219 mm	30.9 MGD	1,245 mm	32.6 MGD
1,372 mm	0.040 (b)	25.6 MGD	1,219 mm	24.2 MGD	1,245 mm	25.6 MGD
1,219 mm	0.072 (a)	25.0 MGD	1,067 mm	22.7 MGD	1,090 mm	24.1 MGD
1,219 mm	0.070 (b)	24.6 MGD	1,067 mm	22.4 MGD	1,090 mm	23.7 MGD

(a) - Maximum slope on project.
(b) - Minimum slope on project.
(c) - Based on fiberglass/plastic pipe with $n = 0.010$.
(d) - Based on actual pipe provided on project; Hobas centrifugally cast fiberglass pipe.

to the quantities of wastewater being transported in these large sewers, by-passing is a very costly procedure.

To compare the selected slipline rehabilitation alternative to the total replacement option, the following criteria or factors were evaluated:

1. Cost,
2. Constructability,
3. Corrosion resistance,
4. Durability, and
5. Disturbance.

A cost comparison developed between the two options found that, in both cities, sliplining of fiberglass liner pipe was a more cost-effective alternative than total replacement. For example, in Jacksonville it was found that total replacement would cost approximately 70% more than the sliplining. This was attributed to several factors:

o   By-passing costs are either substantially reduced or not required during the sliplining project. By-pass requirements during the replacement project involve continuous operation of large electric or diesel pumps and maintenance of temporary force mains 24 hours a day for several weeks or months.

o   Compared to sliplining, excavation work during replacement is very extensive. Excavation for a 1,372- or a 1,829-mm diameter pipe requires trenches up to 2.7 to 3.7 m wide for the entire length and depth of the installation. The excavation work required for the sliplining project is limited to 3.7 m by 9.1 m insertion pits located every 229 to 305 m along the length of the project. Therefore, sheeting, dewatering, and backfilling costs are also substantially reduced for sliplining.

o   Utility relocation costs for potable water, underground, electric, telephone, cable television, gas, storm, fiber optics, etc., is much greater for the replacement option due to the extensive excavation work along the pipeline route. Collector sewer reconnection to the new sewer is also another factor that pushes the replacement cost higher.

The constructability factor of a project is an attempt to assess the complexity of the construction or rehabilitation project. Each alternative involves the performance of a number of sequential tasks which, when successfully completed, result in the finished product.

The potential for undetected quality or workmanship problems is also included in this factor. The sliplining option was considered to be the least complex procedure. The rationale for this is that sliplining has less potential for installation problems since the construction activities are limited to the insertion operation and grouting of the pipe annulus. Replacement involves significantly more construction tasks, e.g., excavation and preparation of subgrade, pipe bedding compaction, which can have a significant effect on the successful construction of this alternative.

Corrosion resistance was probably one of the most important considerations in the evaluation process because of the importance of providing a system with long-term protection against hydrogen sulfide and other corrosive acids. The fiberglass/plastic liner pipe used in the sliplining work is manufactured from fibers or resins which are resistant to specific corrosive environments. Since the fiberglass/plastic pipe can also be considered as an alternative for direct burial, the two alternatives were rated equal. In the case of the installation of new reinforced concrete pipe with mechanically bounded PVC, the sliplining option would rate somewhat higher.

Durability, as it relates to life expectancy or the endurance characteristics of a material or structure, is a concern. Since the fiberglass/plastic pipe could be considered for direct burial, the replacement and sliplining alternatives were considered equal.

Disturbance, such as noise, traffic problems, public inconvenience, etc., occurs on either replacement or sliplining. Conventional construction or installation procedures would create significantly more traffic disruptions and public inconvenience than sliplining. In the congested urban areas of Tampa and Jacksonville, sliplining requires a minimum interruption of vehicular and emergency traffic. Also, since the installation work is being performed in these congested residential, commercial, and industrial rights-of-way, relocating other utilities (water, gas, fiber optics, etc.) to accommodate this new line is almost impossible. Therefore, it was determined that the sliplining alternative would result in the least disturbance to these communities.

When all these factors (cost, constructability, corrosion resistance, durability, and disturbance) were considered in the evaluation of the sliplining or total replacement options, the sliplining method was found to be the most viable alternative.

DESIGN PHASE

Since the structural integrity of the large diameter sewers was less than acceptable in both municipalities, the potential for sewer collapse has been a major concern of these cities for a number of years. Therefore, the design criteria for the project had to be written to ensure that the sliplined pipe was designed and manufactured to be corrosion resistant, structurally sound, and installed properly so that the newly rehabilitated system would perform for many decades. The municipalities not only wanted to provide a continuation of sewer service, they wanted assurances that the public's health and welfare was protected for the present and into the future. Both municipalities, aware of the hazards associated with the failure of large diameter sewers, fully expected the completion of these rehabilitation projects to provide long-term "peace of mind."

The design criteria for the manufacturing of the liner pipe used on the project was established as follows:

1. No consideration of structural load-carrying support was given for the existing concrete pipe. Liner pipe must be designed as a total structural replacement.

2. Pipe must withstand all dead loads such as soil weight and live loads such as H20-44 highway loading or Cooper E80 railroad loading. Weight of soil was given as 120 lbs/cu.ft. and depths were as detailed on the drawings.

3. Water table (hydrostatic pressure) was established as being one (0.3 m) foot below finished grades.

4. Liner pipe must withstand a minimum of five psi grouting pressure.

5. Pipe material must be corrosion resistant and unaffected by hydrogen sulfide and other corrosive gases normally found in domestic wastewater streams.

6. Pipe must withstand all jacking loads during installation.

7. Minimum pipe stiffness, in accordance with ASTM D-2412, was established as 18 psi for Jacksonville and 36 psi for Tampa.

8. Pipe must be manufactured and tested in accordance with applicable nationally recognized AWWA and ASTM standards. Centrifugally cast fiberglass pipe installed on the project was manufactured in accordance with AWWA C950.

9. A minimum factor of safety of 2.5 was required on all design calculations.

In order to ensure that the liner pipe manufacturer strictly adhered to the project design criteria, each Bidder, as a requirement of the contract documents, had to provide a Design Submittal with his bid proposal. This procedure allowed the Engineer to review all design calculations from the manufacturer so that product compliance could be confirmed prior to award of the contract. By analyzing the pipe product (at least from a design standpoint) and locking into a particular manufacturer prior to the bid award, meant that the cities would know, precisely, the pipe product to be utilized before contracts were signed and would also benefit from the best overall pipe price. Many times after the lowest bidder has received the contract, the bidder, by using the leverage of a signed contract, may shop around for a lower pipe price which would not benefit the municipality.

The specifications developed for the project included procedures for cleaning and preparing the existing concrete pipe, installing the liner pipe and performing the grouting work. The key to the success of the installation of the new liner pipe is the proper cleaning and preparing of the deteriorated concrete sewers. Specifications were written to ensure that all sand, sludge, deteriorated rebar, and debris were removed from the sewer line. Project guidelines also detailed that all obstructions and leaks were removed or repaired prior to attempting the slipline activity.

Since the pipe installation guidelines as published by ASTM or AWWA for the fiberglass/plastic pipe are written around direct-burial installation only, specifications had to be developed for the slipline operation. In lieu of detail procedures directing the contractor as to how to install the pipe, certain guidelines were established to ensure proper installation. These guidelines were

1) Optimum location of pits to control length of jacking operation;
2) Homing marks installed on all pipe to ensure proper joint installation;
3) Limitations on pipe deflection and joint separation; and
4) Limitations on length of pipe in order to minimize use of pipe joints.

The actual liner pipes utilized on the 16th Street Trunk Sewer Rehabilitation - Phase III and Main Outlet Interceptor were HOBAS centrifugally cast fiberglass pipe. These pipes were manufactured in accordance with ASTM D3262 and AWWA C950. The pipe, including fabricated bends and shorts, were designed to withstand all dead and live loads, external hydrostatic pressure and grout pressures. The pipe was designed to resist buckling in accordance with AWWA C950. The buckling analysis accounted for a combination of dead, live, and hydrostatic loads and a modulus of soil reaction (E) of 2000 psi.

CONSTRUCTION PHASE

The construction phase activities related to sliplining vary substantially from conventional direct burial pipeline installation. First and foremost is the minimum amount of excavation work that is required to complete the slipline project. Sliplining involves only periodic excavations along the route compared to a continuous excavation for direct-bury. This is a tremendous advantage in the congested city rights-of-ways of today. Generally, the sliplining procedure is conducted as follows:

o   Construction of insertion pits,
o   Preparatory cleaning and pipe condition verification,
o   Insertion of new liner pipe,
o   Liner pipe closure, and
o   Grout fill annular space between liner and existing concrete pipe.

The insertion pits used on the two projects were approximately 3.7 by 9.1 m except at certain pipe bends which were 3.7 by 12.2 m or 3.7 by 18.3 m. The construction of these pits included steel sheeting, excavation, utility relocation, removal of crown of RCP, backfilling and compaction. Pits were located at various intervals to facilitate the pushing of the new liner or to allow the installation of a prefabricated bend.

The preparatory cleaning phase of the project is the removal of solids, sludge, rocks or dislodged pipe materials (concrete and rebar) to allow the designated sewer section to be rehabilitated. This operation also includes the removal, dewatering and proper disposal of the sewer waste. Due to the huge volume of debris in these large diameter sewers, the cleaning operation is the most time-consuming component of the project. The contractor can spend 80% of the total contract time in just the cleaning effort alone. Also, the limited availability and effectiveness of conventional sewer cleaning equipment may require the contractor to develop his own equipment and/or procedures. The cleaning effort/procedure is the major reason why sliplining of large diameter sewers is so much more difficult than conventional sliplining of small diameter sewers.

Prior to beginning the sliplining work, the exiting pipe is inspected either visually or by television camera to ensure that no blockages or major leaks are occurring in the concrete pipe. Any obstructions or major leaks are repaired prior to inserting the new liner pipe.

The actual installation of the liner pipe can be a smooth operation if the sewer has been properly cleaned and sufficient annular space is available between the liner and the existing pipe. The centrifugally cast fiberglass pipe utilized on the Jacksonville project has an outside diameter of 1,758 mm which provided an overall clearance of 71 mm in the 1,829-mm pipe. In Tampa, the outside diameter was 1,311 mm and 1,151 mm which provided clearances of 61 mm in the 1,372-mm pipe and 69 mm in the 1,219-mm pipe.

The sliplining begins by lowering the liner pipe into the insertion pit and placing the pipe into the open concrete pipe. The liner pipe is then pushed into the existing sewer. The pipe is bell and spigot and all joining of the pipe sections is completed in the pit area. Depending on flow conditions, the liner pipe can be pushed in both directions.

Prior to installation of any liner pipe on either project, a close inspection was performed on each piece of pipe by the project inspector. The pipe was inspected for defects such as cracks, splits, imperfect bell or spigot, and general defects which would effect the pipe's performance and/or joining ability.

During the installation, certain guidelines were established to ensure that proper installation procedures were utilized. Generally these guidelines were:

1) Existing line is to be properly cleaned prior to installation of line;
2) Bell and spigots shall be as clean as possible and gasket properly placed;
3) Bell and spigots shall be "homed" in accordance with manufacturer's homing marks;
4) Bell and spigots shall be protected from mechanical injury; and
5) Limitation on length of liner pipe.

After sliplining in both directions of the insertion pit is completed, the new line is closed with a prefabricated bend or mechanical coupling. The insertion pits are then closed by pulling sheeting, backfilling/compacting the excavated soil and then restoring surface materials.

The rehabilitation work is completed by completely grout-filling the annular space between the liner and the concrete pipe.

In Jacksonville, the construction project involved the installation of 1,280 m of 1,676-mm nominal diameter HOBAS centrifugally cast fiberglass pipe. Prior to the shipping of the liner pipe to the job-site, the manufacturer performed project verification testing on the new pipe. The testing was performed at the manufacturer's North Florida plant and witnessed by the Engineer. The testing performed on the pipe was

o   Pipe stiffness testing in accordance with ASTM D2412, and
o   Hydrostatic pipe testing of 10 psig for three minutes.

The results of the tests proved that the pipe products met the project design criteria and the pipe was approved for shipment.

On the Phase III project in Jacksonville, the pipe manufacturer and contractor had to overcome several difficult applications. These problem areas were:

1)   Large quantities of sand and dislocated, hanging rebar presented a monumental cleaning task;
2)   Four 1,829-mm ninety-degree bends;
3)   Existing pipe sections under railroad tracks were deemed as critical due to concrete deterioration; and
4)   Approximately 213 m of the trunkline had very serious structural problems. Concrete loss was so severe that only 25 to 51 mm of the original 178-mm thick wall remained at the crown. Since this pipe section had experienced two previous cave-ins, an emergency repair plan had to be implemented in case of failure prior to the completion of the sliplining.

The cleaning work required an enormous effort by the contractor to remove, dewater, and dispose of the wastes in the trunkline. This was complicated by the fact that new debris was constantly entering the trunk sewer through a collector line which tied into one of the upstream sections of the project and that several large rainfall events occurred during the project which transported debris into the construction area. Eventually, the pipe was adequately cleaned and the sliplining work proceeded.

The ninety-degree bends on the project were repaired by constructing insertion pits over the bend area and installing prefabricated centrifugally cast fiberglass pipe elbows. These bends were closed by mechanical couplings when required.

The repair of the critical sections of the concrete pipe was accomplished successfully without having to implement any emergency plan of action. In fact, the contractor installed over 488 m of 1,676-mm diameter fiberglass pipe in the 1,829-mm concrete pipe in less than one week. The sliplining accomplished during this week included the critical sewer sections. Therefore, the completion of sliplining in these areas prevented the occurrence of a serious public safety and health hazard.

The Tampa sliplining project involved the rehabilitation of 3,292 m of 1,372-mm and 1,219-mm diameter pipe by installing 1,219-mm and 1,067-mm nominal diameter HOBAS centrifugally cast fiberglass pipe. Confirmatory pipe product testing was performed at the manufacturer's plant in the presence of the city and the Engineer. These tests were comprised of the parallel plate test or pipe stiffness test in accordance with ASTM D2412 and hydrostatic pipe testing. The tests confirmed the pipe design and the pipe was shipped to the project site. Several items on this project presented potential problems to Kimmins Contracting Corporation, the contractor for the Main Outlet. The single most difficult problem was the cleaning of the interceptor line. Unfortunately, debris in a major portion of the 1,372-mm diameter concrete sewer was at a depth of 457 to 914 mm. This required a 24-hour cleaning operation to effectively clean this reach for sliplining. After many months of cleaning these sewer sections, the contractor was set to begin his sliplining work.

Other potential obstacles were a 53-degree bend with a radius of approximately 366 m and four short radius bends of 20, 22, 51, and 90 degrees. The long radius bend was sliplined by inserting three m sections of fiberglass pipe into the 1,371-mm diameter concrete pipe. This work was completed without any problems due to the smooth outside surface of the fiberglass pipe and the shortened length of the pipe.

The 22- and 90-degree bends were rehabilitated by constructing insertion pits and installing prefabricated bends or elbows. The other bends were completed with short lengths of fiberglass liner pipe.

After the installation of all liner pipe was completed on these two trunk sewer systems at Jacksonville and Tampa, the annular space between the liner pipe and existing reinforced concrete was completely grouted with 3,000 psi concrete. The grouting of the void not only secured the fiberglass pipe in place permanently but prevented any future collapse of the existing deteriorated concrete pipe.

CONCLUSION

The completion of the two rehabilitation projects marked an end to the serious concern each municipality had about the toll that corrosion, age, and wear had taken on their concrete trunk sewer systems. The exceptional corrosion resistance and structural stability of the new fiberglass liner pipe assures the cities that the public welfare, safety, and economy has been protected.

By selecting the sliplining rehabilitation alternative, these municipalities were saved the inconvenience and social costs associated with the conventional direct-bury installation. These social impacts--such as disrupting vehicular, emergency, and pedestrian traffic, and creating noise, vibration, and air pollution during the extensive trench construction; and social costs--such as utility relocation, by-pass pumping, extensive excavation, dewatering, backfill, and pavement replacement associated with conventional pipeline construction--were substantially reduced on these sliplining projects. The success of the rehabilitation program in these two municipalities proves that the "sliplining" option was the most viable and cost-effective alternative for restoring the structural integrity and corrosion resistance of these large diameter sewers.

312   BURIED PLASTIC PIPE TECHNOLOGY

ACKNOWLEDGEMENTS

The author acknowledges with appreciation the help of Mr.
William B. Stanwix-Hay, Public Utilities, City of Jacksonville; Mr.
Andrew T. Cronberg, Department of Sanitary Sewers, City of Tampa;
Mr. Kyle Kovacs, Hall Contracting Corporation, Mr. John Simon,
Kimmins Contracting Corporation and Mr. J.A. Finch, Price Brothers
Composite Pipe, Inc.

Reduction, Reversion, Renovation:R^3

REFERENCE: Svetlik, H. E., "Reduction, Renovation, Reversion:R^3", Buried Plastic Pipe Technology, ASTM STP 1093, George S. Buczala and Michael J. Cassady, Eds., American Society for Testing and Materials, Philadephia, 1990.

ABSTRACT

Polyethylene pipes for industrial and municipal uses have been installed since 1960, thirty years ago. Insert Renewal of gravity flow sewers is a widely accepted process using high density polyethylene pipe (ie: a pipe in a pipe). The placement of high density polyethylene (HDPE) liner in a high pressure metal pipe began in the early 1960's. That 60's technology was advanced during the 1980's using an expanded liner approach based on the ductility, toughness, and durability of HDPE, as well as using more effective installation equipment, product design, and construction methods. The 1980's innovations were directed towards longer distance, higher pressure, transport pipelines in sizes of 2" (51mm.) through 24" (610mm.) diameter. During the last decade, a new proliferation of HDPE lining products and processes have evolved to meet the challenges of the 1990's. This paper briefly evaluates the requirements and effect of each design and installation approach on the HDPE liner. The growing industry of HDPE lining of pipes has expanded to oil patch downhole tubulars, transport pipelines (oil, gas, and product), industrial/municipal water and gas distribution mains, and in some cases, municipal sewers. The four generic processes considered herein are:

1. Liner Expansion
2. Liner Rolldown Reduction
3. Liner Hot Swage Reduction
4. Liner Visco-elastic Reduction

KEYWORDS: high-density polyethylene, liner, liner expansion, rolldown, swage reduction, visco-elastic reduction, rehabilitation.

Harvey Svetlik is the Senior Technical Representative for Phillips Driscopipe Inc.'s industrial, municipal, and defense pipe products located at 2929 North Central Expressway; Richardson, Texas 75083.

GENERAL HDPE CHARACTERISTICS

Inert polyethylene is used as a liner due to its material characteristics, chemical resistance, excellent "barrier" properties, toughness, ductility, durability, handability, and acceptable economic cost.

Polyethylene is a visco-elastic thermoplastic material. In pipe grade polyethylene there are four "ingredients" that give each polyethylene its specific characteristics: copolymer chemical composition and morphology, density, weight average molecular weight, and molecular weight distribution. Generally, the high density higher molecular weight polyethylenes are tougher, ductile, fatigue resistant and more "elastic" compared to lower density lower molecular weight polyethylene which tends to be more visco-elastic, with reduced toughness. Although all pipe grade polyethylene resins are generally characterized as being tough, ductile, and visco-elastic; specific resins exhibit differing degrees of elasticity, visco-elasticity, and toughness.

Polyethylene exhibits both elastic and visco-elastic characteristics. For elastic materials, stress is said to be a function of strain only:

$$\sigma = E \cdot \varepsilon$$

Where: σ = Stress (psi)
E = Modulus of Elasticity (psi)
ε = Strain (in/in)

For viscous materials, the stress generated in the material depends not only on the strain but also on the rate at which strain is applied:

$$\sigma_{Total} = \sigma_{Spring} + \sigma_{Dashpot}$$

$$\sigma_T = E\varepsilon + \frac{\varepsilon \eta}{t}$$

$$\sigma_T = E\varepsilon + \dot{\varepsilon}\eta$$

Where: η = Coefficient of Elasticity

t = time
$\dot{\varepsilon}$ = Strain Rate

By endowing a material with an instantaneous elastic response together with a time-dependent displacement, one can obtain a reasonable approximation of the behavior of a visco-elastic material. The simplest model having such characteristics is the combination in series of a linear spring and a linear dashpot.

A force applied to the combined series develop a uniform stress: Total = Spring + Dashpot. However, total strain is the sum of the spring and dashpot displacements: $E_T = E_S + E_D$. Dividing the strain by the time duration over which it developed, the <u>strain rate</u> is known. The spring component is considered to be the elastic characteristic and the dashpot the visco-elastic characteristic, as illustrated in Figure 1, A through D.

$$\varepsilon_T = \varepsilon_S + \varepsilon_D$$

$$\frac{\varepsilon_T}{t} = \frac{\varepsilon_S}{t} + \frac{\varepsilon_D}{t}$$

$$\dot{\varepsilon}_T = \dot{\varepsilon}_S + \dot{\varepsilon}_D$$

$$\dot{\varepsilon}_T = \frac{\dot{\sigma}}{E} + \frac{\sigma}{\eta}$$

$$- O R -$$

$$E\,\dot{\varepsilon} = \dot{\sigma} + \frac{\sigma E}{\eta}$$

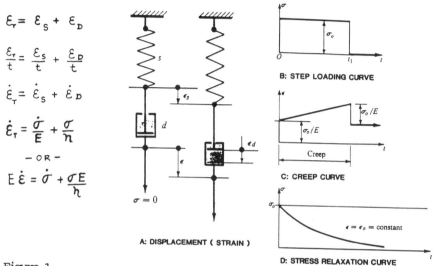

B: STEP LOADING CURVE

C: CREEP CURVE

D: STRESS RELAXATION CURVE

A: DISPLACEMENT (STRAIN)

Figure 1
Representations of Viscoelastic Stress:VS:Strain

The modulus of elasticity and the coefficient of elasticity are strongly temperature dependent.

Each of these four pipeline renovation processes use the elastic and visco-elastic properties of pipe-grade HDPE liner resins in different ways to effect pipe rehabilitation. Specific hardware is necessary in each case to manipulate the PE material properties in a predetermined way.

<u>Expanded Liner</u>

The expanded HDPE liner approach is a proven concept with a 30 year history of performance. Original installations were made in Oklahoma's oil-patch flow lines, downhole tubulars, and brine disposal systems.

Fundamentally, a custom diameter liner of relatively thin wall is extruded from a high performance pipe grade resin so as to provide a 3% to 8% annular clearance between the liner OD and metal pipe ID. As pressure is applied interior to the liner, it strains within its ductile, proportional limit to contact the metal pipe I.D. The pressure maintains the expanded condition while the HDPE resin begins to visco-elastically stress-relieve itself and accept the larger metal pipe ID as its own new permanent OD. The liner becomes molded to the metal pipe I.D. with time.

In all cases, the liner is in radial compression due to the flow stream operational pressure.

In some cases the "molding" or visco-elastic strain of the liner to the larger metal pipe ID can be accelerated by pressure testing the liner at a pressure which exceeds the compressive yield strength of the liner material, or, in the cases of lower test pressures, by warming or heating the liner to lower the compressive yield strength of the liner to within the test or operating pressure of the lined pipeline. All three methods are practiced.

Hence, expanded liner is installed undersize and is visco-elastically compressively molded to the host pipe I.D.

The obvious impact of this process is that ambient temperature flow stream pressures usually greater than 100 psi are generally needed to initiate expansion, visco-elastic strain, and molding of the liner to the pipe ID. Use of hot water or low pressure steam to accelerate compressive molding of the liner works effectively but adds to installation cost.

Over 800 miles of expanded liner have been installed in the last decade with installation lengths averaging 2500 ft to more than a mile. The expanded liner design and installation is cost effective, especially when evaluated on the basis of an actual 20 to 30 year service life, with the expectation of more than 50 years of serviceable life.

Liner Rolldown Reduction

For gravity flow or low pressure pipelines, when an internal corrosion liner is needed, the expanded liner concept cannot be applied. In the mid 1980's a semi-elliptical, dual roller machine was developed which diametrically reduces standard OD high-density polyethylene pipes to an OD sufficiently small to allow insertion of the liner into a metal, cast iron or concrete pipeline.

This was useful in two ways. First, the liner could actually be made with a thicker wall which then could function as a "snug" pipe within a pipe offering maximum flow cross-sectional area and the pressure containment capability of a pipe; and which could then be "tapped" as a custom diameter pipe to distribute water or gas. Second, if a thinner wall HDPE liner were used, it also offered maximum flowstream area while sealing leaking, mechanical joints, gaskets, or pit cell corrosion sites. The thinner liner provides corrosion protection and leak sealing but is usually not "tapped".

In the process of rolling down the liner with a pair of dual rollers, each pair oriented 90° to the other, the liner is "instantaneously" reduced in O.D. as it passes across the tangent point of each roller pair.

Looking at the elastic recovery and the visco-elastic recovery, the liner has to be squeezed significantly to force a visco-elastic diametral reduction which would last sufficiently long to allow liner installation. This significant squeeze is needed due to the very short duration of roller contact. Simultaneously, the percentage of diametral reduction must be sufficiently large to provide a small percentage of permanent undersizing. Hence, by diametrally squeezing the pipe with rollers, the liner is made smaller. Part of this reduction is permanent, part is immediately recoverable (elastic), and a small part is near term recoverable (visco-elastic). If the liner thickness, roll configuration, and other factors are not carefully controlled, the liner may end up only "snug" not tight; and/or it may recover too quickly and lock-up in the host pipeline halfway through the segment being lined.

Axial tension on the HDPE liner being processed through the rollers will tend to hold the diametral reduction by reducing the rate of elastic and visco-elastic recovery as related by Poission's Ratio to the axial length. When the axial tension is released, the diameter recovers and the length shortens (ie: like the "Chinese Finger") with elastic and visco-elastic recovery. This generally leaves this liner in radial compression and axial tension relative to the host pipe I.D.

If the liner is driven or compressively forced through the rollers by means of a circumferential hydraulic and mechanical clamp, the roller reduction generates slight axial elongation and some degree of radial wall thickening. Care is exercised to develop an exact roller configuration such that the roller's compressive force does not cause wall crushing at the horizontal spring line separating the two rollers. Wall crushing at the spring line would result in non-uniform wall thickening and the possibility of "wrinkling" in the pipe wall.

The tangent velocity of the dual rollers is different dependent upon the point of contact with the roller. At the crown of the pipe and at the horizontal spring line of the roller, there may exist up to a 50% difference in rotational velocity resulting in non-uniform slippage of the pipe through the roller depending upon the tangent point, such that roller pair at the horizontal spring-line tends to "work" the liner surface-layer by slippage resulting in some surface layer deformation.

The end user should receive test data showing the effective SDR is maintained in the reduced liner, that satisfactory wall uniformity is maintained; and that the horizontal spring line is as strong as the remaining liner or pipe wall. If the roll reduction method is to be used to form a downsized pipe in a pipe, a stress life curve showing pressure and service life equal to the original pipe might be requested.

If several schedules of steel pipe are used in the host pipeline, the liner might not be tight or even snug, but "loose" where diametral recovery was insufficient. In the loose areas, the liner would then function as an expanded liner under higher pressure, or as a pipe at lower pressure.

If the loose liner did not contact the host pipe ID, at lower pressure it would be subject to circumferential tension due to pressure, and as such, may be acting as an over-pressurized thin wall plastic pipe subject to the long term hydrostatic strength of the HDPE pipe grade resin. Hence, this "liner" portion would be a pipe and the tensile stress life curve would apply. The tensile stress life curve does not apply to liner in radial compression.

To insure this loose liner situation does not occur, a strong pressure test of the liner and pipeline is usually done to re-stretch the previously compressed HDPE molecular structure, which should put the liner in contact with the host pipe ID.

Hence, the rolldown liner reduction method processes HDPE tubular profile both as an insert pipe or a liner. But there appears to be some limitations to its processing of HDPE pipe/liner. The graphs at the end of this paper (which are explained in the next two sections) apply to rolldown and the other liner reduction methods.

Liner Hot Swage Reduction

As previously noted, the modulus of elasticity and the coefficient of elasticity are strongly dependent upon temperature. The colder it gets, the swaging force increases. As the temperature gets hotter, the swaging force and magnitude of elastic recovery decreases.

Hence, there is an optimum temperature range within which to "work" the polyethylene liner.

Hot swaging of the HDPE liner consists of introducing pre-fused continuous HDPE liner (with external fusion beads removed) into a pre-heating chamber. Residence time in the heating zone is affected by the pipe wall thickness, hot air ambient temperature, and initial pipe wall temperature. The pipe is heated above 200°F (93°C) for a predetermined time until the pipe warms beyond a critical temperature on the pipe ID, thought to be about 140°F (60°C).

Hence, there is a thermal gradient in the pipe wall.

But the strength modulus is significantly reduced and the coefficient of elasticity begins to decline. At elevated temperature and strains of 10% to 18%, some permanent strain (diametral reduction) is easily induced. Elevated temperature reduces: the force needed to push the liner into the swaging die; the pull force needed to keep the liner straight; and the tension on the line to minimize elastic and visco-elastic recovery. Even so, the average length of installations to date in the USA averages 500 ft to 1000 (200m to 300m) ft using hot swage reduction.

Although the swaged liner system is intended for use with standard IPS polyethylene pipe, because of the variables in ID and wall thicknesses for steel, cast iron, clay, and cement pipe, it appears some non-standard HDPE liner OD's may be required to provide for insertion with complete diametral recovery of the HDPE liner to all ID's of the pipeline being lined.

Running at 8 ft to 10 ft per minute (3 meters/min.), this process takes about an hour to insert a 500 ft long liner. Colder temperatures, rain and thicker wall liners may affect the rate of insertion.

In order to keep the hot liner at processing temperature, the sewer flowstream or product flowstream must be stopped or by-passed.

When the air temperature gets to about 30°F (0°C), roll reduction does not work well because the coefficient of elasticity approaches nearly zero thus locking in the mechanical roller's reduction strain with virtually no elastic recovery.

Conversely, even in freezing weather, the hot swage lining technique creates an artificial temperature controlled environment. This temperature range provides a zone for material workability while allowing sufficient time for insertion prior to diametrical recovery.

Once inserted, the hot swage liner will diametrically enlarge and axially foreshorten somewhat. After thermal and dimensional stabilization occurs, the ends are usually locked into position by one of several means. Some of the end treatments are internal stiffners which internally, stiffen and lock the end in position; externally applied clamps with inserts; and an imported electrofusion device. Pre-fabricated plastic flanges are not currently used. Allow the manufacturer/installers to describe additional closures.

After installation with end treatments, custom radius side - wall taps might be made if the liner were to be used as a pipe. If the liner were used in a sewer conduit, lateral portals would be cut in 6" and larger diameter sewers. Laterals are usually located by electronic or video detection from the house lateral through the liner wall.

Current projects and equipment cover sizes from 4" (114mm) to 24" (610mm) pipe, with plans for 24" (610mm) to 48" (1220mm) sizes.

Liner Visco-Elastic Reduction:

This process systematically reduces the OD of standard polyethylene pipe sizes to make the pipe temporarily undersize to facilitate its insertion into a host pipeline. It mechanically works the polyethylene to quickly force a diametrical reduction, but then holds the size reduction (compressive strain) for an appropriate time to allow time for the visco-elastic component of the PE material to stress relax under fixed compressive strains.

This reduces the severity of the diametral reduction required to produce undersizing lasting an extended duration. It provides for a greater degree or percentage of recovery from the visco-elastic portion of the dimetrical size reduction. And it preserves the effective SDR of the liner or pipe.

The combined effect is to induce a smaller percentage of permanent deformation; to handle the natural elastic, quickly recoverable deformation, and to provide a greater degree of recoverable deformation from the visco-elastic memory of the high molecular weight polymer (recoverable creep).

The primary size reduction device consists of several staged multi-roller sets. Each roller set consists of three or more roller segments dependent upon liner size. Multiple segmented roller sets provides uniform diametral reduction with uniform wall thickening to maintain the liner SDR with minimal axial elongation, while avoiding wall cracking, crushing and wrinkling/distortion of the liner O.D.

The secondary size/maintenance device consists of a set of proprietary tooling of required length and internal diameter to hold the liner to its reduced diameter for an appropriate duration such that elastic recovery is prevented, and visco-elastic creep is forced to occur in the liner wall.

Upon exiting the machine, the liner exhibits about 4% to 6% elastic recovery very quickly; followed by a larger degree of recoverable creep over a longer period.

Dependent upon the pipeline size, size reductions (strains) of 9% to 19% have been used. With 4% to 6% immediate elastic recovery, recoverable creep of 5% to 15% may be achieved dependent upon the specific liner material used and the process temperature.

From a phenomenological viewpoint, for individual pipe grade high molecular weight high density polyethylenes, the stress relaxation, deformation and creep recovery curves must be considered fundamental properties of a specific resin and must be determined experimentally. Important parameters to be defined for a given resin, pipe size, and wall thickness are illustrated in the attached graphs.

Figure 2 illustrates three different percentage diametral reductions. The greatest reduction does not allow the liner to fully recover hence making it a "snug" pipe. The least reduction shows the slight permanent deformation of the reduction process. The middle reduction illustrates the optimum reduction with elastic recovery and visco-elastic recovery sufficiently slow (1 hr to 5 hrs) to permit installation followed by diametrical contact having an interference fit leaving the liner in radial compression and slight axial tension.

Figure 3 is a study of recoverable creep versus % diametrical reduction at given temperatures. For a given material and a host pipe ID, the liner size and % reduction can then be engineered.

Figure 4 avoids yielding of the liner in tension or compression based upon diametral reduction and temperature. If the liner were to be pulled through the processing machine and pipeline, loss of the pulling head would be catastrophic. Setting tension limits avoids this.

Figure 5 predicts the expected working time available for liner insertion and hence the maximum workable pipeline segment length based on the speed of the reduction hardware.

Figure No. 2

Diametral Reduction
    versus

Working Time .... To Achieve Interference Fit.

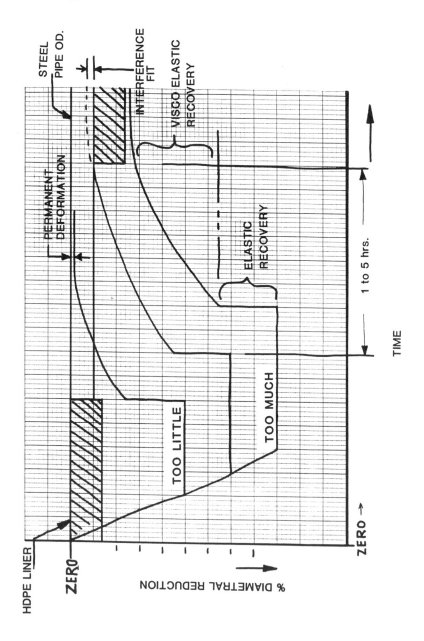

Figure No. 3

Recoverable Deformation
versus
Diameteral Reduction.

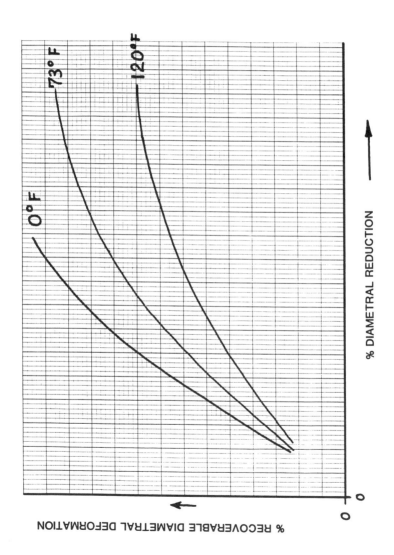

Figure No. 4

Axial Forces versus Diameteral Reduction.

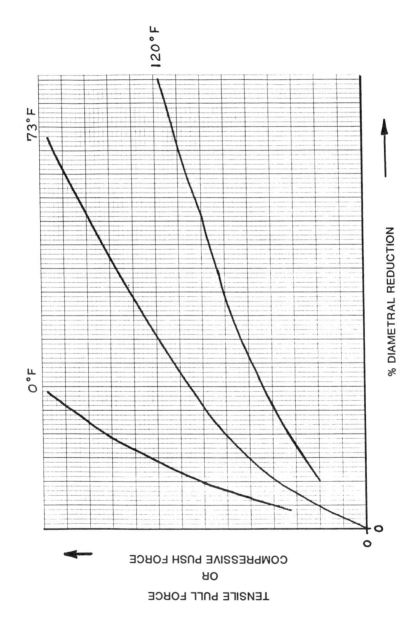

Figure No. 5

Working Time versus Diameteral Reduction.

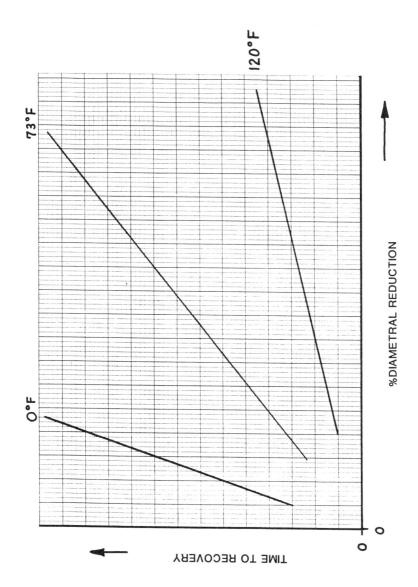

Figure No. 6

Dimensional Changes versus Diameteral Reduction.

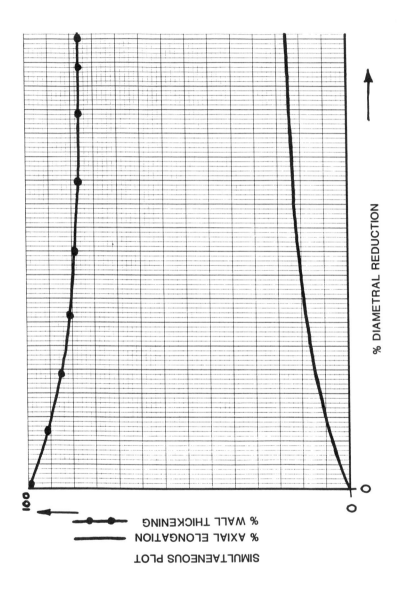

Figure No. 7

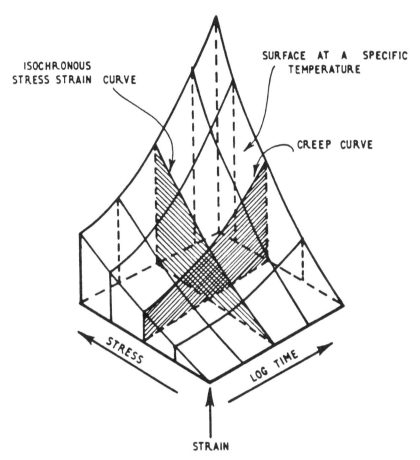

A three-dimensional plot of stress-strain and time. (From *Structural Design with Plastics*, B. S. Benjamin, Van Nostrand Reinhold, 1964)

Figure 6 should illustrate the effect of the process or
machinery on the liner OD, wall and length. With the liner
volume being conserved, transference of diametral reduction into
radial wall thickening with minimal elongation is the most
desirable condition.

Figure 7 illustrates the basic, three dimensional "surface"
relating stress, strain, and time at a given temperature for a
specific polymer. The relationship of the isochronus
stress-strain curve and creep curve is also shown.

The visco-elastic reduction method incorporates the features and
benefits of the hot swage process while gaining production speed
and lengthening working time, thus providing for longer
insertion lengths.

End treatments seem to include pre-fabricated plastic flanges,
thermally formed flanges and mechanical terminations similar to
other processes. The current size range is 2", 3" and 4"
(51mm-114mm) pipelines and tubulars. Future plans may include
6" to 12" followed later with sizes to 24" (168mm-610mm).

Conclusion:

Polyethylene tubular profile can be used as pipe subject to hoop
tension, or, as liners in radial compression. Expanded liners
are usually used in higher pressure pipelines. Expanded liners
are put into service immediately and become molded to the pipe
ID over a period of time or by the use of high pressure and hot
water in a short time. Expanded liners can be inserted in very
long pipeline segments.

Compressed liners may be used in low pressure service or high
pressure service upon installation. Molding to the pipeline ID
is immediate. Insertion lengths are moderately long.

Each process is unique. Allow each to speak for itself. But,
all processes use the elastic and visco-elastic properties of
HDPE in an engineered manner to provide an HDPE lined, corrosion
and abrasion resistant pipeline or downhole tubular.

Philip M. Hannan

"CURED-IN-PLACE PIPE: AN END USER ASSESSMENT"

REFERENCE: Hannan, P. M., "Cured-In-Place Pipe: An End
User Assessment," Buried Plastic Pipe Technology, ASTM STP
1093, George S. Buczala and Michael J. Cassady, Eds.,
American Society for Testing and Materials, Philadelphia,
1990

ABSTRACT: Trenchless technology is rapidly entering the
North American marketplace as demand creates an increasing
need for cost-effective pipe renovation. One of these
technologies is Cured-In-Place Pipe (CIPP). Generally,
these lining processes consist of a resin impregnated
flexible tube, inserted into the existing pipeline by winch,
air pressure or hydrostatic head, and cured to final
structural properties through the application of heat. CIPP
technology will be reviewed and its position in the marketplace discussed. Factors affecting selection of this type
of process will be assessed. Performance criteria and
installation requirements will also be discussed.

KEYWORDS: cured-in-place pipe, liner, inversion, sewer
reconstruction

The deteriorating infrastructure, in the United States as
well as worldwide has been well documented in countless journals
and magazines. Wastewater collection system requirements amount
to billions of dollars over the next decade in this country alone.
To address that need, more cost-effective technology is required
to extend the benefit of the limited monies available to utility
owners.

Trenchless technology is the most rapidly expanding market
of infrastructure renewal. This is evidenced by the many
specialized conferences and symposiums that are devoted to this
aspect of construction practice. Numerous technologies are being
developed here while others are imported to position themselves
for a share of the North American market. Companies are scramb-
ling to meet the growing recognition that reconstruction efforts
must proceed even in times of tight money. Many of these firms
and their technology are represented at this conference and the

Philip M. Hannan, P.E. is the Maintenance Reconstruction
Division Head at the Washington Suburban Sanitary Commission,
4017 Hamilton Street, Hyattsville, MD 20781.

329

subject of presentations and papers. One segment of this trenchless technology market is the cured-in-place pipe or CIPP. Generally, these lining processes consist of a resin impregnated flexible tube, inserted into the existing pipeline by winch, air pressure or hydrostatic head, and cured to final structural properties through the application of heat.

The Washington Suburban Sanitary Commission, the seventh largest water and wastewater utility in the United States, has over 10 years of experience in utilizing this rehabilitation technology. In presenting this topic, the point of view will be primarily that of an owner and end user. CIPP technology will be reviewed and its position in the marketplace discussed. Factors affecting selection of this type of process will be assessed. Performance criteria and installation requirements will also be discussed.

CIPP OVERVIEW

The allure of CIPP, as with any trenchless technology, is the minimization of disruption to the urban community and environment. Those same advantages are also what contribute to the cost-effectiveness of these lining techniques. Through the elimination of excavation, significant costs are deferred for the associated equipment, labor, and paving restoration.

There are a number of characteristics common to CIPP products that influence the selection and application of this technology over other reconstruction methods.

Materials

The fundamental product is a felt liner, sized to the length and diameter of the existing sewer, impregnated with a thermoset resin system to provide structural capabilities after curing in the host pipe. Patent considerations have led to a variety of distinctions in the liner material, resin formulation, and installation characteristics among the manufacturers.

All of the techniques available today utilize existing man-hole access for insertion. "Soft liner" had been used by some to describe these techniques because of the flexible nature of this material prior to curing. This "soft" aspect is what permits the liner to bend and maneuver through the manhole frame, channel and existing pipe conditions to achieve its' final position in the sewer. This also provides the desirable characteristic of one continuous, liner membrane from manhole to manhole.

Because the structural capability is achieved through the interaction of the liner tube and resins, the overall wall thickness of these piping systems is very small with respect to the existing pipe diameter. In many collection system applications, liner thickness of 5-15mm are common. For a 200mm inside diameter pipe, a 6mm close fit liner still maintains nearly 94% of the existing cross-sectional area. An improved flow friction factor on the face of the liner serves to counteract the diameter reduction for sustaining pipe capacity.

The same flexible characteristics that permit insertion of the liner from manholes also allow for line conditions to influence the final product. Poor line and grade in the host pipe will be reflected in the CIPP. A belly, sag, or offset is mirrored in the renovated pipe. Missing pipe will appear as a bulge in the liner. If those characteristics are undesirable in the renovated pipe, they need to be excavated and repaired prior to lining, decreasing the cost-effectiveness of this type of renovation.

Service Connections

A primary advantage of CIPP is the ability to reopen service laterals from inside the pipe. Cutters of various designs are capable of removing the coupon covering over the service connection. This addresses key economic issues in comparison to techniques requiring excavations, particularly when
    o there are frequent taps in the line or
    o when excavations would be very disruptive and costly
    as in a commercial business district or arterial road. Conversely, this is of no advantage if the renovated line section has few or no taps or is in an open area where excavations are of limited economic impact and restoration costs low.

In most CIPP, there is limited interaction between the liner and service connections. Because the cured liner is relatively discrete, connections that are defective before lining will remain essentially unchanged. If leakage existed around the tap or thimble at the point of connection, the leakage will likely resume when the coupon is removed. For an infiltration reduction application, therefore, CIPP is most effective when existing connections are sound or are limited in number. it is ideal for excluding groundwater if the liner is continuous or unbroken the full length from manhole to manhole.

Protruding taps are another item that must be addressed in preparing a host pipe to accept a liner. Thimble connections or pipe to pipe appurtenances are susceptible to shifting over time resulting in an encumbrance on the cross sectional area of the sewer main. All CIPP applications must have significant protrusions removed by cutting, grinding, or excavating before lining.

Execution

Since CIPP involves lining the entire length of sewer and
temporarily blocking the lateral connections, the CIPP process
has been tailored by manufacturers to be completed in a tolerable
time for customers to be without sewer service. For most
techniques, this involves loss of service for 8-12 hours before
coupons are removed and service to the customer re-established.

Since curing the liner in a controlled environment is
required, the host pipe must be isolated by plugs. Bypass pumping is required to move existing sewage flow around the isolated
section for the duration of the operation.

WSSC EXPERIENCE

The Washington Suburban Sanitary Commission, serving 1/3 of
the population of the State of Maryland, has been utilizing CIPP
since 1978. The first application involved an 18" diameter sewer
passing under a heavily traveled railroad corridor. Since that
installation, the Commission has gone on to renovate nearly 126
miles (203 km) of sanitary sewer by this technique. An annual
outlay of $4 million has been established to address continuing
reconstruction needs in our 4500 mile (7,240 km) system.

The smallest diameter lines completed have involved 100mm
sewer lateral pipes and the largest an 1800mm interceptor. Both
projects were performed to explore the applicability and
economics of the process at the two extremes of diameter ranges.
The majority of the lining has been in the 200-400mm category.

Why does WSSC utilize CIPP so extensively? The key elements
lie in making use of those inherent advantages of CIPP. It's
ease of application in urbanized areas with minimal disruption to
the environment was important. Our service area is composed of
well informed and politically savvy communities who demand the
technological conveniences CIPP affords. The location of the
sewer mains and high frequency of taps generally limit the
economic advantage of alternatives such as sliplining. Leaking
service connections have been addressed by other non-excavation
rehabilitation techniques developed with industry in partnership
with WSSC. A local CIPP source was established with volume
pricing influencing the cost available to the Commission. Until
recently, there have also been few alternatives to enter the
marketplace.

CIPP TECHNIQUES

There are currently four potential CIPP products available
in the United States marketed for sewer applications.

o Insituform
o Paltem
o In-Liner
o Insta-Pipe

   The techniques differ by method of insertion, liner material, resin formulation, curing process, and final properties. The pioneer technique of the group is Insituform. It is the oldest (approximately 17 years) and has the longest record of experience in the United States. Paltem and In-Liner are recent licensed techniques from Japan and Germany respectively. Insta-Pipe is a domestically developed product here in the States. A summary of the key characteristics of these four technologies are found in Table 1.

Table 1:   Comparison of CIPP Characteristics

Product

Liner Parameter	Insituform	Paltem	In-Liner	Insta-Pipe
Insertion	inversion using water head	inversion using air pressure	winched in- to place	floated & winched into place
Materials	non-woven tube materials & thermoset resin	woven & non-woven tube materials & thermo- set resin	non-woven tube materials & thermoset resin	woven & non- woven tube materials & epoxy thermo- set resin
Curing Process	circulating hot water	circula- ting hot steam	circulating hot water	circulating hot air

DESIGN & TESTING

   Although Insituform has been in the United States for a decade, the CIPP field is still relatively young. As such, each of the manufacturers has been promoting design formula and test- ing standards. As an end user, the owner of the project is required to sort through these competing approaches to establish a standard specification and minimum performance criteria.

   One movement has been underway to accomplish an industry standard through American Society for Testing and Materials (ASTM). ASTM Committee F17 on Plastic Piping Systems has an ASTM document F1216-89 entitled "Standard Practice for Rehabilitation of Existing Pipelines & Conduits By the Inversion and Curing of a Resin-Impregnated Tube." This is a standard practice that is applicable for the present to only Insituform.

One of its accomplishments is to attempt to define a series of design considerations in specifying performance requirements. It establishes formula for specifying thickness of the liner for partially and fully deteriorated gravity pipe. These take into account groundwater, soil, and live loads. There are also equations for pressure pipe applications and references for chemical resistance testing.

A key element in utilizing these equations is the selection of the structural values for long term modulus of elasticity and flexural strength. This requires industry supplied values or proceeding on a path of independent testing.

In order to define our requirements and address objectively the changing market conditions, the Commission entered into a partnership with WRC Inc./Carmen F. Guarino Engineers, Ltd. to develop a framework for evaluating sewer renovation systems. [1] The report included a basis of strategy for renovation studies, a review of the techniques currently available, and a generic review of material properties and principles of design. This effort included a classification of CIPP to determine appropriate performance criteria for the liner.

In evaluating CIPP, WRC has determined several performance standards. They emphasize the relationship between the design requirement and the product specification to achieve the desired performance. Of key interest to assessing CIPP performance are the following properties;
        o long term flexural strength
        o short term and long term flexural modulus
        o long term tensile strength
        o long term tensile modulus
        o compressive strength

If the CIPP technique meets the minimum criteria established for these properties, the process can be considered for WSSC approval. Many were developed from British Standards where a great deal of research has been performed.

SUMMARY

As a public water and sewer utility established to serve our ratepayers, the optimal situation for WSSC is to move to identify all innovative technologies that can meet minimum performance criteria, understand the inherent strengths and weaknesses that affect installation and applicability, and then devise site specific scopes of work that take advantage of those strengths. Competitive bidding should then develop a low price approach toward utilizing any approved new technologies.

CIPP technology is positioned in the marketplace to cost-effectively satisfy many of the reconstruction needs of utilities. It's "trenchless" lining ability enable CIPP to structurally renew sewer pipe without disruption and long term impact to the community. With proper application, it should continue to be an effective tool in addressing the growing infrastructure rehabilitation market.

REFERENCE

[1] WRC, Inc., "The Development of a Framework for the Evaluation of Sewer Renovation Systems," Report to the Washington Suburban Sanitary Commission, Hyattsville, MD, 1989

Lars-Eric Janson

REVIEW OF DESIGN AND EXPERIENCE WITH
THERMOPLASTIC OUTFALL PIPING

REFERENCE: Janson, L.-E., "Review of Design and
Experience with Thermoplastic Outfall Piping",
Buried Plastic Pipe Technology, ASTM STP 1093,
George S. Buczala and Michael J. Cassady, Eds.,
American Society for Testing and Materials,
Philadephia, 1990.

ABSTRACT: Flexible pipes of thermoplastics such as
polyethylene and polypropylene are excellently suited
for the construction of submarine outfall systems
because they can be extruded in long sections, towed
fully equipped with anchoring weights to the outfall
site from the production base, and sunk directly on to
a seabed with a minimum of underwater work. When lying
on the seabed they can yield to extreme wave and current forces and conform to the seabed when underscour-
ed without failure. This factor allows a light structure, a short implementation time and a cost-effective
outfall system.

KEYWORDS: ocean outfall, submarine pipeline, flexible
piping, polyethylene pipe.

1. Introduction

In this paper are presented the principle topics of
the flexible plastic submarine piping concept. This concept has been applied for a large number of submarine
pipelines mostly in Europe since the early 1960's. The
experience today comprises large diameter pipes made of
high density polyethylene (HDPE), medium density polyethylene (MDPE) and polypropylene (PP). The largest pipe
diameter is 1.6 m, while the most frequently used diameter
range is from 0.4 m to 1.2 m, [1].

The problems related to ocean outfall are connected
with heavy sea conditions and unstable seafloor. In the
case of using conventional non-flexible pipe materials
such as concrete or steel, the pipeline has to be buried

Dr Sc. M. ASCE. VBB Consulting Ltd
Head of Research and Development
P.O. Box 5038
S-102 41 STOCKHOLM, Sweden

well below the seafloor thus being protected from wave and current forces and from effects of sand transport and seabed motion. Such construction may require expensive and time consuming sheet piling and trenching. The main purpose with introducing the flexible plastic piping concept is to solve these problems in a safe way but at a significantly reduced installation cost. (In many cases it has been proved that plastic ocean outfalls have a cost which is only half the one valid for conventional outfalls made of steel or concrete.)

Another topic to be dealt with concerns the acceptable amount of temporary movement of the pipeline during heavy storm conditions, as the design stability of a flexible submarine pipeline is normally chosen for a significantly lower wave height than the statistically largest wave during the service life of the structure.

To the special problems to be treated when designing a flexible outfall belong also the internal hydrodynamics. A particular question concerns the risk of free air coming into the pipe during certain transient flow conditions. This is connected with the fact that the pipeline is normally not loaded as to prevent floating when air filled. To the topic belongs also the control of underpressure inducing buckling.

2. Behaviour of a flexible submarine pipeline exposed to wave and current forces

The external forces affecting a submarine pipeline are caused primarily by wave and current actions but often also by an unstable seabed. This type of forces requires large safety factors in the structural design, as the exact magnitude of the forces is always difficult to predict. This means also high costs for conventional submarine pipes made of concrete, cast iron or steel, as such pipes can afford only small deformation or strain before burst or leakage. Consequently, these pipelines have to be heavily loaded or buried down in the seabed sediments.

The flexible submarine piping concept implies instead that the high safety factor, which is required by the uncertainty concerning design load, is replaced by a certain movability of the flexible pipeline in case of extraordinary forces coming up. This means that the flexible pipeline can normally be placed directly on the seabed without any trenching and only lightly loaded. Moreover a mimimum of pipe bed preparation is required as the pipeline can easily follow the bottom configuration independent upon whether the seabed is stable or mobile.

It is primarily the very high strainability in combination with the stress relaxation ability of the

polyolefin materials, which create the basis for this new
type of design philosophy for submarine pipelines, [2].
The flexibility and the continuity in the longitudinal
direction with only few pipe joints, combined with the
light weight, imply also that a simple and economic
submersion technique can be applied. In this way long
ocean outfalls or transmission pipelines can be submerged
in a short time even during heavy sea conditions. By using
polyolefin pipes many submarine pipe projects have been
accomplished, which would not have been economically
realistic with conventional pipe materials.

The following shows examples of how the flexiblity and
the strainability of the plastic pipe is a basis for the
flexible submarine piping philosophy:

Wave forces on pipes close to seabed consist of three
components: a horizontal drag force, a vertical lift force
and a horizontal inertial force. The first two components
vary with the wave induced velocity, the latter with the
wave induced acceleration. The latter is therefore phase-
shifted from the other components and the magnitude of the
maximum combined force is therefore smaller than the sum
of the components.

The lifting force on a pipe resting close to the sea
bed and caused by wave action is significantly greater
than on a pipe which is placed at a certain distance from
the bed. This means that concrete weights designed so that
they give an open space between the pipe and the seafloor
will give rise to smaller lifting and horizontal forces on
the pipeline than, for instance, saddle-shaped weights
loosely placed over the pipe. When the wave induced forces
on a pipe, distanced from the seabed by its weights,
exceeds the seabed resistance, then the pipe will start to
move forward and backward on the seabed by the combined
action of the horizontal drag force and the inertial force
without the lifting force being able to actually lift the
pipe and weight.

However, even if the loading weights are designed in
such a manner that the pipe is placed at a certain distance from the seabed, the pipe may, as a consequence of
settlements and erosion, gradually sink and come to rest
on the bed. When, in such a case, the maximum design wave
force is exceeded, the pipe will (if not arrested by the
soil suction) be heaved to a level, where the loading
weights balance the lifting force from the wave. At the
same time the lateral drag force will move the pipe horizontally. The pipe has, however, now reached the basic
situation described above.

Furthermore, as the pipeline can normally be placed so
that the wave forces do not attack the pipe along its
whole length simultaneously (wave refraction turns the
wave direction towards the coast normal direction which

also is the preferred direction of the pipe) the longitudinal continuity and the axial and torsional stiffness
of the pipeline imply that neighbouring parts of the pipe
are capable of moving the displaced part of the pipe more
or less back to its original position.

By using computer simulation it is easy to predict the
movement of the structure for various relations between
pipe loading and wave forces, securing that the pipeline
dislocation is kept within acceptable limits also for a
wave height recurrency period of 100 years or more.

3.   Installation and operation

One important idea of the flexible submarine piping
concept is that the installation of the pipe shall be possible to be performed from a stage where the long pipeline
is floating on the sea surface. This means that the pipeline which is filled with air in this first stage cannot,
from a practical towing and submerging point of view, be
weighted to more than about 65 % of the pipe displacement.
Fortunately, according to the flexible design philosophy
described above, normal sea conditions close to the shore
will not require weighting above 70 % of the displacement
and in most cases a weighting of 25 to 50 % is quite
sufficient. The submerging can then easily be accomplished
by venting the pipe as described below.

Loading weights of reinforced concrete are commonly
used, symmetrically designed in relation to the centre of
the pipe, and so constructed that the pipeline is placed
at a distance from the sea floor corresponding to at least
one-quarter of the pipe diameter. In such cases when the
sea conditions are so heavy that a weighting above 65 % is
needed, it is preferable to mount additional weights after
completion of the submerging.

The weights can in many cases also be used for strengthening the pipe during operation. Thus a change in pumping or a sudden flow stop will give rise to transient
pressures. A pressure drop will propagate along the pipeline producing an ovalization of the pipe. This ovalization must be limited if buckling of the pipe is to be
prevented. Here the weights, if properly distanced, can
work as ring stiffeners. It has been shown that also the
flexibility of the plastic pipe wall itself significantly
reduces the propagated underpressure as compared with a
pipeline of rigid material.

The submersion of the air filled pipeline is effected
by filling it with water from the shore and outwards
whereby it gradually sinks to the bottom. In order to
control the submersion procedure, arrangements are made
for compression of the air of the air-filled part of the
pipeline which has not yet sunk. In this connection the

air pressure must be carefully controlled and adjusted
with regard to the load and depth of the bottom so that
the bending radius at the time of sinking does not become
smaller than is permissible both with regard to the
strength of the material as well as with regard to the
risk of buckling of the circular cross-section of the
pipe. This technique also makes it possible to lift the
pipeline in whole or in part if it proves necessary during
the submerging operation to conduct repairs or corrections
of the position.

It has to be considered that the pipe material is viscoelastic, which means that the buckling criteria are both
time and temperature dependent. Therefore, long stoppage
of the submerging procedure must be avoided with the
pipeline bended. Instead measures have to be taken in
beforehand so as to allow a continuous submerging. A sinking velocity of about 500 m/hour is normally recommended.
Using numerical simulation, it is possible to predict the
behaviour of the pipe during the total submerging procedure.

As mentioned above, an easy submerging procedure requires the loaded and air filled pipeline to weigh less
than the water displaced by the pipe. Consequently, it is
essential in such cases that air is prevented from entering the outfall during operation so that the pipeline does
not refloat or looses its lateral stability. The risk of
getting air in the pipe is greatest in the case of ocean
outfalls, for which consequently the design of the shore
devices is of great importance. If, for example, the land
section of the pipeline is joined directly to the submerged section, too rapid a pump start may result in the
formation of an air bubble which may then transfer out
into the submerged section, causing it to rise. If the
rate of pumping is suddenly reduced, a hydraulic jump may
form above the water line with a strong entrainment of air
as a result. In the case of a partial or total stoppage of
the pumps the kinetic energy in the water head may be so
great that the interface between water and air in the pipe
may proceed down into the underwater sections.

These undesirable conditions can be avoided by inserting on the shore a surge chamber which should be sufficiently deep to prevent the water level in the chamber
sinking below the entrance of the outfall pipe in the
event of pump stopping.

4.   Quality control of the pipe material

With the ambition to secure a plastic material which
can really sustain the particular strength properties
needed for fulfilling the advanced impact on a flexible
submarine pipeline described above, some important quality
control tests have to be performed. First of all the long

term strength of the material must be checked. A minimum requirement is that the accelerated long term hydrostatic internal pressure test at +80°C gives a time to failure exceeding 1,000 hours when the ring tensile stress is 4 MPa. Simultaneously the linear relation between the tensile strain in the pipe wall and the logarithmic loading time shall show a rectilinear relationship. The same long term tensile strength shall be proved valid for butt welded joints.

Another important quality control concerns the pipe manufacturing process in order to prevent thermal oxidation to occur at the internal surface of the pipe wall. The most simple way to prevent such an event is to require that the pipe shall be internally loaded with an inert gas, such as nitrogen or carbon dioxid during the extrusion. This is an experienced practice in Europe, particularly for large diameter pipes or thick-walled pipes.

In addition to these main requirements, actual national or international standards for polyethylene pressure pipes have to be referred to.

5. Technology assessment

For all structures there is a risk of failure which, however, can be minimized if the design and specifications and construction control are based upon established experience.

Assisting when evluating such risks could be first to identify the various impacts the structure has to withstand, and then try to assess to which degree the technique is sufficiently known for making it possible to disregard the consequences of these impacts as more or less insignificant. In this case the evaluation can be systemized by treating separately the impact, which the structure will be exposed to, of physical (mechanical, hydraulic, thermal), chemical and biological nature. Guiding for the evaluation should of course be the requirements which have to be fulfilled and which can be related to the user's demand of operational safety and functional stability of the structure.

Concerning the physical impact on a submarine pipeline, it is an established fact that both the mechanical and hydraulic load can much easier be taken up by a viscoelastic material as polyethene than by conventional rigid materials. This is due to the very large strainability of polyethylene (5-10 %). In spite of the creep property of the viscoelastic material, the knowledge about how to chose safe stress and strain levels is well established since more than 30 years for standard specified polyethylene pipes. A part of the structure of particular concern is, however, butt welded joints which must be

perfectly performed under experienced supervision, shall
the axial strength of the joints reach the same level as
valid for the pipe itself.

The thermal impact on polyethylene has always implied
a weakness concerning the long term durability, but only
at temperatures above +50°C. In most cases the temperature
will not reach that limit.

Concerning the impact of chemical nature, it is a
well-known fact that such corrosive environment which is
dangerous for conventional pipe materials is totally safe
for polyethylene. No accelerating degradation of the
stabilization system of the polymer material has been
noticed as a consequence of the impact of muncipality
waste water or of waste water from cellulose industries.
Nor is any biological degradation of the pipe material
stored in the ocean referred to in the literature, neither
of macro or micro nature.

Based upon the statements above, it may momentarily be
believed as there is no impact on a submarine polyethylene
pipeline, which can be described as negative (in the sense
of risky). Of course this is not the case. Hence there is
an obvious and well-known risk which, however, primarily
has to be classified as caused by the human factor. As an
example, many pipe failures have occurred due to neglec-
tion of an accurate quality control of the pipe delivery
and of the butt welding procedure. Other types of failure
(floating up to the sea surface, clogging, buckling during
submerging, etc.) have been caused by pure design faults
or by lack of understanding of the specific conditions
valid for the flexible viscoelastic pipe material. The
purpose of a proper design is to eliminate these types of
negative impact.

One impact which in addition has to be mentioned, and
which has to be classified as exceptional, is the one
which may be caused by big ships. In emergency cases, such
as non-steerable situations, a ship can be forced to drop
anchor near the coast. If the anchor catches the pipeline
and the ship's drag force is big enough, the pipeline will
be dislocated and probably spring at leak. This is a type
of failure which cannot be disregarded despite good knowledge and experience. That it can be accepted is of course
due to the fact that the probability for the failure to
occur, i.e. the risk, is very small. Hereto belongs also
that appropriate technique and methods have been developed
for rather rapid and simple repair of the polyethylene
pipe in situ. Such effluents which can cause negative
impact on the environment in the case of leakage, have of
course to be monitored with special care. An experienced
aid, to minimize the negative impact of uncontrolled
discharge, is to store easily accessible equipment for
adequate repair preparedness. A maintenance manual for the

ocean outfall shall perferably describe the equipment and how to use it in case of such failures.

REFERENCES

[1]   Janson, L.-E., "The Utilization of Plastic Pipe for Submarine Outfalls. - State of the Art". Water Science Technology Vol. 18, No. 11, 1986, pp. 171-176.
[2]   Janson, L.-E., "Plastic Pipes for Water Supply and Sewage Disposal". - Published by Neste Chemicals, Stenungsund, Sweden. Stockholm 1989.

Ian D. Moore and Ernest T. Selig

USE OF CONTINUUM BUCKLING THEORY FOR EVALUATION OF BURIED PLASTIC PIPE
STABILITY

REFERENCE: Moore, I.D. and Selig, E.T., "Use of Continuum
Buckling Theory For Evaluation of Buried Plastic Pipe
Stability", Buried Plastic Pipe Technology, ASTM STP 1093,
George S. Buczala and Michael J. Cassady, Eds., American
Society for Testing and Materials, Philadelphia, 1990.

ABSTRACT: A buckling theory for design of buried plastic pipes
is described, which combines linear shell stability theory for
the structure with elastic continuum analysis for the
assessment of ground support. The theory provides stability
estimates which are superior to those generated using 'spring'
models for the soil, predictions of phenomena such as
long-wavelength crown buckling without the need to pre-guess
the deflected shape, and rational assessment of the influence
of shallow cover and the quality and quantity of backfill
material. As well as describing the continuum buckling theory,
the literature is briefly reviewed, buckling as a performance
limit for buried plastic pipe is discussed, and the selection
of appropriate soil and polymer moduli for use in the theory is
also considered.

KEYWORDS: Buckling, Buried Pipe, Design, Stability

INTRODUCTION

Currently a variety of procedures are being used for the design
of buried plastic pipe, depending on the pipe product and its country
of origin. It is widely recognised that these compressed flexible
cylinders can become elastically unstable. Buckling may be caused by
external soil pressures (e.g., Molin [1], Carlstrom [2]) or fluid
pressures associated with groundwater or internal vacuum (e.g.,
Carlstrom [2], Taprogge [3], Heierli and Yang [4]), and it can be
influenced by shallow burial (e.g., Greatorex [5]). However, there is
some confusion about the mechanisms involved, with phenomena such as

Dr. Moore is Senior Lecturer in the Department of Civil Engineering
and Surveying, University of Newcastle, N.S.W., 2308, Australia;
Dr. Selig is Professor, Department of Civil Engineering, University of
Massachusetts Amherst, MA, 01003, U.S.A.

"local" buckling (Schluter [6], Jeyapalan and Bolden [7]), "flattening" and "curvature reversal" (Hurd [8], Jeyapalan and Bolden [7]), "ring" buckling (Chambers and McGrath [9], Schluter [6]) and "upward" buckling (Greatorex [5]) being mentioned in the literature.

A number of different procedures have been proposed for buckling strength assessment. The Levy [10] solution for the stability of an unsupported circular ring under external pressure is sometimes used (e.g., Taprogge [3]) and various versions of the Luscher [11] theory for a pipe supported by elastic springs are employed (e.g., Carlstrom [2], Heierli and Yang [4], Schluter [6], Jenkins and Kroll [12]). Correction factors are sometimes employed to adjust for the effects of burial depth and fluid loads (AWWA [13]) as well as out of roundness (Taprogge [3], Jenkins and Kroll [12]). The issue is further complicated as "excessive deflection" is often treated as a stability rather than serviceability criterion (e.g. Jenkins and Kroll [12], Watkins, Dwiggins and Altermatt [14]).

Recently Gumbel [15] and Moore [16] have reported that the combination of linear shell theory for the pipe with elastic continuum analysis for the assessment of ground support provides superior estimates of buried pipe buckling strength. This "continuum buckling theory" has been applied to the design of long span corrugated metal culverts, Moore, Selig and Haggag [17]. The model can be used to examine the effect of shallow burial (Moore [18]), nonuniform hoop thrust (Moore and Booker [19]), noncircular pipe shape (Moore [20]) and nonuniform ground support (Moore, Haggag and Selig [21]).

This paper begins with a brief discussion of the mechanism of buried pipe buckling. The benefits of the continuum soil model compared to the Winkler or "elastic spring" model are then outlined. A design procedure based on the continuum buckling model is described, for use in prediction of hoop thrusts which destabilise buried plastic pipe compressed by the surrounding soil. The influence of shallow burial, nonuniform ground support, nonuniform thrust distribution and the soil modulus are discussed. The selection of pipe modulus is examined through reference to an example problem (this is an important issue for polymers where modulus is a function of time). Finally, observed plastic pipe buckling phenomena are interpretted in a discussion of performance limits for buried plastic pipes.

BURIED PIPE BUCKLING

Column Buckling

One important concern with flexible structures is the effect of thrust acting in the plane of the structure on the bending response. For example, the familiar "Euler buckling" problem is shown in Figure 1a, where at some critical value of axial thrust N an initially straight column bends freely even when lateral pressures are only small. Now if that column is provided with lateral supports, Figure 1b, the wavelength of the deformations ("buckles") is reduced and much higher thrusts N can develop before lateral stability is affected. The case of a ground supported structure is quite similar, Figure 1c, where lateral support is distributed along the structure leading to similar reductions in buckle wavelength and increases in thrust capacity. In each case, the thrusts influence the flexural response, because this force produces bending when lateral pipe deformations provide eccentricity. For those readers who want this buckling phenomenon expressed in theoretical terms, the structure buckles when a "critical" distribution of thrust develops so that the structure can move from one equilibrium position to an adjacent position with no net energy input (the equilibrium state becomes "elastically unstable").

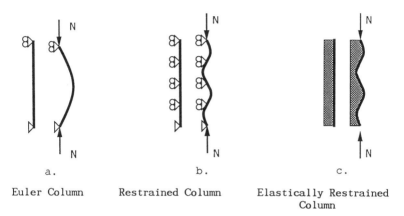

        a.   b.   c.

    Euler Column   Restrained Column   Elastically Restrained
                                                   Column

        Figure 1. Buckling of a Straight Column

Elastically Supported Buried Pipe Buckling

        Let us now consider the case of a flexible pipe buried in massive ground, Figure 2. When the pressures applied to the pipe across the interface are large enough, the hoop thrusts N which develop produce wave-like flexural deformations, Figure 2a, (at the locations of maximum thrust or over the crown where ground support is reduced). The thrusts N affect equilibrium in two ways:

a.   simple statics indicates that a net lateral force develops – this is obviously resisted by the action of the interface pressure due to soil weight p_w, Figure 2b.

b.   local bending is induced as the buckling deformations provide eccentricity for the thrust – this is resisted by the structure which has a small amount of bending stiffness and, significantly, the nonuniform earth pressures p_r which develop as the soil is deformed, Figure 2b.

        The wavelength of the "buckling" deformations and the ability of the pipe to support thrust, depend on the flexural stiffness of the pipe and the magnitude of the ground support. For a pipe immersed in a fluid, the nonuniform lateral pressures p_r cannot develop so that thrust capacity is low and the pipe deforms into an oval shape. Where the pipe is buried in stiff ground it has high thrust capacity and short wavelength buckles occur, since the soil actively resists the pipe deformation, and large earth pressures p_r develop.

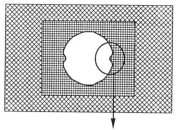

a. Pipe Close to Buckling Failure

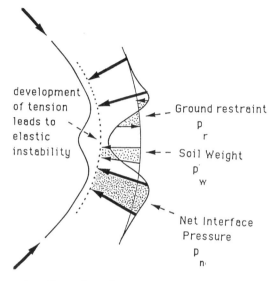

development
of tension
leads to
elastic
instability

Ground restraint

p_r

Soil Weight

p_w

Net Interface
Pressure

p_n

b. Interface Stresses p_n, p_w and p_r

Figure 2. Buried Pipe Buckling

Separation of Soil and Structure

The net interface pressures $p_n = p_w + p_r$ are also shown in Figure 2b. We have already shown how initially as pressures p_w increase pipe equilibrium is maintained by the action of soil support pressures p_r. These are able to act while the pipe and the soil surrounding it remain in contact. Eventually, however, when pipe deformations become large enough, the net pressures p_n at some location dwindle to zero and the pipe and ground separate. At that location the ability of the soil to resist local bending is affected and the thrust capacity of the pipe suddenly decreases. If earth pressures p_w are maintained the pipe is elastically unstable and a catastrophic failure results, with the pipe "unzipping" further from the soil as it deforms into a long wavelength deformation pattern, Figure 3a [15,16,22] (if the moment capacity of the pipe wall is fully developed, the large pipe deformations which occur produce local yield and "plastic hinging"). This sudden increase in buckle wavelength and decrease in thrust capacity is similar to that which occurs when a straight column has one of its lateral supports removed, Figure 3b. Energy is released and rapid "snap-through" buckling is observed.

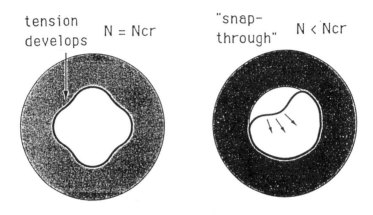

a. Buried Pipe: Destabilised by Separation of Pipe From Ground

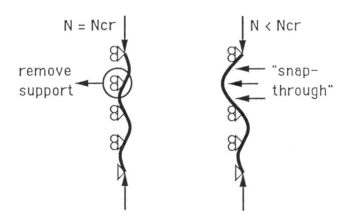

b. Column: Destablised by Removal of Lateral Support

Figure 3. Snap-Through Buckling After Removal of Support

Deflection Control

It is worth noting, in relation to this type of pipe failure, that pipe shape does not reliably indicate the extent of pipe stability, and that pipe stability cannot be assured through shape control. The pipe collapses at the point of separation from the ground, and this can occur before or after it flattens. Naturally, pipe flattening is of concern to the engineer responsible since it is a sign that pipe buckling strength may be barely adequate. However, distortions such as flattening and reverse curvature can, for example, be the result of concentrated load from a rock in contact with the pipe, or construction induced deformations. In such cases the potential for buckling collapse may be minimal and the wave-like deformations observed should not be regarded as "buckles".

SELECTION OF PREFERRED BUCKLING THEORY

There are two approaches available for estimating the nonuniform pressures p_r which occur as the soil resists lateral pipe deformations: (i) the spring model, and (ii) the continuum model.

Spring Model

The use of the Luscher [11] buckling theory is common in the design of buried plastic pipes (e.g. AWWA [13]). This approach employs the elastic spring (or "Winkler") model to characterise the soil support. For ground support with $E'R^3 > 8EI$, the critical hoop thrust is given by an expression of the form

$$N_{ch} = 2 (EI)^{1/2} (E')^{1/2}/(R)^{1/2} \qquad (1)$$

where EI is flexural rigidity of the pipe, E' is the elastic spring stiffness, and R is the pipe radius.

The spring model represents an over-simplified analysis of the soil-structure system, since it ignores shear deformations in the soil. Except through the structure, each point on the internal soil boundary is artificially separated from the other points (no deflection occurs when loads are applied at the other points). A severe penalty results – the elastic spring stiffness is a function of the wavelength of the pipe deformation which occurs, Duns and Butterfield [23]. There is no unique relationship between spring stiffness and elastic soil modulus. Empirical spring estimates obtained from static pipe deformation studies (e.g., Howard [24]) cannot be employed in buckling strength assessment since static pipe response has totally different mechanics. Also, the task of empirically determining the effect on spring stiffness of complexities such shallow pipe burial and nonuniform ground support, is formidable indeed.

Continuum Model

Also available are analyses which use elastic continuum theory to determine ground support (e.g., Forrestal and Herrmann [25]). For a flexible pipe supported by uniform elastic ground of modulus E_s the

continuum buckling theory suggests that where $E_s R^3 > 10EI$, critical
hoop thrust takes the form

$$N_{ch} = 1.2 \ (EI)^{1/3} \ (E_s)^{2/3} \tag{2}$$

In contrast to the spring model, continuum theory employs
modulus parameters with real physical meaning. Using numerical
analysis, ground support can be determined for complex pipe burial
conditions. Also note that according to equation (2) pipe radius does
not influence the critical thrust for deeply buried pipes in uniform
soil. This is perhaps the most significant difference between the
continuum and elastic spring buckling solutions. As we shall see in
the following section, pipe radius does affect the manner in which
nonuniformities in ground support influence the buried structure. Of
course, even for deeply buried pipes where thrust capacity is
independent of radius, the pipe size significantly influences the
thrusts that develop, and therefore the factor of safety.

Comparision of Models

A number of studies have been performed over the past two
decades comparing experimental data with theoretical predictions of
buckling strength. For example, Figure 4 shows normalised maximum
thrust $N_{max} 8R^3/EI$ data from tests on deeply buried pipes (see Moore
[16] for further details) as well as predictions of critical thrust
from linear continuum buckling theory and the theory based on elastic
springs (for the latter the common but incorrect assumption has been
made that spring stiffness E'is given by $E_s/2(1-v_s)$, where v_s is
Poisson's ratio of the ground). The horizontal axis is normalised
ground modulus $8E_s R^3/EI$, where the wavelength of the buckles steadily
decreases as normalised ground stiffness is increased.

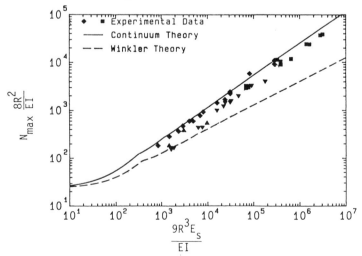

Figure 4.   Maximum Thrust N_{max} From Theory and Experiment [16]

For these deeply buried pipes, the scatter of test results is such that both models need some calibration before they provide reliable estimates of stability. Examining the line of best fit to the experimental data, Moore [16], we find a slope of 0.647. This is very close to the 2/3 relationship associated with the continuum model of equation (2), but not the 1/2 relationship given for the spring model in equation (1). This illustrates the point made earlier that it is incorrect to assume spring stiffness is independent of buckle wavelength. Clearly interaction is occuring through the ground similar to that predicted using elastic continuum analysis. When this problem is taken together with those described before concerning the spring model, the performance of the continuum model is clearly superior.

DESIGN OF BURIED PLASTIC PIPES FOR BUCKLING

Design Equation

A design approach based on the continuum buckling theory is now described. A similiar approach has recently been reported for corrugated metal culverts, Moore, Selig and Haggag [17].

Critical hoop thrust N_c is given by

$$N_c = \phi \, N_{ch} \, R_h \qquad (3)$$

Here, the critical thrust for a pipe deeply buried in elastic ground N_{ch} is modified using the calibration factor ϕ and the correction factor R_h.

For typical buried plastic pipes the flexural stiffness EI is less than $E_s R^3 / 10$, and under these cirumstances the critical thrust N_{ch} for deeply buried pipe in uniform ground is given by equation (2).

Otherwise, N_{ch} can be found by minimising the expression

$$(n^2 - 1) \; EI/R^2 \; \frac{E_s a}{1.82n + 0.52}$$

with respect to integer $n \geq 2$ (see Moore and Booker [26] for further discussion).

Calibration

Discrepancies between linear continuum buckling theory and the experimental data occur, Figure 4, probably as result of inelastic and nonlinear soil response as well as geometrically nonlinear structural and interface behaviour. A simple arithmetic calibration is used to allow for these effects.

An examination of the available experimental data indicates that the calibration factor ϕ takes the value 0.55 for granular materials, Moore [16]. Lower values should probably be used if fine grained backfill materials are employed, although at present there is insufficient information for a specific recommendation.

Nonuniform Hoop Thrust

The use of linear stability analysis for the pipe structure together with the elastic continuum soil model for a pipe with nonuniform thrust, Moore and Booker [19], indicates that buckle waves develop at the locations of maximum thrust, and the buckle wavelength is almost equal to that for the same buried pipe subjected to uniform thrust. It is convenient and conservative to estimate critical thrust using stability analyses based on uniform thrust distribution, predicting factor of safety F by dividing that critical thrust N_c found using (3) by maximum thrust expected N_{max}: $F = N_c/N_{max}$.

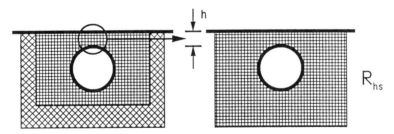

a. R_{hd}: for burial within ring of stiff soil within poor soil

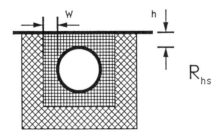

b. R_{hs}: for shallow burial within uniform soil

c. R_{hs}: for shallow burial within nonuniform soil
Figure 5 Ground support models for buried pipe

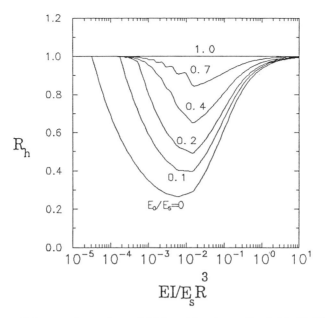

Figure 6 R_{hd} values for stiff soil ring of width W=R/2, [21]

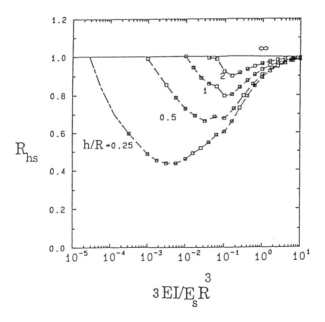

Figure 7 R_{hs} values for shallow burial in uniform soil, [18]

Nonuniform Ground Support

The factor R_h corrects for the effects of pipe burial close to the ground surface, and for the influence of poorer quality soil beyond the zone of engineered backfill.   Such nonuniformities in ground support affect the buckling strength of the pipe crown and pipe walls and invert differently.

a.   Walls and Invert: The buckling strength of the pipe walls and invert is not significantly affected by the presence of the ground surface (excepting the influence of burial depth on earth pressures, and therfore soil modulus).   The correction factor $R_h = R_{hd}$ is employed, Figure 5a.   The factors R_{hd} have been found using a closed form buckling solution which considers the hoop thrust capacity of a deeply buried circular conduit surrounded by a ring of backfill material of modulus E_s, which is in turn surrounded by poorer quality material of modulus E_o, Moore, Haggag and Selig [21].   One set of solutions for R_{hd} are shown in Figure 6, where the ring of backfill has width $W = R/2$.   Results are given for $E_o/E_s = 0, 0.1, 0.2, 0.4, 0.6, 0.8, 1.0$ and 1.5.

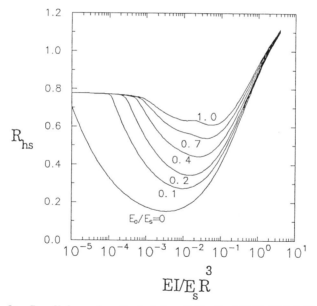

Figure 8: R_{hs} Values for Burial $h = R/2$, Backfill Width $W = R/2$

b.   Crown: The buckling strength of the crown can be influenced by both the burial depth h and modular ratio E_o/E_s.   Firstly, finite element buckling analysis of the system Moore [18] reveals that reductions in buckling strength associated with long wavelength buckling across the pipe crown can occur when the pipe burial depth is less than R.   For pipes buried at

greater depths, the crown can be analysed in the same manner as
the walls and invert, as described above, Figure 5a. Secondly,
where structural backfill extends a long way from the pipe
(more than one pipe diameter), $R_h = R_{hs}$ solutions for pipes
shallow buried in uniform ground Figure 5b can be employed,
Figure 7 (see also Moore [18]). Finite element analysis can
also be used to consider the case where the backfill zone
stretches only a short way from the pipe, Figure 5c. For
example, Figure 8 gives values of $R_h = R_{hs}$ for a pipe buried a
distance R/4, with good quality backfill stretching W = R/2
from the pipe.

In all cases, reductions in critical thrust are influenced by
the normalised flexural stiffness $EI/E_s R^3$. Generally, buckle
wavelengths decrease as $EI/E_s R^3$ is reduced, ground deformations
associated with pipe buckling attenuate more rapidly away from the
pipe, and the influence of nonuniform ground support is reduced.

Choice of Soil Modulus

To make use of the design equation (1), it is necessary to
estimate the soil modulus. Selig [27] discusses this issue in detail
and readers are referred to that work for further discussion.

Choice of Pipe Modulus

During the selection of pipe modulus, creep effects for the
polymer concerned need to be considered carefully. For example, let
us consider a high density polyethylene with an instantaneous modulus
of 1200 MPa, a 24 hour modulus of 800 MPa and a long term modulus of
only 200 MPa. Which of these values, if any, should be used in the
assessment of pipe flexural rigidity EI? (Modulus values are also
influenced by stress and temperature, but these issues are not
considered here).

The process of buried pipe buckling was discussed earlier in
some detail. The importance of pipe separation from the ground was
noted, where this localised loss of ground restraint leads to a sudden
reduction in buckling strength and therefore collapse. Hence pipe
stability is a function of the pipe deformations, and for a
viscoelastic material this implies a dependence on the manner in which
the pipe is loaded over time (the "load path"). A correct
understanding of the pipe collapse mechanism is important here, since
incorrect conclusions about controlling pipe modulus are drawn if
linear buckling theory based on small, rapid deformations of the pipe
from its undeformed position is used as the framework for assessing
the influence of pipe creep.

To illustrate the influence of load path let us examine an
example structure of 2 m diameter, of thickness 40 mm, composed of
high density polyethylene with time dependant modulus as quantified in
the preceeding paragraph and Poisson's ratio 0.5. The pipe is buried
a distance of 4 m in soil of modulus 13 MPa, Poisson's ratio 0.3 and
unit weight 20 kN per cubic metre.

The hoop thrusts which develop depend on the time spent in backfilling the pipe. For example, using the static solutions of Haggag [28], if the pipe is buried rapidly then using the 24 hour modulus of 800 MPa a hoop thrust of about 80 kN/m is estimated for the pipe immediately at the end of burial. If pipe burial is extremely slow, then the long-term modulus 200 MPa is employed and thrust reduces to 60 kN/m.

Now the pipe deformations and therefore the thrust capacity also depend on the manner in which loads are applied. For rapid loading, pipe deformation and therefore collapse is controlled by 24 hour modulus, and using this in the buckling theory (3) yields a critical hoop thrust of 127 kN/m. For very slow loading, pipe deformations and therefore stability depend on long term modulus, which yields 82 kN/m when used in equation (3).

Overall stability may be quantified using factor of safety defined as $F = N_c/N_{max}$. For rapid loading we obtain $F = 127/80 = 1.6$, while for very slow loading $F = 82/60 = 1.3$, so this implies an overall decrease in factor of safety with reduction in load rate. Note however, that these are the two simplest load paths, and that more realistic paths (involving, say, initial pipe burial followed by loading to failure at some later time) will give different values of both thrust and thrust capacity which may be more critical. This load path dependence merits further study.

SUMMARY AND CONCLUSIONS

a.   A design equation for estimation of critical hoop thrust has been described for ground supported plastic pipes. The equation can be used to determine thrust capacity for a range of ground support and burial conditions. The equation has been calibrated using available experimental data. Stability can be estimated by comparing thrust capacity with expected thrust.

b.   For pipes subjected to external earth (rather than fluid) loads, the distinction between "ring" and "local" buckling mechanisms is unnecessary, since both are multi-wave buckling phenomena. The buckle waves simply become more localised when the distribution of hoop thrust becomes less uniform, (observers will notice "local" buckling at the locations of maximum thrust). In both cases, continuum buckling theory can be used to determine critical thrust.

c.   The continuum buckling theory has been modified using linear finite element buckling analyses of shallow buried pipes (Moore [18]). Such analyses show how reductions in the quantity of ground material reduce ground support, and therefore buckling strength. The buckle wavelength increases, and a long wavelength "upward" buckle develops at the crown [5] together with long-wavelength "downward" buckles across the shoulders.

d.   Modifications have also been incorporated to consider the case of pipe surrounded by a limited zone of granular backfill. The effect on critical hoop thrust of the poorer soil beyond the granular backfill has been considered both for deeply buried and shallow buried pipes. This poorer soil can reduce critical thrust substantially.

e.   Buckling collapse can occur at both very large ($>$ 20%) and very small ($\ll$ 5%) deformations [24, 29, 30], so that deflection control does not necessarily prevent instability nor can the degree of stability be inferred from the deformed shape of the pipe.

f.   The linear buckling solutions for earth supported pipe are unsuitable for assessing the stability of pipes subjected to large external fluid pressure. Nonlinear solutions for single-wave buckling have the potential to provide stability estimates for fluid pressure buckling. No such solution is yet available where continuum theory has been used to model the soil, and further work is needed to investigate this case.

g.   Data is available concerning the modulus of typical backfill materials [27]. For polymer materials which creep, the pipe modulus used in the design equation depends on the rate of loading and is, in general, a function of the load path. More work is needed to investigate this issue further.

REFERENCES

[1]   MOLIN, J., "Calculations Principles for Underground Plastic Pipes", Svenska Vatten-Och Avloppsverksforeningen, VAV P16, January, 1971.

[2]   CARLSTROM, B.I., "Structural Design of Underground GRP Pipe", International Conference on Underground Plastic Pipe, New Orleans, 1981, American Society of Civil Engineers, pp. 56-78.

[3]   TAPROGGE, R.H., "Large Diameter Polyethylene Profile-Wall Pipes in Sewer Applications", International Conference on Underground Plastic Pipe, New Orleans, 1981, American Society of Civil Engineers, pp. 175-190.

[4]   HEIERLI, W. and YANG, F.L., "The Static Analysis of Buried Sewer Pipes", Europipe 1982 Conference, Basel, Switzerland, Paper 13, pp. 127-142.

[5]   GREATOREX, C.B., "The Relationship Between the Stiffness of a GRP Pipe and Its Performance When Installed", International Conference on Underground Plastic Pipe, New Orleans, 1981, American Society of Civil Engineers, pp. 117-129.

[6]   SCHLUTER, J.C., "Large Diameter Plastic Pipe Design", International Conference on Underground Pipeline Engineering, Wisconsin, 1985, American Society of Civil Engineers.

[7]   JEYAPALAN, J.K. and BOLDEN, B.A., "Performance and Selection of Rigid and Flexible Pipes", Journal of Transportation Engineering, American Society of Civil Engineers, Vol. 113, No. 3, 1986, pp. 315-331.

[8]   HURD, J., "Field Performance of Corrugated Polyethylene Pipe Culverts in Ohio", Transportation Research Record 1087, 1986, pp. 1-6.

[9]   CHAMBERS, R.E. and McGRATH, T.J., "Structural Design of Buried
        Plastic Pipe", International Conference on Underground Plastic
        Pipe, New Orleans, 1981, American Society of Civil
        Engineers, pp. 10-25.

[10]   LEVY, M., "Memoire sur un nouveau cas integrable du probleme
        de l'elastique et l'une de ses applications", Journal de Math
        Pure et Applied, Series 3, No. 7, 1884, p 5.

[11]   LUSCHER, U., "Buckling of Soil-Surrounded Tubes", Journal of
        Soil Mechanics and Foundation Engineering, American Society of
        Civil Engineers, Vol. 92, No. 6, 1966, pp. 211-228.

[12]   JENKINS, C.F. and KROLL, A.E., "External Hydrostatic Loading
        of Polyethylene Pipe", International Conference on Underground
        Plastic Pipe, New Orleans, 1981, American Society of Civil
        Engineers, pp. 527-541.

[13]   AMERICAN WATER WORKS ASSOCIATION, "AWWA Standard for Glass
        Fibre Reinforced Thermosetting-Resin Pressure Pipe", ANSI/AWWA
        C950-1981, American Water Works Association, Colorado.

[14]   WATKINS, R.K., DWIGGINS, J.M. and ALTERMATT, W.E., "Structural
        Design of Buried Corrugated Polyethylene Pipes", Presented to
        67th Annual Meeting, Transportation Research Board,
        Washington D.C., 1981.

[15]   GUMBEL, J.E., "Analysis and Design of Buried Flexible Pipes",
        Ph.D. Thesis, Department of Civil Engineering, University of
        Surrey, U.K., 1983.

[16]   MOORE, I.D., "Elastic Buckling of Buried Flexible Tubes - A
        Review of Theory and Experiment", Journal of Geotechnical
        Engineering, American Society of Civil Engineers, Vol. 115,
        No. 3, 1989.

[17]   MOORE, I.D., SELIG, E.T. and HAGGAG, A., "Design Procedure for
        Estimating the Strength of Flexible Metal Culverts",
        Transportation Research Record, Washington D.C., 1988.

[18]   MOORE, I.D., "The Elastic Stability of Shallow Buried Tubes",
        Geotechnique, Vol. 37, No. 2, 1987, pp. 151-161.

[19]   MOORE, I.D. and BOOKER, J.R., "The Behaviour of Buried
        Flexible Cylinders Under the Influence of Nonuniform Hoop
        Compression", International Journal of Solids and Strucures,
        Vol. 21, No. 9, 1985, pp. 943-956.

[20]   MOORE, I.D., "Elastic Stability of Buried Elliptical Tubes",
        Geotechnique, Vol. 38, No. 4, 1988, pp. 613-618.

[21]   MOORE, I.D., HAGGAG, A. and SELIG, E.T., "Buckling of
        Cylinders Supported by Nonhomogenous Elastic Ground", Internal
        Report No. TRB, 1988-346I, Geotechnical Engineering,
        University of Massachusetts, U.S.A.

[22]   FALTER, B., "Grenzlasten von einseitig elastisch gebetten
        kreiszylindrrischen konstructionen (Critical Loads for Circular
        Cylindrical Structures With One-Sided Elastic Support",
        Bauingenieur, Vol. 55, No. 10, 1981, pp. 381-390.

[23]   DUNS, C.S. and BUTTERFIELD, R., "Flexible Buried Cylinders – Part III: Buckling Behaviour", International Journal of Rock Mechanics and Mining Sciences, Vol. 8, No. 6, 1971, pp. 613–627.

[24]   HOWARD, A.K., "Laboratory Load Tests on Buried Flexible Pipe", Journal of the American Water Works Association, Vol. 64, No. 10, 1972, pp. 655–662.

[25]   FORRESTAL, M.J. and HERRMANN, G., "Buckling of a Long Cylindrical Shell Surrounded by an Elastic Medium", International Journal of Solids and Structures, Vol. 1, 1965, pp. 297–310.

[26]   MOORE, I.D. and BOOKER, J.R., "Simplified Theory for the Behaviour of Buried Flexible Cylinders Under the Influence of Uniform Hoop Compression", International Journal of Solids and Structures, Vol. 21, No. 9, 1985, pp. 929–941.

[27]   SELIG, E.T., "Soil Properties for Plastic Pipe Installations", Buried Plastic Pipe Technology, ASTM STP 1093, George S. Buczala and Michael J. Cassady, Eds., American Society for Testing and Materials, Philadelphia, 1990.

[28]   HAGGAG, A., "Structural Backfill Design for Corrugated Metal Buried Structures", Ph.D. Thesis, Department of Civil Engineering, University of Massachusetts, Amherst, MA, 1989.

[29]   GUMBEL, J.E. and WILSON, J., "Interactive Design of Buried Flexible Pipes – A Fresh Approach from Basic Principles", Ground Engineering, Vol. 14, No. 4, 1981, pp. 36–40.

[30]   CRABB, G.I. and CARDER, D.R., "Loading Tests on Buried Flexible Pipes to Validate a New Design Model", Transportation and Road Research Laboratory, Supplementary Report 204, Crowthorne, U.K., 1985.

Applications/End Users

Robert J. Bailey

INSTALLATION OF FIBERGLASS PIPE ON SEWER FORCE MAIN
PROJECTS

REFERENCE: Bailey, R. J., "Installation of
Fiberglass Pipe on Sewer Force Main Projects,"
Buried Plastic Pipe Technology, ASTM STP 1093,
George S. Buczala and Michael J. Cassady, Eds.,
American Society for Testing and Materials,
Philadelphia, 1990.

ABSTRACT: The stringent design criteria set for
large diameter sewer force mains encompass many
factors. These mains must resist external loading
from backfill and live loads prior to the line
being pressurized. Once pressurized, the pipe must
withstand the combined effects of both the internal
pressure and external loads. In addition, the pipe
must be corrosion resistant to the interior flow
and, in many cases, to severe external corrosive
environment. If thrust blocking is not provided,
the pipeline must be capable of resisting the
unbalanced forces at bends, wyes, tees, etc. By
meeting all of these criteria, the pipe should
perform satisfactorily over the design life of the
project.

KEYWORDS: Fiberglass pressure pipe, sewer force
mains, corrosion, thrust restraint, burial design.

During the mid-1980's the Sewerage and Water Board
of New Orleans (S & WB) was faced with the decision to
upgrade an older treatment plant to meet the increasingly
stringent U.S. Environmental Protection Agency (EPA)
requirements or to divert the flow to a more modern plant
several miles away. Based on a detailed economic analysis,
the decision was made to phase out the older Michoud sewage
treatment plant and build a new sewer line system to link
the affected service areas to the more modern East Bank
sewage treatment plant. The service area involved is the
only area of New Orleans that's not fully developed.

Mr. Bailey is a product manager at Price Brothers
Company, P. O. Box 825, Dayton, OH 45401.

The flat terrain of the service area, the high
water table and long pipeline lengths dictated that the new
sewer line system be a pressure system. In planning the
new system, the New Orleans S & WB Engineering Design
Department was requested to consider non-ferrous pipe.
This was made necessary due to the internal corrosion
threat of hydrogen sulfide and the external corrosion
potential of the swampy landfill areas involved.

Besides the vital need to be corrosion resistant,
the pipe, it's jointing systems, fittings and specials had
to be capable of functioning as a force main conduit with
an internal pressure of 50 psi (345 kPa). Pipe with
nominal inside diameters of 18 in., 30 in. and 48 in.
(450mm, 750mm and 1200mm) were needed to satisfy the
hydraulic flow and pumping criteria that were established.
Construction of the new sewer line system was broken down
into two separate contracts, as follows:

Contract No. 3492

30 in. and 18 in. (750mm and 450mm) sewer force
main.

Major pipe items were: 18,700 linear ft. (5700m)
of 30 in. (750mm) diameter pipe and 2,845 ft.
(867m) of 18 in. (450mm) diameter pipe. Also
included were 50 ft. (15m) of 30 in. (750mm)
diameter pipe within 48 in. (1200mm) diameter
casing. Contract bid date was November 18, 1987.

Contract No. 3504

48 in. (1200mm) sewer force main.

Major pipe item was: 11,800 linear ft. (3600m)
of 48 in. (1200mm) diameter pipe. Contract bid
date was February 26, 1988.

In selecting pipe materials to meet the demanding
requirements of the projects the City of New Orleans'
efforts centered on corrosion resistant pipes. Pipe with
inherent corrosion resistant properties, rather than
resistance supplied by supplemental coating and lining
operations, were deemed preferable. Three types of plastic
pressure pipe that were determined to be capable of meeting
the corrosion requirements as well as the internal pressure
and external loading requirements were: fiberglass pipe,
polyvinyl chloride (PVC) pipe, and high density
polyethylene (HDPE) pipe. On Contract No. 3504 only the
fiberglass pipe was available to meet the 48 in. (1200mm)
diameter requirement at the pressures required. Therefore,
the contract specifications included two types of ferrous
pipe, steel and ductile iron, with supplemental corrosion
requirements considered necessary for interior and exterior

Table 1 - Pipe Highlights (Contract No. 3492)

Type of Pipe	Specification Requirements
Fiberglass	Manufactured in accordance with AWWA C950 [1] Type II: Centrifugally cast Grade 4: RPMP polyester Liner D: Non-reinforced thermoset resin O.D. of 30 in. pipe = 32.00 in.; O.D. of 18 in. pipe = 19.50 in. Pipe stiffness = 72 psi Min. liner thickness of 0.04 inches of non-reinforced polyester resin.[a] Min. coating thickness of 0.03 inches of polyester resin and sand.[a] Joints: Exterior coupling with an elastomeric membrane with dual function sealing fins on each side of a center stop with the membrane overwrapped with a filament wound glass fib er reinforcement sleeve. Pressure class = 100 psi All pipe to be hydrostatically tested in the plant to 200 psi.
High Density Polyethylene (HDPE)	Manufactured in accordance with ASTM F714[2] having a cell classification of PE 345434-C per ASTM D3350 [3]. O.D. of nominal size 34 in. pipe = 34.00 in. O.D. of nominal size 20 in. pipe = 20.00 in. DR of pipe = 21[b] Wall thickness of 34 in. nominal size = 1.619 in. Wall thickness of 20 in. nominal size = 0.952 in. Pipe stiffness - 61 to 89 psi for DR = 21 and cell classification specified. Joints: Made in the field by the butt fusion technique. Pressure class = 80 psi for DR - 21 and cell classification specified.
Polyvinyl Chloride (PVC)	Manufactured in accordance with UNI-BELL PVC Pipe Association recommended standard specification UNI-B-11[4][c] O.D. of 30 in. pipe = 32.00 in.; O.D. of 18 in. pipe = 19.50 in. DR of pipe = 32.5[b] Wall thickness of 30 in. pipe = 0.985 in. Wall thickness of 18 in. pipe = 0.600 in. Pipe stiffness = 57 psi for DR = 32.5 and modulus of elasticity of 400,000 psi per UNI-B-11. Joints: Shall be of push-on type utilizing elastomeric gaskets. Pressure class = 125 psi All pipe to be hydrostatically tested in the plant to 250 psi.

[a] Liner and coating are integral part of pipe manufacture and are not added as supplemental operations.

[b] DR = Dimension ratio = outside diameter/wall thickness

[c] The provisions of UNI-B-11 have basically been incorporated in AWWA C905 [5]

1 in. = 25.4 mm
1 psi = 6.9 kPa

corrosion protection. Highlights of the pipe alternates
on Contracts No. 3492 and 3504 are given in Tables 1 and
2. Tabulation of bid results for Contracts No. 3492 and
3504 are given in Tables 3 and 4.

The low bidding contractors on each project elected
to use fiberglass pressure pipe from among the possible
alternates. Both contractors chose as their pipe supplier,
Price Brothers Composite Pipe Inc. Centrifugally cast
fiberglass pressure pipe and fittings, marketed under the
tradename HOBAS[R] pipe were supplied from manufacturing
facilities in Florida.

There were many unique aspects of the design,
specifications, and installation of the two projects that
should be of interest to those responsible for large
diameter sewer force mains that will be discussed.

HYDRAULICS:

The hydraulic capacity and pumping capacity designs
were based on having pipe of full 18 in., 30 in. and 48 in.
(450mm, 750mm, and 1200mm) pipe diameters with friction
coefficients appropriate to plastic interior surfaces.
Hazen-Williams friction coefficients of C = 140 were used
in the design. Pipe geometry including pipe inside
diameters for Contracts No. 3492 and 3504 are given in
Tables 5 and 6.

SOILS:

Core borings along the length of both projects
indicated a wide range of soil conditions awaited the
contractor. The top 1 to 2 feet (0.3 to 0.6m) often
consisted of fill containing miscellaneous debris, shell,
concrete, brick and wood. Soil in the foundation and
backfill zone of the pipe was logged as being: very soft
to soft clay, medium stiff clay, clay with shell and sand,
soft clay with organic and wood, sand, soft brown humus,
silt, silt and sand.

The different types of soils provided varying degrees
of bottom and side trench wall support to the pipe as well
as varying degrees of corrosion potential to pipe that
could be susceptible to exterior corrosion.

CONSTRUCTION:

All of the plastic pipe alternates were specified to
have a foundation and side backfill of a mixture of 65%
clam shell or crushed reef shell and 35% river sand that
is to be mechanically batch mixed prior to installation.
As an alternate to clam shell, the contractor could
substitute a Class 1 angular material (1/4 in. to 1/2 in.)
(6 to 13mm) per ASTM D2321[10]; (i.e. coral, cinders,

Table 2 - Pipe Highlights (Contract No. 3504)

Type of Pipe	Specification Requirements
Fiberglass	Same as Table 1 except: O.D. of 48 in. pipe = 50.80 in.
Steel	Manufactured in accordance with AWWA C200[6]. I.D. of 48 in. pipe = 48.00 in.; I.D. of 30 in. pipe = 30.00 in. Wall thickness of 48 in. and 30 in. pipe = 3/8 in. Lining: Liquid two-part chemically cured rust inhibitive epoxy primer and one or more coats of a two-part coal-tar epoxy with a total thickness of 16-20 mils in accordance with AWWA C210[7]. Coating: Coated and wrapped outside with "Prefabricated Multilayer Cold-applied Polyethylene Tape Coating" in accordance with AWWA C214[8]. Total thickness of coating shall be a minimum of 50 mils consisting of primer, 20 mil inner layer for corrosion and 30 mil outer layer for mechanical protection. Cathodic Protection System: Install fifty (50) pound magnesium anodes with No. 6, 600-volt black, standard copper wire at designated locations. Joints: Field welded in accordance with AWWA C206[9].
Ductile-Iron	Manufactured in accordance with AWWA C151. Diameter Class 50 O.D. of 48 in. pipe = 50.80 in.; O.D. of 30 in. pipe = 32.00 in. Wall thickness of 48 in. pipe = 0.51 in. Wall thickness of 30 in. pipe = 0.39 in. Lining : Coal-tar epoxy lining the same as for steel pipe, including sand blasting. Coating: Bituminous coating plus installing a sealed polyethylene tubular wrap, 8 mils thick, covering all pipe, fittings and joints. Joints: Shall be of push-on type utilizing elastomeric gaskets.

1 in. = 25.5mm

crushed concrete or crushed stone).

The shell-sand bedding and backfill was to be placed in layers not exceeding 9 in. (230mm) and compacted by a mechanical vibrating compactor until a minimum of 90% standard Proctor density is attained for the full width of the trench. The specified limits of the foundation and bedding for all three 30 in. (750mm) plastic pipe alternates on Contract No. 3492 are given in Fig. 1. On Contract No. 3504 the backfill was specified to extend up approximately 60% of the O.D. of the 48 in. (1200mm) diameter fiberglass pipe.

The minor deflections recorded on the installed pipes has led the New Orleans' S & WB engineers to feel that on future projects the extent of the shell-sand backfill will be lowered, possibly to the springline of the pipe as proposed in ASTM D2321.

Table 3
Tabulation of Bids - Contract No. 3492   ($1,000)

	Contractor				
Bid Item	I[a]	II	III	IV	V
Section "A" (Fiberglass Pipe)	1,876	2,112	N.B.	2,829	N.B.
Section "B" (HDPE Pipe)	N.B.[b]	N.B.	N.B.	N.B.	N.B.
Section "C" (PVC Pipe)	N.B.	N.B.	2,537	N.B.	3,132
Section "D" (Remaining Contract Items)	720	526	450	659	615
Total for comparison of bids	2,596	2,638	2,987	3,488	3,747

[a] Low bidder: CFW Construction Company, Fayetteville, TN.

[b] N.B. = No Bid

Table 4
Tabulation of Bids - Contract No. 3504   ($1,000)

	Contractor				
Bid Item	I[a]	II	III	IV	V
Section "A" (Fiberglass Pipe)	1,744	1,778	1,801	2,076	2,022
Section "B" (Steel Pipe)	N.B.[b]	N.B.	N.B.	N.B.	N.B.
Section "C" (Ductile Iron Pipe)	N.B.	N.B.	N.B.	N.B.	N.B.
Section "D" (Remaining Contract Items)	426	450	472	343	431
Total for comparison of bids	2,170	2,228	2,273	2,419	2,453

[a] Low bidder: Boh Brothers Construction Company, New Orleans, LA.

[b] N.B. = No Bid

Table 5 - Pipe Geometry (Contract No. 3492)

Type of Pipe	Nominal Diameter (in.)	Outside Diameter (in.)	DR[c]	Wall Thickness (in.)	Inside Diameter[b] (in.)
Fiberglass	30.00[a]	32.00[a]		0.745	30.510
HDPE	34.00[a]	34.00[a]	21[a]	1.619[a]	30.762
PVC	30.00[a]	32.00[a]	32.5[a]	0.985[a]	30.030
Fiberglass	18.00[a]	19.50[a]		0.470	18.560
HDPE	20.00[a]	20.00[a]	21[a]	0.952[a]	18.096
PVC	18.00[a]	19.50[a]	32.5[a]	0.600[a]	18.300

[a] Values given in project specifications
[b] I.D. = O.D. - (2 x wall thickness)
[c] DR = outside diameter / wall thickness

1 in. = 25.4 mm

Table 6 - Pipe Geometry (Contract No. 3504)

Type of Pipe	Nominal Diameter (in.)	Outside Diameter (in.)	Wall Thickness (in.)	Liner Thickness (in.)	Inside Diameter[c] (in.)
Fiberglass	48.00[a]	50.800[a]	1.197		48.406
Steel	48.00[a]	48.750	0.375[a]	0.008[a,b]	47.984
Ductile Iron	48.00[a]	50.800[a]	0.510[a]	0.008[a,b]	49.764

[a] Values given in project specifications
[b] Epoxy liner specified = 8-10 mils (0.008-0.010 in.)
[c] I.D. = O.D. - (2 x wall thickness) - (2 x liner thickness)

1 in. = 25.4 mm

Thrust Restrain and Harnessed Joints: Except for the 18 in. and 30 in. (450mm and 750mm) field butt fusion joint HDPE pipe alternate on Contract No. 3492 and the 30 in. and 48 in. (750mm and 1200mm) field welded joint steel pipe alternate on Contract No. 3504, all pipe was to be designed with thrust restraining joints meeting the following requirements:

"Thrust forces in elbows and bends shall be designed to be resisted only by frictional drag against the soil surrounding the adjoining sections of straight pipe.   Bends shall be harnessed to conform with the minimum requirements of Table 7 unless shown otherwise on the contract drawings."

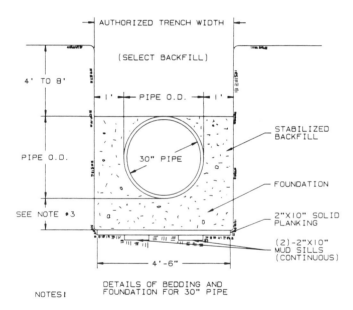

DETAILS OF BEDDING AND
FOUNDATION FOR 30" PIPE

NOTES!

(1) STABILIZED SHELL-SAND FOUNDATION AND BEDDING SHALL BE COMPACTED TO MINIMUM DENSITY OF 90% STANDARD PROCTOR.

(2) PLANKING AND MUD SILL LUMBER FOUNDATION TO BE INSTALLED IF DIRECTED BY THE ENGINEER.

(3) DEPTH OF FOUNDATION VARIES BETWEEN 9" AND 15", DEPENDING ON PROJECT LOCATION

FIGURE 1

Table 7 - Harnessed Joint Lengths (Contracts No. 3492 and 3504)

Degree of Bend	Minimum Length of Harnessed Pipe Required for Each Bend	
	Ft.	(m)
6-30	20	(6.1)
30-45	40	(12.2)
45-55	60	(18.3)
56-90	80	(24.4)

Hydrostatic Test of the Force Mains: The main was to be sealed at each end and filled with water. The contractor shall then apply a hydrostatic pressure of 75 psi (518 kPa) on the force main and shall maintain this pressure for a period of 24 continuous hours. During this period, the total leakage shall not exceed 50 gallons per inch of internal diameter for each mile of pipe (4.63 liters/mm of diameter/km of pipe). If greater leakage than the said quantity is developed, the contractor shall locate the leaks and repair them.

The contract specifications then go on to state: "It is the intent of these specifications and of the contract based thereon, that all pipe joints be water tight under all service conditions and, even though the total leakage of any test is within the permissible limits as stated herein, any and all leaks from improperly laid or defective joints which are discovered during the leakage test or tests shall be repaired by and at the expense of the contractor. Field hydrostatic acceptance tests indicated no leakage.

Restrained Joints

Thrust forces in elbows and bends were to be designed to be resisted only by frictional drag against the soil surrounding the adjoining sections of straight pipe. Bends shall be harnessed together with adjoining straight lengths of pipe to conform with the minimum requirements of Table 7 unless shown otherwise on the contract drawings.

The restrained joint was to be designed for the full thrust of 100 psi (690 kPa) against a dished head (25,400 lbs. for 18 in. pipe, 70,650 lbs. for 30 in. pipe, and 180,800 lbs. for 48 in. pipe) (113 kN for 450mm pipe, 314 kN for 750mm pipe and 804 kN for 1200mm pipe). The manufacturer had to submit certified test results for proof of harness joint design.

The above restrained joint requirements were for the fiberglass and PVC pipe on Contract No. 3492 and the

fiberglass and ductile iron pipe on Contract No. 3504. The
field fusion jointed HDPE pipe on Contract No. 3492 and the
steel pipe with field welded joints on Contract No. 3504
were exempt from the harness joint requirements.

Pipe Design: The fiberglass pipe was designed for the
project burial and service conditions in accordance with
Appendix A of AWWA C950. The design shall be based on a
strain analysis and the corrosion liner shall not be
considered as contributing to the structural strength of
the pipe.

The 18 in. (450mm) and 30 in. (750mm) diameter
fiberglass pressure pipe for Contract No. 3492 was designed
for a minimum cover of 4 ft. (1.2m) and a maximum cover of
8 ft. (2.4m). The 48 in. (1200mm) diameter pipe for
Contract No. 3504 had a minimum cover of 4 ft. (1.2m) and
a maximum cover of 12 ft. (3.7m). All pipe was designed
for a 16,000 lb. (71 kN) wheel live load. Ground water
was assumed to be 1 ft. (0.3m) beneath grade.

Pipe was required to have a minimum pressure class of
100 psi (690 kPa) with a working pressure of 100 psi (690
kPa) and no surge pressure. Pipe designed in accordance
with AWWA C950 is always designed for a surge pressure
equal to 40% of the working pressure.

Per AWWA C950 pipe, long-term deflection shall not
exceed 5%. A deflection lag factor of 1.50 was used with
a soil specific weight of 120 lbs/ft^3 (1922 Kg/m^3).

Bid dates for the two force main contracts required
that the pipe be designed in accordance with AWWA C950-
81, the edition of the standard at the time. The design
appendix, Appendix A, was extensively modified with the
issuance of AWWA C950-88 which became effective in October,
1989. Appendix 1 of this paper gives the calculations for
the 30 in. (750mm) diameter fiberglass force main pipe done
in accordance with Appendix A of the latest issue of AWWA
C950.

ACKNOWLEDGEMENTS

The author acknowledges with appreciation the help of
G. Joseph Sullivan and Wesley L. Busby, both of the New
Orleans Sewerage and Water Board, in the preparation of
this paper.

APPENDIX 1
Conditions and Parameters for Design Example
30" Pipe (Contract No. 3492)

Conditions and Parameters	
Design Conditions	
Nominal pipe diameter, in.	30
Working pressure P_w, psi	100
Surge pressure P_s, psi	0
Vacuum P_v, psi	0
Cover depth H, ft (min.-max.)	4 - 8
Wheel load P, lb	16,000
Soil specific weight γ_s, lb/ft^3	120
Service temperature, °F	40 - 100
Native soil conditions at pipe depth	Medium stiff clay
Groundwater table location	1 ft. below grade
maximum h_w, in.	84
minimum h_w, in.	36

Basis for HDB and S_b	Strain in./in.
Pipe Properties	
Trial pressure class P_c, psi	100
Reinforced wall thickness t, in.	0.6995
Liner thickness t_L, in.	0.0455
Total wall thickness t_t, in.	0.7450
Minimum pipe stiffness F/Δy, psi	72
Hoop tensile modulus E_H, psi	938,000
Hoop flexural modulus E, psi	1,680,000
HDB	0.006615
S_b	0.008800
Mean diameter D, in.	31.3005
Distance between joints L, in.	240
Poisson's ratio γ, in./inc.	
Hoop load γ_{hl}	0.30
Axial load γ_{lh}	0.15

Installation Parameters	
Pipe-zone installaion description	moderately compacted 65% clam shell (or crushed reef shell) 35% river sand
Shape factor, D_f	4.5
Backfill soil modulus E', psi	2000
Deflection coefficient K_x	0.103
Deflection lag factor D_l	1.5

Deflection	
Maximum deflection permitted, Δy_a, %D	4

1. Calculate pressure class P_c from HDB

$$P_c = 100 \leq \left(\frac{HDB}{FS}\right)\left(\frac{2\ E_H\ t}{D}\right)$$

$$\leq \left(\frac{0.006615}{1.8}\right)\left(\frac{(2)(938,000)(0.06995)}{31.3005}\right)$$

$$\leq 154\ psi\ \therefore\ o.k.$$

2. Check working pressure P_w using P_c

$$P_w \leq P_c$$

100 psi $\leq$ 100 psi $\therefore$ o.k.

3. Check surge pressure P_s using P_c

$$P_w + P_s \leq 1.4\ P_c$$

$$100 + 0 \leq 1.4\ (100)$$

$$100\ psi \leq 140\ psi\ \therefore\ o.k.$$

4. Calculate allowable deflection $\triangle y_a$ from ring bending:

$$\mathcal{E}_b = D_f \left(\frac{\triangle y_a}{D}\right)\left(\frac{t_t}{D}\right) \leq \frac{S_b}{F.S.}$$

Substituting

$$4.5 \left(\frac{\triangle Y_a}{31.3005}\right)\left(\frac{0.7450}{31.3005}\right) \leq \frac{0.00880}{1.5}$$

$$0.00342\ \triangle y_a \leq 0.00587$$

$$max.\ \triangle y_a = \frac{0.00587}{0.00342} = 1.716\ in.$$

$$\frac{\triangle Y_a}{D} \leq 0.04$$

Taking the maximum $\triangle y_a$ as the smaller of 1.716 in. or 0.04D

$$0.04\ (31.3005) = 1.252\ in. < 1.716\ in.$$

$$\therefore\ max.\ \triangle y_a = 1.252\ in.$$

5. Determine external loads

$$W_c = \frac{\gamma_s\ H\ (D + t\ )}{144} = \frac{120\ H\ (31.3005+0.6995)}{144} = 26.67H$$

For H =   4 ft. W_c =   106.7 lb/in.
For H =   8 ft. W_c =   213.3 lb/in.

$$W_L = \frac{C_L P\ (1 + I_f)}{12}$$

$$I_f = 0.766 - 0.133H;\ (0 \leq I_f \leq 0.50)$$

For H =   4 ft. I_f =   0.234
For H =   8 ft. I_f =   0.0

C_L from live load coefficient table in AWWA C950

For H =   4 ft. C_L =   0.066
For H =   8 ft. C_L =   0.019

substituting in equation for W_L

For H = 4 ft. $W_L = \dfrac{0.066(16,000)(1+0.234)}{12} = 108.6$ lb/in.

For H = 8 ft. $W_L = \dfrac{0.019(16,000)(1+0.0)}{12} = 25.3$ lb/in.

6. Check deflection prediction Δy

$$\Delta y = \frac{(D_L\ W_c + WL)\ K_x\ r^3}{EI\ +\ 0.061K_A E' r^3}$$

where: $r = D/2 = 31.3005/2 = 15.6503$ in.

$$I = \frac{t^3}{12} = \frac{(0.6995)^3}{12} = 0.02852\ in^4/in.$$

for H = 4 ft.

$$\Delta y = \frac{[(1.50 \times 106.7) + 108.6] \times 0.103 \times 15.6503^3}{(1,680,000 \times 0.02852)+(0.061 \times 0.75 \times 2000 \times 15.6503^3)}$$

=   0.266 in. $\leq$ 1.252 in. $\therefore$ ok

for H = 8 ft.

$$\Delta y = \frac{[(1.50 \times 213.3) + 25.3] \times 0.103 \times 15.6503^3}{(1,680,000 \times 0.02852)+(0.061 \times 0.75 \times 2000 \times 15.6503^3)}$$

=   0.342 in $\leq$ 1.252 in. $\therefore$ ok

7. Check combined loading strain ε_c

$$\varepsilon_c = \frac{P_w D}{2E_H t} + D_f\, r_c \left(\frac{\Delta Y_a}{D}\right)\left(\frac{t_t}{D}\right)$$

where: $r_c = 1 - \left|\dfrac{P_w}{435}\right| = 1 - \left|\dfrac{100}{435}\right| = 0.770$

$$\varepsilon_c = \frac{100 \times 31.3005}{2 \times 938,000 \times 0.6995} + (4.5 \times 0.770)\left(\frac{1.252}{31.3005}\right)\left(\frac{0.745}{31.3005}\right)$$

$$= 0.00239 + 0.00330 = 0.00569 \text{ in/in}$$

$\varepsilon_c \leq S_b/1.5;\quad 0.00569 \leq (0.00880/1.5 = 0.00587)\quad \therefore \text{ ok}$

Check that ε_c satisfies following equation

$$\frac{\varepsilon_{pr}}{HDB} + \frac{\varepsilon_c - \varepsilon_{pr}}{S_B} \leq \frac{1}{1.5}$$

where: $\varepsilon_{pr} = \dfrac{P_w D}{2E_H t} = 0.00239 \text{ in./in.}$

$$\frac{0.00239}{0.006615} + \frac{(0.00569 - 0.00239)}{0.00880} \leq 0.667$$

$$0.361 + 0.375 = 0.736 > 0.667 \text{ N.G.}$$

Since ε_c does not satisfy the equation the value of Δya (the allowable long-tem deflection) must be recalculated to satisfy the equation.

Calculation would show that a Δya = 1.020 in. (or 3.26% of D) will satisfy all strain criteria. Verification:

$$\varepsilon_c = 0.00239 + (4.5 \times 0.770)\left(\frac{1.020}{31.3005}\right)\left(\frac{0.745}{31.3005}\right) = 0.00508 \text{ in/in.}$$

$$\frac{0.00239}{0.006615} + \frac{(0.00508 - 0.00239)}{0.00880} \leq 0.667$$

$$0.361 + 0.306 = 0.667 \leq 0.667 \quad \therefore \text{ ok}$$

8. Check buckling

$$q_a \;=\; \frac{1}{FS} \left(\frac{32\ R_w\ B'E'\ EI}{D^3} \right)^{1/2}$$

where:
$R_w = 1 - 0.33\ (h_w/h)$

For H = 4' $R_w = 1-0.33\ (36/48) = 0.752$
For H = 8' $R_w = 1-0.33\ (84/96) = 0.711$

$$B' \;=\; \frac{1}{1 + 4e^{\,-0.065H}}$$

For H = 4' $B' = 0.245$
For H = 8' $B' = 0.296$

For H = 4 ft.

$$q_a \;=\; \frac{1}{2.5} \left[\frac{32 \times 0.752 \times 0.245 \times 2000 \times 1{,}680{,}000 \times 0.6995^3}{12 \times 31.3005^3} \right]^{1/2}$$

= 54.30 psi

For H = 8 ft.

$$q_a \;=\; \frac{1}{2.5} \left[\frac{32 \times 0.711 \times 0.296 \times 2000 \times 1{,}680{,}000 \times 0.6995^3}{12 \times 31.3005^3} \right]^{1/2}$$

= 58.03 psi

with no vacuum pressure present q_a must satisfy the following equation

$$\gamma_w\, h_w + R_w \left(\frac{Wc}{D} \right) + \left(\frac{W_L}{D} \right) \leq qa$$

For H = 4 ft.

$(0.0361 \times 36) + \dfrac{0.752 \times 106.7}{31.3005} + \dfrac{108.6}{31.3005} \leq 54.30$ psi

7.33 psi $\leq$ 54.30 psi ∴ ok

For H = 8 ft.

$(0.0361 \times 84) + \dfrac{0.711 \times 213.3}{31.3005} + \dfrac{25.3}{31.3005} \leq 58.03$ psi

8.69 psi $\leq$ 58.03 psi ∴ ok

REFERENCES

[1] AWWA C950 Fiberglass Pressure Pipe

[2] ASTM F714 - Polyethylene (PE) Plastic Pipe (SDR-PR) Based on Outside Diameter

[3] ASTM D3350 - Polyethylene Plastics Pipe and Fittings Materials

[4] UNI-B-11 - Polyvinyl Chloride (PVC) Water Transmission Pipe (Nominal Diameters 14-36 Inch)

[5] AWWA C904 - Polyvinyl Chloride (PVC) Water Transmission Pipe, Nominal Diameters 14 In. through 36 In.

[6] AWWA C200 - Steel Water Pipe 6 In. and Larger

[7] AWWA C210 - Liquid Epoxy Coating Systems for the Interior and Exterior of Steel Water Pipelines

[8] AWWA C214 - Tape Coating Systems for the Exterior of Steel Water Pipelines

[9] C206 - Field Welding of Steel Water Pipe Joints

[10] ASTM D2321 - Underground Installation of Flexible Thermoplastic Sewer Pipe

Reynold K. Watkins

PLASTIC PIPES UNDER HIGH LANDFILLS

REFERENCE: Watkins, R.K., "Plastic Pipes Under High Landfills",
Buried Plastic Pipe Technology, ASTM STP 1093, George S. Buczala and
Michael J. Cassady, Ed., American Society for Testing and Materials,
Philadelphia, 1990.

ABSTRACT: Plastic pipes have long life because of their
resistance to corrosion and erosion. Consequently, they are
attractive for use under long-term landfills and in aggressive
environments such as sanitary landfills.   But sanitary
landfills are usually high landfills.   Tests at Utah State
University investigated the performance of plastic pipes under
high landfills.   It was found that a plastic pipe can perform
under enormous soil loads -- hundreds of feet -- if an
envelope of carefully selected soil is carefully placed about
the pipe. The creep of plastic materials allows the pipe to
relax and so to conform with the soil in a mutually supportive
pipe-soil interaction.

KEYWORDS: pipes, buried, flexible, structural stability, high
soil cover, landfills

INTRODUCTION

Plastic pipes are an attractive alternative for collection and
transmission of fluids in erosive and aggressive environments.   One
example is the collection of leachate under sanitary landfills.   The
leachate is highly acidic.   Erosive sediment may enter the collection
system. The required service life is over a hundred years. And now comes
a demand for very high landfills. Our throw-it-out generation is running
out of out.

Can plastic piping perform under landfills that are hundreds of ft
high? Since 1984, tests have been performed in the soil cells at USU to
evaluate the performance and limits of performance of plastic pipes under
high landfills. The soil cells are basically large containers into which

Reynold K. Watkins, PhD is Professor of Engineering, Department of
Civil and Environmental Engineering, Utah State University, Logan, UT
84322-4110.

pipes can be buried and then loaded by hydraulic jacks to simulate the loading of high landfills. Various types of plastic pipes were tested. Various soil types and soil densities were used as pipe-zone-backfill.

TEST RESULTS

The capacity of the USU soil cells is vertical soil pressure of about sixteen kips per square ft (766 kPa) which is equivalent to 210 ft (64 m) of landfill at 75 lb per cubic ft (1.2 Mg/m^3). Higher loads were simulated.

The results are conclusive. Structurally, plastic pipes perform adequately under high landfills if the pipe-zone-backfill is of good quality, is carefully placed, and is adequately compacted.

Performance limits for the cross section (ring) are ring compression strength; excessive ring deflection; and, under extraordinarily poor pipe-zone-backfill, incipient ring collapse. Ring compression strength is compressive strength of the wall.

Performance limits for longitudinal (beam) action are excessive longitudinal stress and beam deflection (low spots in the pipeline or sharp bends). Longitudinal performance is basically alignment which is assured by specifications and careful installation.

PRINCIPLES

Following are useful principles for analyzing the structural performance of buried plastic pipes under high landfills.

1. Plastic pipes are flexible. Because flexible pipes can deform, they conform with the soil and relieve the pipe of pressure concentrations. The differences between horizontal and vertical soil forces on the cross section (ring) are reduced. See Fig. 1. Any stresses in the pipe due to hard spots in the pipe-zone-backfill are partially relieved.
2. Arching action of the soil supports vertical load. The soil performs as a masonry arch over the pipe. No cement is needed to hold the arch together because the pipe retains the soil arch. The pipe is a liner for a soil conduit.
3. The flexible pipe ring is held in shape by the soil. The soil-stabilized ring itself can carry substantial load. Without soil support, the ring would collapse under light load.
4. Performance limits are ring crushing and excessive deflection. If performance limit is reached, it usually happens during completion of the landfill (in the short term -- not in the long term).
5. Stresses in the plastic relax. If the soil holds the pipe in a fixed shape, the plastic relaxes over a period of time and relieves itself of part of the stresses in it.
6. Ring deflection is approximately equal to, but not greater than, the vertical strain of sidefill soil due to the weight of the landfill. See Fig. 2 for nomenclature.

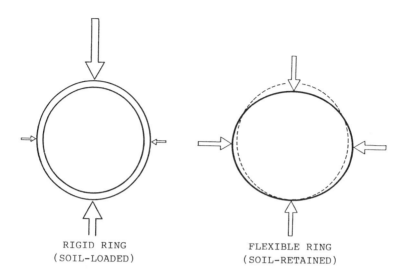

RIGID RING
(SOIL-LOADED)

FLEXIBLE RING
(SOIL-RETAINED)

Figure 1.   Comparison of typical soil loads on a rigid ring and on a flexible ring showing how the flexible ring deflects just enough to equalize horizontal and vertical forces.

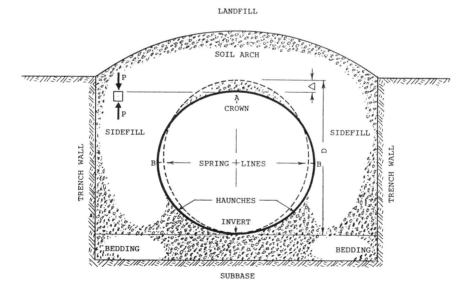

Figure 2.   Nomenclature for the cross section (ring) of a buried flexible pipe showing the deflected ring in a select soil envelope called the pipe-zone-backfill (PZB), within a trench, loaded by a (Ring deflection as defined as d = Δ/D).

7. Collapse of plastic pipe is not common.   Collapse becomes incipient if ring deflection is excessive and if strength of the sidefill soil is inadequate.   "Incipient" does not mean that collapse is inevitable.   It means that the pipe ring itself is not able to carry any additional loads.   Any additional loads, then, must be carried by the soil. Collapse may be progressive soil slippage over a period of time if additional loads do, in fact, occur and are great enough to cause soil slippage.

RATIONALE FOR DESIGN

Successful performance of plastic pipes under high landfills is based on four requirements:
1. limits for ring compression stress,
2. limits for deflection -- both ring deflection and longitudinal beam deflection,
3. stability; i.e. no incipient ring collapse, and
4. intimate contact of the soil against the pipe.

RING COMPRESSION STRESS

The performance limit for ring compression stress is wall crushing at 9:00 and 3:00 o'clock.   It occurs when ring compression stress reaches the yield strength of the plastic.   Ring deflection causes circumferential flexural stresses which are also maximum at 9:00 and 3:00 o'clock. However flexural stresses are compression on the inside of the pipe and tension on the outside.   See Figure 3. Wall crushing can occur only when the pipe wall is at compression yield stress throughout the entire wall thickness.   Therefore flexural stress does not affect wall crushing even though it may cause plastic hinging as discussed below under "STABILITY".   Ring compression stress analysis does not include circumferential flexural stress.

Under high landfills, the effect of surface live loads on the buried pipe is negligible.   Because plastic relaxes under fixed deformation, the vertical load felt by a nearly-circular ring is not greater than $P(OD)$;

where (See Figures 3 and 4):
OD = outside diameter of the pipe
P = γH = vertical soil pressure at the top of the pipe
H = height of soil cover over the top of the pipe
γ = unit weight of the soil overburden
σ = ring compression stress in the pipe wall
D = mean diameter of the pipe = OD-t
t = wall thickness (minimum)
r = mean radius of the pipe
d = ring deflection = Δ/D
Δ = decrease in vertical diameter
DR = dimension ratio = OD/t.   DR is a measure of pipe stiffness for plastic pipes.

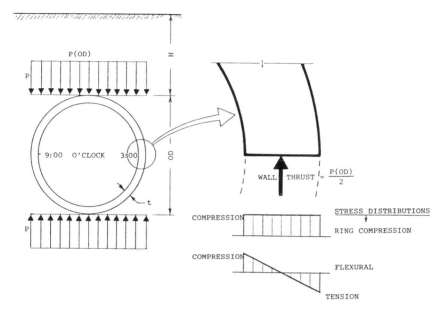

Figure 3.   Ring compression of a circular pipe ring loaded vertically, showing the ring compression stress distribution and the flexural (ring deflection) stress distribution across the wall.

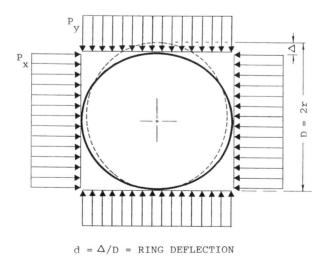

d = Δ/D = RING DEFLECTION

Figure 4.   Ring deflection, d = Δ/D for a flexible pipe cross section.

To avoid wall crushing of a deflected ring, the ring compression
stress, given by the equation,

$$\sigma = 0.5P(1+d)DR \ldots \ldots \ldots \ldots \ldots \ldots \ldots (1)$$

must be less than yield strength of the plastic. Ring deflection, d =
Δ/D, occurs during completion of the landfill, but is often negligible.
If the strength of the pipe-zone-backfill (PZB) is enough to hold the
pipe ring fixed in its buried shape, then the yield strength is short
term; i.e. the time of completion of the landfill. Once the shape of the
ring is fixed, stresses in the pipe wall begin to relax. For a high-
quality plastic pipe under a fixed deformation, the stresses relax faster
than the yield strength decreases. So wall crushing does not occur in
the long term. Long term yield strength is not a performance limit.

Example

A polyethylene pipe is buried under a landfill for which the
vertical soil pressure on the pipe is 280 psi (1.93 MPa). The dimension
ratio for this pipe is DR = 9.2. The ring deflection is not greater than
10% according to a bullet drawn through the pipe. What is the safety
factor for ring compression if the short term yield strength of the
polyethylene is 2300 psi (15.9 MPa)?

From equation 1, the ring compression stress is 1417 psi (9.77
MPa). The safety factor, sf = 2300/1417 = 1.6. After completion of the
landfill, stresses relax if the PZB is good granular soil. The safety
factor increases. 1.6 is an adequate safety factor for ring compression
stress because of stress relaxation.

DEFLECTION

Longitudinal deflection is usually not of concern. With careful
placement of the bedding, the pipe does not sag or hump as a beam. Pipe
manufacturers specify a minimum longitudinal radius of curvature of the
pipe to avoid excessive longitudinal (beam) stress during installation.
Over the less critical long term, longitudinal stresses relax. Excessive
longitudinal bending may cause plastic hinges (creases) to form in the
beam.

Ring deflection, d = Δ/D, is of greater concern. See Figure 4.
Excessive ring deflection can cause leaks at appurtenances and joints;
it reduces flow; it contributes to incipient collapse of the ring. Ring
deflection is usually limited by specification. It can be predicted by
classical or empirical methods.

RING STABILITY

Because plastic pipes are flexible, instability is a performance
limit. Instability is incipient collapse of the buried flexible ring.
The ideal flexible ring is as flexible as a chain-link watch band.

Practically, enough stiffness is built into the flexible ring to hold it in shape during installation and during non-uniform loading. Because soil specifications for buried flexible pipes usually require compacted granular pipe-zone-backfill (PZB), the ring is flexible compared with the soil. It is usually assumed conservatively that shearing stresses are zero between the pipe and the PZB. It is also assumed that the first mode of ring deflection is from circular to elliptical. As the flexible ring deflects under load, it relieves itself of part of the load which is transferred to the soil. The soil forms an arch which supports load.

The conditions for incipient collapse are shown in Figure 5. Analysis starts with a uniformly distributed vertical pressure P on the pipe. For high landfills, $P = \gamma H$. Collapse is incipient when the cube of soil at B shears because of an excessive ratio of horizontal to vertical stresses. Ring deflection changes radii of curvature of the ring. See Fig. 6. For an ellipse, the horizontal radius of curvature, r_x, is minimum. The vertical radius of curvature, r_y, is maximum. The effect of the maximum radius of curvature is to increase the ring compression stress, Pr_y/t, at the crown of the pipe. The effect of the minimum radius of curvature is to increase the soil support requirement at the spring lines. For a known percent of ring deflection, the design engineer can analyze the effects of ring deflection on ring compression at the crown and on soil strength at the spring lines. The procedure follows.

Consider Fig. 5 which shows the deflected flexible ring with vertical pressure P acting on it and with horizontal pressure KP supporting it. It is assumed that:

1. The ring is deflected into an ellipse.
2. The soil is cohesionless, for which,
   ϕ = soil friction angle
   K = $(1+\sin\phi)/(1-\sin\phi)$
   c = 0 = soil cohesion (granular soil)
   γ = unit weight of the soil
3. No shearing stress exists between the ring and the soil.
4. Vertical and horizontal soil stresses are each uniform.
5. The ring has some stiffness called pipe stiffness, F/Δ,

   where, see Fig. 7
   F = diametral line load applied in a parallel plate test
   Δ = decrease in diameter due to the load F
   F/Δ is called pipe stiffness (It is the slope of some initial portion of the F-Δ plot from a parallel plate test.)
6. P is assumed to be the same vertical soil pressure at the crown as at the spring lines. This assumption is justified for high landfills. P is based on maximum soil cover -- not minimum soil cover.
7. It is assumed that the circular pipe cross section deflects from a circle to an ellipse during soil placement. See Fig. 6.
   a = minor semi-diameter = r(1-d)
   b = major semi-diameter = r(1+d)
   d = ring deflection = Δ/D
   D = mean diameter of the originally circular ring = 2r

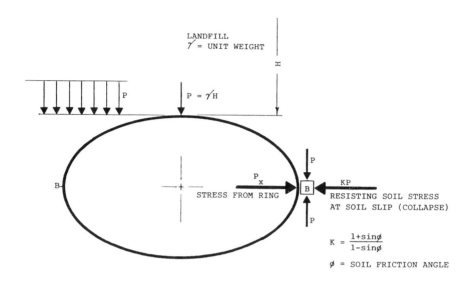

Figure 5.   Free-body-diagram of an infinitesimal cube of soil B at the
spring line, showing the principal stresses acting on it
(for equilibrium, P_x = KP).

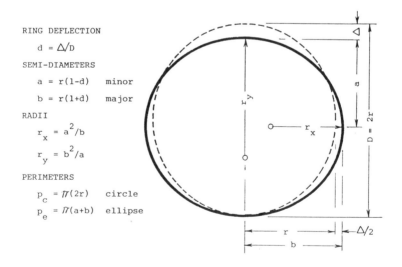

Figure 6.   Pertinent notation for the approximate geometry of an
ellipse.

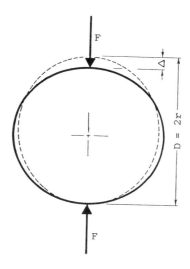

Figure 7.   Diagramatic sketch of a parallel plate test from which a
plot of F vs Δ provides pipe stiffness F/Δ which is the
slope of the plot of line load F as a function of deflection
Δ.

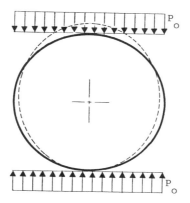

Figure 8.   Diagramatic sketch of the flexible ring carrying a small
load P_0 by the stiffness of the ring only -- without support
from the sidefill soil.

Δ = decrease in diameter due to external pressures
r = mean radius
r_x = horizontal (minimum) radius of the ellipse = a^2/b
r_y = vertical (maximum) radius = b^2/a
r_r = r_y/r_x = $(1+d)^3/(1-d)^3$
r_r = ratio of maximum radius to minimum radius.

Because the flexible ring has some pipe stiffness, F/Δ, it can support a portion P of the vertical pressure as shown in Fig. 8, for which, by the Castigliano equation,

$$P_o = 1.7854(F/\Delta)d$$

P_o is the load carried only by the stiffness of the ring, with no support from the sidefill soil. From Fig. 9, neglecting friction between soil and ring, it can be shown that,

$$P_y r_y = P_x r_x$$

where
$$P_y = P - P_o; \text{ i.e. } P = P_y + P_o$$

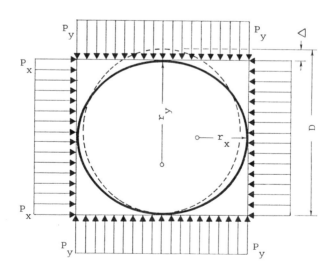

Figure 9.   Deflected flexible ring in equilibrium with uniform vertical pressure, P_y and with uniform horizontal soil reaction P_x.

It follows that the horizontal pressure P_x of the soil against the ring is,

$$P_x = (P-P_o)r_r = [P-1.7854(F/\Delta)d]\ r_r$$

But the soil at the spring lines must be able to provide P_x without soil slip (failure). At soil slip, see Fig. 5,

$$P_x = KP$$

Equating the two expressions for P_x, and solving:

$$P = 1.7854(F/\Delta)d/(1-K/r_r) \quad \ldots\ldots\ldots\ldots (2)$$

where:

P = vertical pressure at soil slip (soil failure)
F/Δ = pipe stiffness from a parallel plate test
d = ring deflection of the ellipse = Δ/D
Δ = vertical decrease in diameter
D = mean diameter of the pipe
K = $(1+\sin\phi)/(1-\sin\phi)$
ϕ = soil friction angle for compacted sidefill
r_r = r_y/r_x = ratio of vertical to horizontal radii
r_r = $(1+d)^3/(1-d)^3$ for an ellipse

From equation 2 it is clear that if $K \geq r_r$, there is no soil slip regardless of soil pressure P. In fact, P becomes negative.

Fig. 10 comprises plots of equation 2. Sidefill soil is compacted. The vertical scale is the dimensionless vertical soil pressure term, $P/(F/\Delta)$ at incipient collapse. The soil friction angle of the sidefill is ϕ. The horizontal scale is ring deflection d. If the soil pressure term, $P/(F/\Delta)$, and the ring deflection term, d, locate a point to the lower left of a soil friction line (ϕ-line), the buried pipe is stable. If the point is located to the upper right of a ϕ-line, collapse is incipient -- not imminent -- but possible, progressively, over a period of time, due to soil dynamics such as earth tremors, wetting and drying, pipe or soil deterioration, etc. As collapse progresses, plastic hinges form at the spring lines. Of course, plastic hinges cause the ring cross section to deviate from an ellipse.

No safety factor is included in either Fig. 10 or equation 2. Soil arching action assures some margin of safety. The $\phi = 0$ line at the bottom is an asymptote for the ϕ-lines, but otherwise is meaningless. If ring deflection is more than 20%, equation 2 loses accuracy because it is derived from various simplifying approximations of elliptical analyses that apply only for small ring deflections. To the right of d = 20%, the ϕ-lines are of little practical value. In fact, pipes with more than 10% ring deflection are usually rejected even though they are structurally stable against collapse. It is noteworthy that if the compacted sidefill soil has a friction angle of at least $\phi = 20°$, stability is assured for ring deflections less than d = 11.8%. Including a safety factor, minimum soil friction angle should be increased --say to $\phi = 30°$.

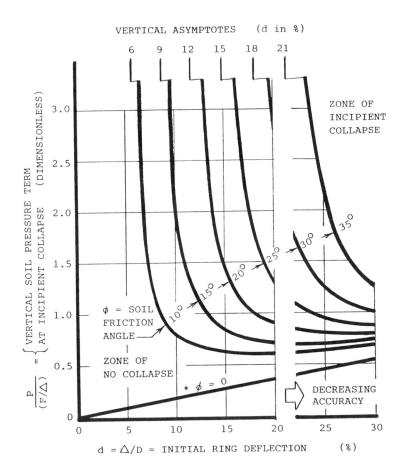

* At ϕ = 0, vertical soil pressure is supported
by the stiffness of the elastic ring only --
no sidefill support.

Figure 10.   Vertical soil pressure term at incipient collapse of
flexible pipes buried with initial ring deflection d, in
soil compacted on the sides where the soil friction angle
is ϕ. (The graphs lose accuracy to the right of d = 20% as
the ring deviates from an ellipse, and as yield point of the
plastic is exceeded.)

Clearly the conditions for stability are assured if good granular soil is carefully compacted in the sidefills, and if the ring deflection is held to less than 10%. Under conditions where mitigation is sought, the height of soil cover can be restricted such that the soil pressure term, $P/(F/\Delta)$, is reduced thereby causing the point, $P/(F/\Delta)$ vs. d, to fall below the ϕ-line where stability is assured.

Example

PVC piping is proposed for drainage under a sanitary landfill to be 600 ft high. Unit weight of the landfill is 75 pounds per cubic ft (1.2 Mg/m^3). Fifteen years are anticipated to complete the landfill. It is to last for at least 100 years.

1. What dimension ratio (DR) is required? DR is the ratio of outside pipe diameter and minimum wall thickness. It is a measure of the pipe stiffness for PVC. Assume that the 15 year yield strength of the PVC is 5000 psi (34.5 MPa). Safety factor is 1.5. From equation 1, ring compression stress is,

$$\sigma = 0.5 \ \gamma H(1+d)DR$$

For a long-term sanitary landfill, it is prudent to hold ring deflection to nearly zero by careful compaction of sidefill. Solving the above equation with d = 0, and with a safety factor of 1.5; DR = 21.3. A good selection is PVC pipe SDR 21(200) ASTM D 2241. SDR is a standard dimension ratio. It is defined the same as DR; i.e. SDR = OD/t.

2. What is the maximum allowable ring deflection? If select PZB is specified, the soil friction angle is no less than $\phi = 30°$. From the UNI-BELL Handbook of PVC pipe, published by the Uni-Bell PVC Pipe Association, 2655 Villa Creek Drive, Ste. 155, Dallas, TX 75234, for SDR = 21, the pipe stiffness is F/Δ = 234 psi (1.6 MN/m/m) for E = 400,000 psi, and F/Δ = 292 psi (2.0 MN/m/m) for E = 500,000. Entering Figure 10 (or equation 2) with $\phi = 30°$, and $P/(F/\Delta) = 600(75)/234(144) = 1.34$; the ring deflection at incipient collapse is about d = 24.4%. Clearly, if ring deflection is held to less than d = 10%, the safety factor against incipient collapse is greater than two which is adequate.

Ring deflection can be controlled by the quality and compaction of the sidefill. From laboratory tests, select crushed stone compacted to 95% density AASHTO T99 (70% relative density) will hold ring deflection to less than d = 5% under 600 ft (183 m) of soil cover at unit weight of 75 pcf (1.2 Mg/m^3).

INTIMATE SOIL CONTACT

Intimate contact of the pipe-zone-backfill against the pipe helps to assure alignment and position of the pipe. Where vertical alignment is critical in order to prevent ponding and consequent sedimentation in the pipe, intimate contact of the bedding must be assured. If groundwater flow could erode channels along the pipe, intimate pipe soil contact would be necessary. Of course, intimate soil contact retains the circular cross section of the pipe ring. By fixing the shape of the pipe ring, stresses in the pipe relax in the long term.

Intimate soil contact is assured by using select granular pipe-zone-backfill (PZB) placed under the haunches of the pipe by shovel-slicing, J-barring, flushing, etc. It is noteworthy that intimate soil contact may not be absolutely essential for adequate structural performance of the pipe. This is true in the case of small diameter plastic pipes of low DR with excellent sidefill. The excellent sidefill can support the landfill by arching action with or without contribution from the pipe ring. The phenomenon is tantamount to boring a tunnel under the landfill and inserting the pipe as a tunnel liner. The pipe only has to support the talus that would fall into the tunnel if the pipe were not there. Consequently, design engineers may mitigate the importance of intimate soil contact specifications under some conditions, even though soil under the haunches may not be as well compacted as the sidefill. It is the sidefill that is most essential to arching action of the pipe-zone-backfill.

TEMPERATURE

The properties of plastics are affected by temperature. Pipe manufacturers can provide the necessary design data. For example if PVC pipe is to be used at a temperature of 120°F (49°C), it may be prudent to assume a modulus of elasticity of E = 400 ksi (2.8 GPa) rather than E = 500 ksi (3.4 GPa), and yield strength of 4 ksi (28 MPa) rather than 5 ksi (34 MPa). Decomposition of the biomass in sanitary landfills generates heat. Temperatures of 120°F (49°C) are not uncommon. Similar adjustments apply to other plastics where temperature is of concern.

Dennis E. Bauer

15 YEAR OLD POLYVINYL CHLORIDE (PVC) SEWER PIPE; A DURABILITY
AND PERFORMANCE REVIEW

REFERENCE: Bauer, D. E., "15 Year Old Polyvinyl Chloride
(PVC) Sewer Pipe; A Durability and Performance Review," Buried
Plastic Pipe Technology, ASTM STP 1093, George S. Buczala and
Michael J. Cassady, Eds., American Society for Testing and
Materials, Philadelphia, 1990

ABSTRACT: A sample of 15 year old 257 mm (nominal 10 inch) diameter Polyvinyl Chloride (PVC) sewer pipe was excavated and tested
in accordance with the ASTM standards to which it was manufactured. Test results for standard requirements such as workmanship,
dimensions, flattening, impact resistance, pipe stiffness, joint tightness and extrusion quality are presented and compared to current requirements of ASTM D 3034, "Standard Specification for Type PSM
Poly(Vinyl Chloride) (PVC) Sewer Pipe and Fittings." This information serves as a basis for review of the physical durability of PVC
sewer pipe.

A substantial amount of initial installation data was available for this
particular sewer project. Information such as bedding and haunching
requirements and initial deflections were retrievable as well as original
plans. Several pre-excavation procedures were completed in an effort
to assess current performance. These included a review of City
maintenance records, measuring depth of flow, televising the line and
pulling a deflection mandrel. Actual in-situ soil classifications and
density measurements were completed as the excavation proceeded.
This information serves as a basis for a review of the performance of
the PVC sewer line.

KEYWORDS: modulus of elasticity, deflection, pipe stiffness, joint
tightness, tensile strength

Numerous testimonials to PVC's superior long-term performance are
available from users throughout North America. However, they have neither the
time, budget nor inclination to dig-up perfectly good PVC sewer pipe to test its
durability. Members of the Uni-Bell PVC Pipe Association thought it prudent to
provide such information in support of the selection of PVC.

A decision was made in 1988 to locate and remove sufficient PVC sewer pipe
for testing in accordance with American Society for Testing and Materials (ASTM)
Standard D 3034, "Standard Specification for Type PSM Poly(Vinyl Chloride)

Dennis Bauer is the Association Engineer of the Uni-Bell PVC Pipe Association,
2655 Villa Creek Drive, Suite 155, Dallas, TX 75234.

(PVC) Sewer Pipe and Fittings." The pipe was to be one of the earliest gasketed SDR 35 sewer pipes manufactured in accordance with this standard.

ASTM D 3034 was originally published in 1972. Therefore, the oldest pipe, if installed in 1972 and excavated in 1988, would be 16 years old. Certainly an adequate period of time to formulate opinions on long-term durability.

There were prerequisites, in addition to it being gasketed SDR 35, established upon which a specific site selection would be based. One of the requirements was that the line chosen be at least 10 years old. It should also be a sanitary sewer application, with typical slopes, depth of cover and operating conditions. A final prerequisite was that the project selected be one for which initial installation documentation was available. This requirement would insure an ability to objectively review the changes with time in the pipe performance as well as durability.

SITE SELECTION

The records for older PVC sewer locations were examined from a number of cities. Because Dallas, Texas, has used PVC sewer pipe for many years, a number of sites could be considered. Information about the initial deflections and the installation requirements for their early PVC pipe installations had been documented.

Further review led to a 254 mm (10 inch) diameter gasketed SDR 35 PVC sewer line on the north side of Dallas, which had been installed in May of 1973. The City had plans and profiles of the project with embedment description as well as a report on initial deflections.[1] Depths of burial ranged from 2.4 to 3.0 meters (8 to 10 feet) with a planned slope of 0.80 percent. By tapping the collective memory of local manufacturer's representatives, we learned that a complete installation report[2] had also been prepared immediately following the pipe's installation. This site had everything we required.

The City of Dallas willingly cooperated on this research project which required the removal of an active PVC sewer line from beneath a city street. The Dallas Water Utilities Department believed the research would greatly benefit their information base as well as the user community in general. The City of Dallas Water Utilities Department acted as the contractors for the project.

PRE-EXCAVATION

Before excavating the line and sending it to the laboratory for testing, information was gathered relative to how well it was currently operating. City crews reported that there were no recorded instances of required maintenance on this line in its 15 years of service. The following five steps summarize the pre-excavation protocol.

- Depth of flow measurement
- TV the line
- Clean the line
- Pull a deflection mandrel through the line
- Re-TV the line

Dallas Water Utilities had set up a mobile recording device to measure depths over a seven day period. Depth of flow over the seven days ranged from a low of approximately 25.4 mm (1 inch) to a high of 76.2 mm (3 inches) of depth.

The Dallas utility crew, which televised the line, explained that they no longer televise their PVC sewer lines, as a part of scheduled maintenance, because experience had shown that for PVC sewers, such periodic visual examinations were not necessary. They pulled the camera through the 74.7 meters (245 foot) long section of 254 mm (10 inch) PVC while it was in service. The video revealed that the interior of the pipe was fairly clean. It had what appeared to be very light and inconsistent residue throughout its length. No heavy build-ups were found.

The line was then cleaned with a water jetting device before the deflection mandrel was pulled. Both the City's and manufacturer's post-installation test reports indicated that deflections ranged from two to five percent over the length of the line. They both had used deflectometers to measure the entire length. The City crew attempted to pull a 5 percent deflection mandrel (a go/no go testing device) through the line. Due to the heavy build-up of concrete at the outlet and inlet structures, of the upstream and downstream manhole inverts, respectively, they were unable to introduce the mandrel into the line. With much effort and manipulation, a 7.5 percent deflection mandrel, which is ASTM's published recommended allowable deflection limit, was wedged into the upstream manhole outlet structure. The 254 mm (10 inch) PVC sewer line passed the mandrel without any hang-ups.

Because the interior of the pipe was relatively clean to begin with, the re-televising, after the jet cleaner had been pulled through, revealed no significant changes.

EXCAVATION

In December of 1988, Dallas Water Utilities' personnel began excavation of the pipe. Before removal of any overburden, a device was placed in the pipe to lock-in in-situ deflection over a two foot length. By maintaining in-situ deflection, long-term structural properties of the material could later be determined, in particular, long-term modulus.

The asphalt was removed and a nuclear densometer, operated by a Dallas utility consultant, was calibrated and used to measure soil density just below the surface and then again three feet down. The densities ranged from 82 percent to 92 percent along the length of the trench.

A Dallas based engineering consultant specializing in geo-technical evaluations was retained to classify and define the native trench soil as well as the embedment material. Their soils' report[3] revealed the following. The final backfill material, which was placed from approximately four to six inches above the pipe to about one foot below the pavement, was classified as CH material, described as "Dark gray to tan clay with calcareous modules and limestone fragments."

The initial backfill, which was placed from the springline of the pipe to four to six inches above, was classified as SM and described as "Tan silty fine sand with trace fine gravel." The City crew knew we were getting close to the top of the pipe when the backhoe operator struck this sand.

The City requirements called for a "Class B" embedment material. Class B material basically consists of sand and gravel. The material which was removed from the bottom of the pipe and from the sides to the springline was classified as SC and described as tan and grayish brown clayey fine to coarse sand with some fine gravel. Because the invert was approximately 2.4 meters (8 feet) deep and the water table was slightly above the crown of the pipe, some of the native clay materials washed into these samples during the excavation process. Their embedment requirements were sufficient to maintain deflections within the allowable limit.

The water table being above the pipe made for nasty working conditions which were aggravated by rainy weather. This didn't deter the Dallas utility crew. They brought in pumps to lower the water level and uncovered the pipe. Once uncovered, they cut holes in the crown and let the pipe assist in draining the trench.

A total of approximately 12.4 meters (40 feet) of pipe was required to test in accordance with ASTM D 3034. The 254 mm (10 inch) PVC was removed in two sections, one 10.1 meters (33 feet) in length and the other 2.1 meters (7 feet). A house lateral prevented us from removing one contiguous piece. As Figure 1 reveals, the pipe

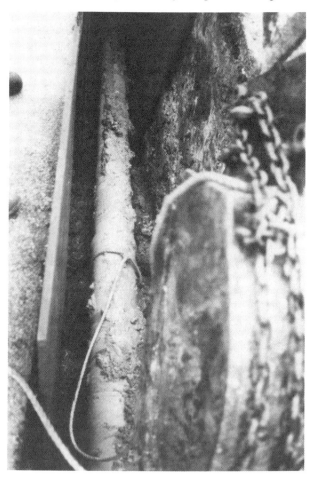

Figure 1. The 257 mm (10 inch) PVC was removed from Dallas.

was covered with mud upon removal. Even so, many of the crew members commented that the pipe looked practically new. The mud was washed off, the interior was rinsed and the pipe was cut in lengths convenient for shipment.

TESTING

The pipe was sent to Utah State University's (USU) mechanical engineering department in Logan, Utah. USU is a recognized pipe research facility.

USU tested the received pipe samples in accordance with ASTM D 3034, which has requirements for workmanship, diameter dimension, wall thickness dimension, flattening, impact resistance, pipe stiffness, joint tightness and extrusion quality.

Workmanship

ASTM D 3034 requires that the pipe and fittings shall be homogeneous throughout and free from visible cracks, holes, foreign inclusions or other injurious defects. The pipe shall be as uniform as commercially practical in color, opacity, density and other physical properties.

After 15 years, the PVC pipe passed these requirements. In fact, the report from USU[4] states that, "after cleaning with soapy water the appearance was almost like new pipe."

Diameter Dimension

The average outside diameter when measured in accordance with ASTM D 2122, "Standard Method of Determining Dimensions of Thermoplastic Pipe and Fittings," was found to be two thousandths of an inch under the tolerance.

This minor variation in diameter would not affect pipe performance.

Wall Dimension

The wall thicknesses exceeded the ASTM D 3034 requirements when measured in accordance with ASTM D 2122. The samples had average wall thickness ranging from 7.80 to 7.98 mm (0.307 to 0.314 inches). After 15 years of service, the wall thickness, even in the invert, was greater than that required by the standard for new pipe. Domestic sewage is not normally very abrasive, but confirmation of the pipe invert wall thickness was felt to be important. Previous studies have confirmed PVC's high resistance to abrasion.

Impact Resistance

ASTM D 3034 requires that 254 mm (10 inch) diameter pipe withstand 298 joules (220 ft-lbf) when tested in accordance with ASTM D 2444, "Standard Test Method for Impact Resistance of Thermoplastic Pipe and Fittings by Means of Tup (Falling Weight)." This standard requires that pipe specimens be able to withstand a blow from a missile-shaped falling weight called a tup. The tup is 9.1 kg (20-lbs), therefore, a 3.4 meter (11 foot) drop height is used. This test is a quality control requirement and was never intended as a requirement for field installation. All samples passed, providing further evidence that no embrittlement occurs over time with buried PVC pipe.

Pipe Flattening

ASTM D 3034 states that there shall be no evidence of splitting, cracking or breaking when pipes are flattened to 40 percent of their outside diameter. By flattening to 40 percent, the pipe is deflected 60 percent.

The 15 year old PVC pipe passed the test. There was no embrittlement with time. See Figure 2.

Figure 2. Fifteen year old PVC pipe subjected to 60% deflection, without splitting or cracking.

Pipe Stiffness

A minimum of 317 kPa (46 psi) is ASTM's requirement for SDR 35 PVC sewer pipe. All samples were tested in accordance with ASTM D 2412, "Test Method for Determination of External Loading Characteristics of Plastic Pipe by Parallel Plate Loading." The average pipe stiffness of the 15 year old pipe was 433 kPa (62.8 psi).

Joint Tightness

Within D 3034 is a requirement that the elastomeric gasket joints perform in accordance with ASTM D 3212, "Standard Specification for Drain and Sewer Plastic Pipe Using Flexible Elastomeric Seals." This performance standard requires that pipe, with gasketed joints, undergo both an internal pressure and vacuum requirement when axially deflected. The 15 year old joint passed the test. (See Figure 3.)

After 15 years of service, the joints met the same requirements as those of new pipe. Meeting these requirements insured that the pipe could still comply with a cost-effective allowable infiltration/exfiltration requirement of 50 gallons per inch of internal diameter per mile per day.

Figure 3. Fifteen year old PVC passes original joint requirements.

Extrusion Quality

ASTM D 3034 requires that the pipe not flake or disintegrate when tested in accordance with ASTM D 2152, "Standard Test Method for Degree of Fusion of Extruded Poly(Vinyl Chloride) (PVC) Pipe and Molded Fittings by Acetone Immersion." Samples were submerged in anhydrous acetone for the required time and inspected. All of the 15 year old pipe samples successfully passed.

Structural Properties

Within a limited group of users and non-users of PVC there lingers a question concerning long-term structural properties. They would suggest that properties such as tensile strength and pipe stiffness (modulus of elasticity) decrease with time.

When the 15-year-old PVC pipe was tested in both the circumferential and longitudinal direction for tensile strength and modulus of elasticity, the following results were obtained.

These values are typical of newly manufactured PVC sewer pipe.

Table 1 - Circumferential Direction

Specimen Number		Tensile Strength MPa[a] (psi)	Modulus MPa[a] (psi)
1		51.05 (7410)	2591 (0.376 x 10⁶)
2		52.36 (7600)	3038 (0.441 x 10⁶)
3		53.05 (7700)	2777 (0.403 x 10⁶)
4		<u>52.85 (7670)</u>	<u>2963 (0.430 x 10⁶)</u>
	AVG.	52.36 (7600)	2839 (0.412 x 10⁶)

Table 2 - Longitudinal Direction

Specimen Number		Tensile Strength MPa[a] (psi)	Modulus MPa[a] (psi)
1		55.05 (7990)	3094 (0.449 x 10⁶)
2		54.98 (7980)	3011 (0.437 x 10⁶)
3		56.08 (8140)	2976 (0.432 x 10⁶)
4		<u>55.53 (8060)</u>	<u>3156 (0.458 x 10⁶)</u>
	AVG.	55.40 (8040)	3059 (0.444 x 10⁶)

[a]MPa = psi x 0.00689

ADDITIONAL TESTING

To remove any doubt about the validity of the long-term structural properties, a locking brace was placed inside one of the pipe samples prior to its excavation. That device served to maintain the in-situ, 15 year deflection, allowing for incremental load and deflection measurements. At 5 percent deflection, pipe stiffness was determined to be 449 kPa (65.1 psi). The corresponding modulus of elasticity was determined to be 3405 MPa (494,220 psi). The 15 year old, buried PVC sewer pipe had not lost any of its stiffness when compared with the ASTM D 3034 requirement.

CONCLUSIONS

Clearly PVC pipe is providing excellent performance. After 15 years of service there have been no required maintenance calls, deflections were held below recommended limits and the joints met the tightness requirements of new pipe.

The results of testing in accordance with ASTM D 3034 reveal that no measurable degradation of any sort took place in the course of 15 years. There was no embrittlement, no loss of wall thickness, no decrease in pipe stiffness and no decrease in modulus.

The PVC pipe's ability to perform has not changed over 15 years and all indications suggest it will not change in the foreseeable future.

REFERENCES

[1] Chandler, R. W., P.E., "Evaluation of In-Place Polyvinyl Chloride (PVC) Sewer Pipe," Dallas Water Utilities, Dallas, 1974.

[2] Morrison, R. S., "A Report on Deflection Measurements of Johns-Manville SDR 35 PVC Gravity Sewer Pipe in the Dallas/Fort Worth Area," Johns-Manville, Dallas, 1973.

[3] Grubbs, B. R. and Oswald, T. W., "Results of Classification Tests PVC Sewer Pipe Project," Report DR-8088, TERRA-MAR Consulting Engineers, Dallas, January 1989.

[4] Moser, A. P. and Shupe, O. K., "Testing Fifteen Year Old PVC Sewer Pipe," Buried Structures Laboratory, Utah State University, Logan, Utah, March 1989.

Author Index

403

Subject Index

A

Additives, 79
Aging, 7, 21
American National Standards
  Institute, 79
American Water Works
  Association, 40, 185
Arching, 266
ASTM standards
  D 1784: 7, 159
  D 2321: 281
  D 2412-87: 7
  D 2487: 245
  D 3034: 159, 393
  D 3839: 281

B

Bend test, 92
Buckling, 57, 141, 266
  continuum theory, 344
Bulk modulus, 141

C

Certification, third party, 79
Chemical resistance
  polyethylene, 21
  polyvinyl chloride, 7
CIPP (cured-in-place) pipe, 329
Classification, soils, 245
Compaction, 141, 281
Compressive wall crushing, 266
Concrete pipe, reinforced, 297
Constrained modulus, 141
Continuum buckling theory, 344
Crack damage, 40
Crack growth, slow, 21
Creep testing, 159
Culverts, 281
Cured-in-place pipe, 329

D

Deflection, 171, 281, 393
  analyzing, 141, 233

estimating, 185
Iowa formula, 266
monitoring, 125, 217
Deformation, 171
Design, 40, 57, 217
  buckling theory, 344
  liner, 297, 313
  ocean outfall, 336
  sewer, 363
  strain, 159
Drinking water, 79
Durability, 7, 21

E

Elastic continuum analysis, 344
Elasticity, modulus of, 233, 393

F

Fatigue behavior, 101
Fiberglass pipe, 40, 281, 297, 363
Fiberglass reinforced plastic
  pipe, 185, 217
  environmental effects, 217
Fittings, pipe, 101
Flexible pipe, 7, 57, 336, 379
  deflection, 125, 141, 171,
    185, 281

G

Gas distribution, pipes for, 233
Glass fiber reinforced plastic
  pipe, 185, 217
Gravity flow pipes, 57, 159, 313

H

Health effects, 79
Hydrogen sulfide, 297

I

Installation
  D 2321: 281
  D 3839: 281